Stability of Parallel Flows

APPLIED MATHEMATICS AND MECHANICS

An International Series of Monographs

EDITORS

F. N. FRENKIEL

Applied Mathematics Laboratory
David Taylor Model Basin
Washington, D. C.

G. TEMPLE

Mathematical Institute
Oxford University
Oxford, England

Stability of Parallel Flows

Robert Betchov

AEROSPACE CORPORATION
EL SEGUNDO, CALIFORNIA

William O. Criminale, Jr.

DEPARTMENT OF AEROSPACE AND
MECHANICAL SCIENCES
PRINCETON UNIVERSITY
PRINCETON, NEW JERSEY

ACADEMIC PRESS   New York · London · 1967

PHYSICS

Cat as Sep

ACADEMIC PRESS INC.
111 Fifth Avenue, New York, New York 10003

United Kingdom Edition published by
ACADEMIC PRESS INC. (LONDON) LTD.
Berkeley Square House, London W.1

LIBRARY OF CONGRESS CATALOG CARD NUMBER: 66-30120

PRINTED IN THE UNITED STATES OF AMERICA

Preface

This book was written with the dual purpose of combining a text with a monograph. In this way a source is provided for the uninitiated to the subject along with an up-to-date review of the state of the art for the more knowledgable researcher.

In planning the book it became apparent that the timing of the writing meant that the digital computer, be it of very large size or only of moderate capacity, offered new possibilities for the study of flow instability both in the classroom and in the research institution. After we had explored this point further, we found that the influence of the computer was already so greatly established in the general sense (even to the point that a new breed of scientist is developing) that it seemed desirable to make its presence better known in this field. At the same time, however, we recognized that, by introducing the computer with any degree of importance, the classical philosophy and style of presentation that are known and established would be noticeably affected. As a result of this fact, we have purposely avoided the traditional practice that is found in a treatise on mathematical physics. Many common detours have been replaced by straightforward numerical procedures, for example, but never in any case has a transformation or substitution been made at the expense of the physics involved or to the understanding of the problem.

Of course, computers must be used with caution, lest they produce numerical asphyxia. We hope that the reader will share with us the view that the computer, by providing specific examples of solutions, does indeed facilitate the physical interpretation of the situation. This doctrine has been admirably expressed in the words of Hamming (1962) in his book on numerical methods: " The purpose of computing is insight, not numbers." And perhaps a word should be said about the economical aspects of nu-

v

merical machine calculations. For several years it has been our experience in associations with various academic and industrial organizations that efficient research requires approximately equal funds for the programming and operations of the digital computer as for the upkeep of a principal investigator's gray matter. Thus, the cost of computing should be comparable to the cost of providing a senior experimental man with laboratory assistants, overhead, glass blowing service, etc., for his investigations.

Because of the goals, this book is naturally organized into two principal subdivisions. In Part One we follow a gradual path leading to an understanding of such essential concepts as the inviscid oscillations of jets and wakes and the resistive instability of boundary layers. As the road crosses over the well-known critical layer, we have tried to give a detailed and intuitive description so that this obstacle can be overcome without permanent trauma.

In Part Two, we have aimed at greater rigor and at establishing a broad review of the various manifestations of fluid instability. It is at this stage that we consider three-dimensional effects, compressibility, magnetohydrodynamics, and a variety of other ramifications.

Finally, analytic methods suitable for solutions to stability problems are condensed in an appendix. Such tools are quite naturally needed as aids to understanding any numerical output or as bases for planning and interpretation of a program. The general problem is complex enough and is best explored by means of a suitable balance between analysis and computation. We offer this material within the framework of what is now deemed inner and outer expansions.

In addition to the references cited in the course of the text, we have included a comprehensive bibliography of the research on the treated subject.

June, 1967   ROBERT BETCHOV
                                                WILLIAM O. CRIMINALE, JR.

Acknowledgments

Before making acknowledgments for specific assistance in our effort, we would like to recognize the general and exemplary influence in our minds of the publications of Professor C. C. Lin of the Massachusetts Institute of Technology.

To Professor O. M. Phillips of the Johns Hopkins University, we are very grateful for giving us valuable time in reading the manuscript, making detailed corrections, and offering valuable suggestions for improvements.

At a later stage of development, Professors G. M. Corcos of the University of California at Berkeley and W. C. Reynolds of Stanford University were kind enough to provide critical appraisals of the work and make recommendations for improvement.

Dr. L. M. Mack of the Jet Propulsion Laboratory and Professor R. E. Kaplan of the University of Southern California have made available to us not only material of limited distribution but also a wealth of supplementary information.

Some sections could not have been written without the guidance of Professor M. T. Landahl of the Massachusetts Institute of Technology and Mr. P. S. Klebanoff at the National Bureau of Standards and of Dr. W. B. Brown of The Northrop Aircraft Company. We are indebted for their special aid.

The Bibliography was compiled by the capable office of Mr. J. K. Lucker, Assistant Librarian for Science and Technology at Princeton University. We express to him our sincere gratitude for his patience and complete competence. The result is an extremely valuable and necessary asset to this work.

We must recognize several colleagues who kindly provided us with material for figures: Professor F. H. Abernathy of Harvard University

(Figure 31.4); Dr. J. E. Fromm of Los Alamos (Figure 30.3); Dr. F. R. Hama of the Jet Propulsion Laboratory (Figure 30.1); Dr. J. M. Kendall of the Jet Propulsion Laboratory (Figure 31.1); and Dr. A. Michalke, Institut für Turbulenz Forschung, Berlin (Figure 20.1).

Acknowledgment is made to the Mechanics Division of the Air Force Office of Scientific Research for support of one of us (WOC) in many aspects of research which lead to this work.

It is now with sincere gratitude that we direct the attention of our readers to a special and somewhat forgotten group of collaborators, the programmers. Throughout a long period, the following persons have given us the benefit of their talents: Miss Nathalie Gamberoni of the Douglas Aircraft Company; Mr. Paul Firnett, Miss Heather Malcom, Mrs. Sylvia Porjes, and Mr. G. Takata of the Aerospace Corporation; Mrs. Marilyn Taynai of Princeton University; and Mr. T. S. Englar, Jr. now with Bellcomm, Inc., formerly RIAS, Baltimore.

We acknowledge the diligent and efficient typing and editing activities of Mrs. S. Leyton at Princeton University.

Finally, our thanks is given to Academic Press who constantly tolerated our delays and provided us with every assistance we needed.

Contents

ix

Chapter IX   Magnetohydrodynamic Effects

Chapter X   Additional Topics and Complexities

Key to Notation

x, y, z	Cartesian coordinates
t	Time
u, v, w	Eulerian components of the fluid velocity
p	Pressure (often on basis of unit density)
U, V, W	Mean velocity components
P	Mean pressure
r	True vorticity; a scalar in two dimensions
$\vec{r}$	The vorticity vector in three dimensions
R	Mean vorticity
S	Mean entropy
T	Mean temperature
$\vec{J}$	Mean electric current
g, h, k	Cartesian components of the magnetic field (commonly in units such that the Alfvén velocity is directly given by the magnetic field)
G, H, K	Mean magnetic field components
$\tilde{u}, \tilde{v}, \tilde{w}, \tilde{p}, \tilde{r}, \tilde{g}, \tilde{h}, \tilde{k}$	Fluctuations in respective quantities; functions of x, y, z, t and real
$\hat{u}, \hat{v}, \hat{w}, \hat{p}, \hat{r}, \hat{g}, \hat{h}, \hat{k}$	Complex fluctuations; functions of x, y, z, t
$\mathbf{u}, \mathbf{v}, \mathbf{w}, \mathbf{p}, \mathbf{r}, \mathbf{g}, \mathbf{h}, \mathbf{k}$	Complex amplitudes; functions of y only: all boldface quantities are complex and indicate the amplitude and phase of the corresponding function.
$c = c_r + ic_i$	Complex phase velocity
A, B, C, D	Generally used as integration constants; often complex
f, g, h	Generally auxilliary functions, except for magneto-hydrodynamics
τ_{ij}	True stress tensor (on basis of unit density)

l	Constant reference length (used in study of thin viscous regions, critical layers, etc.)
$Z, \mathscr{F}$	Special complex functions, describing viscous effects near a wall (see Eqs. (9.12) and (9.16))
$\mathscr{V}$	Volume of integration
$h, \mathbf{h}$	Displacement of a wall or an interface
L	A characteristic length, related to the width of a layer
$\mathscr{R}$	Reynolds number of the mean flow, based on free stream velocity and a thickness such as δ or L
a	Speed of sound
c_v, c_p	Specific heats at constant volume and pressure
$\mathscr{P}\imath$	Prandtl number
M	Mach number
α	Wave number in x direction; sometimes complex
β	Wave number in z direction; real
$\omega = ac$	Complex frequency
ν	Kinematic fluid viscosity
η	Distance used in the y direction within a viscous region; in special units
δ	Abcissa of the outer edge of a boundary layer or shear layer (it is larger than that defined on 99% of the free stream velocity, for Blasius, δ is four times δ^*, the displacement thickness)
T, τ	Fluctuation in temperature; function of x, y, z, t
Υ	Mean fluid density
$\theta, \boldsymbol{\theta}$	Density fluctuations; function of x, y, z, t
τ_{xy}, etc.	Stress tensor fluctuations; functions of x, y, z, t
μ_j	Various productions of energy (see Eq. (2.13))
Ψ	Stream function
$\sigma, \boldsymbol{\sigma}$	Fluctuating entropy
λ	Thermal expansion, by volume, at constant pressure
γ	Ratio of specific heats
—	Overbar denotes average
x, y, z, t	In subscript, partial derivative
$', d/dy, D$	Ordinary derivative
$*$	In superscript, complex conjugate

Introduction

Stability can be defined as the quality of being immune to small disturbances. In general the disturbances need not necessarily be infinitesimally small in magnitude, but the *concept of amplification is always implicit.* Thus, we can examine the stability of mechanical, astronomical, or electrical systems which have only a few discrete degrees of freedom. In the study of continuous media, however, the number of degrees of freedom is infinite and the question becomes more difficult. In particular, the basic equations take the form of nonlinear partial differential equations. In spite of the additional complications, though, progress toward an understanding of such systems is made with the use of linearized approximations and by extending the theory developed with discrete systems.

The first major contribution to the study of hydrodynamical stability can be found in the theoretical papers of Helmholtz (1868). Previous to this time, many scholars had certainly become aware of the question but their efforts did not progress beyond the stage of description. In particular, the drawings of vortices by Leonardo da Vinci (Fifteenth century!) should be mentioned as well as the experimental observations of Hagen (1855).

About twelve years after the findings of Helmholtz, the combined efforts of Reynolds (1883), Kelvin (1880, 1887a, 1887b) and Rayleigh (1878, 1880, 1887, 1892a, 1892b, 1892c, 1895, 1911, 1913, 1914, 1915, 1916a, 1916b, 1917) produced a rich harvest of knowledge. Reynolds discovered the first experimental evidence of "sinuous" motion in water and is generally credited for a first description of random or "turbulent" flow. By dimensional analysis he uncovered the all important number now bearing his name. Reynolds pointed out that the disorder begins when this pure number exceeds a critical value, and that special stresses must be taken into consideration. The founder of hydrodynamical stability is

Lord Rayleigh, who, as cited above, published between 1878 and 1917 a great number of papers on this subject. He was thirty-six years old when the stability of flames attracted his attention, leading to a first paper on jets. The contributions that we now regard as most important cover many topics, such as the importance of inflection points in the velocity profile and the instability of rotating flows between cylinders.

At the age of seventy-two he examined nonlinear effects, a subject which was to remain inactive for the next thirty-seven years. In comparison, the work of Kelvin is of less importance, but nevertheless remarkable.

Around 1907, it became apparent that the existence of a critical Reynolds number could not be explained easily and that the problem involved both the effect of the second derivative of the mean flow and of the viscous forces. The key equation was arrived at independently by Orr (1907) and by Sommerfeld (1908). At this stage, of course, it was generally believed that the study of infinitesimally small perturbations would lead to some definitive answers to the problem of turbulence. This Orr-Sommerfeld equation remained unsolved for twenty-two years, until Tollmien (1929) calculated the first neutral eigenvalues and obtained a critical Reynolds number. This remarkable success was greatly facilitated by the thesis of Tietjens (1925) and the analysis of Heisenberg (1924). Tietjens studied diffusion in a flow of constant shear, computing an important function. The work of Heisenberg is more abstract, but points toward the possibility of resistive instability.

During the intermediate period we find, in addition to the works of Prandtl (1921–1926, 1929, 1935), a first confirmation of a linear theory by experiments: the work of Taylor (1923) on vortices between concentric rotating cylinders being the principal and best known contribution. Indeed this was a dual effort where theory and experiment were matched simultaneously.

The initial success of Tollmien was due to the application of newer mathematical techniques. Soon, the same road led Schlichting (1932a, 1932b, 1933a, 1933b, 1933c, 1934, 1935) to further evaluations of the critical Reynolds number and amplification rates of disturbances.

Later, Lin (1944, 1945) reviewed and improved the mathematical procedure and laid the foundations for a general expansion of the stability analysis. At this time the stability of Poiseuille flows had become a particularly controversial issue and Lin put it in order by his newer and more general analysis. Any additional doubts with respect to this system were finally settled by the first use of a digital computer in hydrodynamical stability. The results of Heisenberg and Lin were in fact found to be correct.

After this success, it became apparent that the critical Reynolds number marked only the threshold of " sinuous " motion and not that of turbulence. This was made perfectly clear by the experimental results of Schubauer and Skramstad (1943), who used a vibrating ribbon to impress a controlled perturbation in the study of boundary layer stability. As an instrument of diagnostics, the hot-wire anemometer was now available which had been perfected by Burgers and Dryden over a period of several years. Incidentally, the perturbations considered here were small but finite.

Experiment and theory agreed, as far as eigenvalues and eigenfunctions were concerned. Yet turbulent transition was not understood, and it still remains an enigma.

The attack on compressible flows was initiated by the work of Landau (1944) and Lees (1947) and continued by Dunn and Lin (1955). Magnetic, gravitational, and convective effects were examined by Bénard (1901) and Chandrasekhar (1961), to cite but two of a great number of references concerning this aspect.

In 1955 appeared the comprehensive monograph of Lin, which gave a systematic evaluation of the entire field of hydrodynamical stability. Although this volume deals essentially with asymptotic power series, it already contains the seed of future developments. It also settled many controversial questions that had been built up over the years.

Also, at about this time, enough information was available about the various manifestations of instability in a continuous medium to formulate three main categories. These are, first, the oscillations of parallel flows, or nearly parallel flows: channel flows, boundary layers, jets and wakes. Second, there is the class with curved stream lines, such as vortices between rotating cylinders or boundary layers along curved walls. In the third category we place those cases where the mean flow is truly zero: Bénard cells and convective instabilities being the chief examples.

The first category contains the classical examples of inviscid instabilities which are generally damped when friction becomes significant. However, it also contains outstanding examples of resistive instability. It is well known that a resistance can cause instability in a mechanical device or an electrical circuit, but a similar process in a fluid was somewhat surprising to the early investigators.

After establishing the instability of Poiseuille flow, the computer gradually made further impacts. Of note are the works of Brown (1959, 1961a, 1961b, 1962, 1965), Mack (1960, 1965a, 1965b), and Kaplan (1964), among others.

Slowly, the theory of nonlinear processes was set on its way by Meksyn and Stuart (1951). The experiments of Klebanoff, Tidstrom, and Sargent (1962) revealed unexpected vortices which were explained by Benney and Lin (1960).

Recently, some simple nonlinear problems have been successfully treated by Fromm and Harlow (1963). This work used a purely numerical method and demonstrated the resources of modern computers.

A surge of interest was created by the finding that the stability of a flow is enhanced when the wall is endowed with negative damping. There is no evidence that dolphins take advantage of this other manifestation of resistive instability.

This book is devoted to the stability of parallel flows. Indeed, resistive instability is the salient aspect of our work. There is a short excursion into the topic of curved streamlines, namely, the conclusions that can be reached when dealing with a boundary layer over a curved wall and the role of secondary instability. A wealth of information on convective instability, all classes of curved streamline problems, fluid stratification, and magnetic processes can be found in the exhaustive treatise of Chandrasekhar (1961).

Stability problems which illustrate resistive instability can be found in many fields. Examples occur in mechanics, electronics, naval and aeronautical engineering, oceanography, administration, and economics. Slowly, this phenomena is being uncovered in magnetohydrodynamics and plasma physics.

It appears that the destabilizing effect of a resistance is often simply a matter of delay. Let us consider an oscillator whose restoring force is simply retarded by a constant time τ. The basic equation for this case is

$$\ddot{x}(t) + x(t - \tau) = 0.$$

An expansion in powers of τ, limited to the first term, gives the equation of an oscillator with negative damping, namely,

$$\ddot{x}(t) - \tau \dot{x}(t) + x(t) = 0.$$

Delayed effects have been detrimental to the performance of ship stabilizing machinery (cf. Minorsky, 1947) or advantageous to the design of electronic beam amplifiers (Birdsall *et al.*, 1953). As we wrote this book, we were very aware of the existence of a general interest for resistive instability in the mechanics of continuous media.

We have made an effort to improve the terminology. Thus the " inviscid Orr-Sommerfeld equation " is properly labeled as the Rayleigh equation. The supersonic waves in the free stream are designated as acoustic waves.

We have not hesitated to replace traditional, but obscure, notations such as the many Z functions of compressible flow stability theory by more intuitive symbols.

As a conclusion, we wish to compare our state of affairs with the conditions existing when Rayleigh started his work. He was interested in flame stability, but found that he should first examine the behavior of a very simple type of jet. Today we barely understand the Kármàn row of vortices, but our ambition is already focused on the design of leakproof magnetic bottles.

Note added in proof: The oscillations so often encountered in parallel flows are, by the historical reasons outlined, generally termed Tollmien-Schlichting waves. It should be noted, however, that Tollmien-Schlichting waves strictly correspond to those waves where *friction is critical* and do not exist in the absence of viscosity. In other words, this designation is only used when, in the limit of infinite Reynolds number, the flow is stabilized. Moreover, insofar as is known, Tollmien-Schlichting waves have only been found along walls.

Initiation to
Two-Dimensional Problems

Chapter I

General Equations and Inviscid Cases

1. General Equations

In Part One we present the essentials to a good fundamental under-standing. We limit our interest to two-dimensional incompressible flows and write the general equations along with certain derived equations that have special physical meaning. In this way the bases are provided for a discussion of the oscillations of uniform flows, shear flows away from walls (jets and wakes), and finally of shear flows along a wall. A shear flow along a wall is generally called a boundary layer and is of special interest to aeronautical engineers. The oscillation of a boundary layer has played a large role in the historical development of stability theory because it lends itself relatively easily to good experimental measurements and observations. The oscillations of boundary layers are first analyzed with maximum clarity and limited rigor in Chapter II, and then with the powerful resources of numerical analysis in Chapter III.

The general equations of a two-dimensional incompressible flow are

$$u_x + v_y = 0. \tag{1.1}$$

$$u_t + uu_x + vu_y + p_x = v(u_{xx} + u_{yy}), \tag{1.2}$$

$$v_t + uv_x + vv_y + p_y = v(v_{xx} + v_{yy}), \tag{1.3}$$

where $u(x, y, t)$ is the actual fluid velocity parallel to the Cartesian x-axis and $v(x, y, t)$ is the actual velocity parallel to the y-axis. More precisely, u and v are the Eulerian velocity components. It is convenient to refer to the x-axis as the "horizontal" axis and to the region $y > 0$ as being "above" the region $y < 0$; this refers to the usual way of drawing axes and does not imply the existence of gravity forces. The function $p(x, y, t)$ is the kine-matic pressure, that is, the ordinary pressure divided by the density.

Because the ordinary pressure is a force per unit area, the kinematic pressure has the dimensions of the square of a velocity. Equation (1.1) expresses the conservation of matter. Equation (1.2) states that the horizontal acceleration is produced by a pressure gradient and by a viscous force proportional to v, the coefficient of kinematic viscosity. Equation (1.3) plays the same role for the vertical acceleration.

It is assumed that v is a constant, and in those cases in which v varies with the temperature, location, or any other variable it may become necessary to add special terms.

The coordinates x, y and the time t are independent variables and the functions u, v, and p are the dependent variables, functions of x, y, t. Note that (1.2) and (1.3) are nonlinear, which is precisely the cause of the trouble now before us.

We now derive the equations that control the small oscillations of a parallel and steady mean flow. This is done in three steps: separation of fluctuations, linearization, and recourse to complex functions.

(a) SEPARATION OF FLUCTUATIONS

We begin by assuming that in the absence of oscillations the flow is laminar and can be specified as follows:

$$u = U(y), \qquad v = 0, \qquad p = P(x). \tag{1.4}$$

Note that in a channel flow P varies linearly with x and that the gradient of P balances the term vU_{yy}. For a boundary layer P is constant and vU_{yy} is balanced by $UU_x + VU_y$. Such terms, however, are neglected in the approximation of (1.4). We now assume that the flow oscillates about these values and use the tilde superscript to indicate a fluctuation. Thus we assume that

$$u = U(y) + \tilde{u}(x, y, t), \tag{1.5}$$

$$v = \tilde{v}(x, y, t), \tag{1.6}$$

$$p = P(x) + \tilde{p}(x, y, t). \tag{1.7}$$

If these equations are introduced into (1.1), (1.2), and (1.3), we obtain the set

$$\tilde{u}_x + \tilde{v}_y = 0, \tag{1.8}$$

$$\tilde{u}_t + U\tilde{u}_x + U_y\tilde{v} + \tilde{p}_x + (\tilde{u}\tilde{u}_x + \tilde{v}\tilde{u}_y) = v(\tilde{u}_{xx} + \tilde{u}_{yy}), \tag{1.9}$$

$$\tilde{v}_t + U\tilde{v}_x + \tilde{p}_y + (\tilde{u}\tilde{v}_x + \tilde{v}\tilde{v}_y) = v(\tilde{v}_{xx} + \tilde{v}_{yy}), \tag{1.10}$$

where the term νU_{yy} no longer appears because the mean flow is a solution in the absence of oscillations.

(b) LINEARIZATION

In these equations the terms containing products such as $\tilde{v}\tilde{u}_y$ correspond to an effect of a fluctuation on another fluctuation. If the fluctuation has a frequency ω, the coupled terms will have frequency 0 or 2ω. Therefore they will either modify the nonfluctuating flow or introduce higher harmonics. Such difficulties disappear if we assume that the fluctuations and their derivatives have small amplitudes. The terms in parentheses in the left-hand sides of (1.9) and (1.10) can be neglected in comparison with the other terms and we obtain the equations

$$\tilde{u}_x + \tilde{v}_y = 0, \tag{1.11}$$

$$\tilde{u}_t + U\tilde{u}_x + U_y\tilde{v} + \tilde{p}_x = \nu(\tilde{u}_{xx} + \tilde{u}_{yy}), \tag{1.12}$$

$$\tilde{v}_t + U\tilde{v}_x + \tilde{p}_y = \nu(\tilde{v}_{xx} + \tilde{v}_{yy}). \tag{1.13}$$

These equations are linear. Indeed, if $\tilde{u}_1$ is a solution in combination with functions $\tilde{v}_1$ and $\tilde{p}_1$ and if $\tilde{u}_2$, $\tilde{v}_2$, and $\tilde{p}_2$ form another solution, any linear combination such as $3\tilde{u}_1 + 2\tilde{u}_2$, $3\tilde{v}_1 + 2\tilde{v}_2$, $3\tilde{p}_1 + 2\tilde{p}_2$ will be another solution. The same fundamental property of linearity occurs in accoustics, electromagnetics, and ordinary quantum mechanics, in which it guarantees that the various oscillations will proceed independently. In thermodynamics, ferromagnetism, electronics, and fluid dynamics nonlinear equations must often be retained. With some luck we may find two solutions, but these two solutions cannot be combined to give many others.

(c) RECOURSE TO COMPLEX FUNCTIONS

We shall immediately take advantage of linearity and seek solutions in terms of complex functions. In this way we will be able to reduce our system of partial differential equations (1.11) to (1.13) to ordinary differential equations, producing an obvious facility in the analysis. Thus, we can hope to find solutions of the type

$$\hat{u}(x, y, t) = \mathbf{u}(y)e^{i\alpha(x - ct)}, \tag{1.14}$$

$$\hat{v}(x, y, t) = \mathbf{v}(y)e^{i\alpha(x - ct)}, \tag{1.15}$$

$$\hat{p}(x, y, t) = \mathbf{p}(y)e^{i\alpha(x - ct)}, \tag{1.16}$$

where the superscript ˆ indicates a preliminary complex solution which will lead to a real solution marked with the tilde. In (1.14) we identify α as the wavenumber and c as the wave velocity. We consider α as a real positive number and c as a complex number. The boldface symbols $\mathbf{u}$, $\mathbf{v}$, $\mathbf{p}$ are used to indicate a complex function of y only. The three functions defined in (1.14), (1.15), and (1.16) must obey (1.11), (1.12) and (1.13), and, after operating with the derivatives in x and t and eliminating the exponential factors, we find

$$i\alpha\mathbf{u} + \mathbf{v}' = 0, \tag{1.17}$$

$$i\alpha(U - c)\mathbf{u} + U'\mathbf{v} + i\alpha\mathbf{p} = \nu(\mathbf{u}'' - \alpha^2\mathbf{u}), \tag{1.18}$$

$$i\alpha(U - c)\mathbf{v} + \mathbf{p}' = \nu(\mathbf{v}'' - \alpha^2\mathbf{v}), \tag{1.19}$$

where we use the prime to indicate the derivative with respect to y. These equations no longer contain x or t; from the partial differential equations (1.11), (1.12), and (1.13) we have reduced the system to ordinary differential equations as desired.

We now observe that if $\mathbf{u}$, $\mathbf{v}$, and $\mathbf{p}$ are solutions of (1.17), (1.18), and (1.19) another set of solutions is given by the complex conjugates defined by three relations, such as

$$\hat{u}^*(x, y, t) = \mathbf{u}^*(y)e^{-i\alpha(x - c^*t)}. \tag{1.20}$$

This property stems from the fact that there are no complex quantities in the original equations (1.11), (1.12), and (1.13). If both $\hat{u}$ and $\hat{u}^*$ are solutions, we can construct the following purely real solutions:

$$\tilde{u} = \tfrac{1}{2}(\hat{u} + \hat{u}^*), \tag{1.21}$$

$$\tilde{v} = \tfrac{1}{2}(\hat{v} + \hat{v}^*), \tag{1.22}$$

$$\tilde{p} = \tfrac{1}{2}(\hat{p} + \hat{p}^*). \tag{1.23}$$

As an example, let us suppose that we have the solution

$$\hat{p} = (20y + i3y^2)\exp\{i7\alpha[x - (0.7 + i0.2)t]\}. \tag{1.24}$$

It leads to

$$\tilde{p} = \{20y\cos[7\alpha(x - 0.7t)] - 3y^2\sin[7\alpha(x - 0.7t)]\}e^{1.4t}. \tag{1.25}$$

Note that the imaginary part of c leads to an exponential growth in time, that the real part of $\mathbf{p}$ gives the amplitude of the cosine component, and that the imaginary part of $\mathbf{p}$ gives the amplitude of the sine component except for a minus sign.

The advantage of using complex quantities should now be evident; in general, an oscillation has an amplitude and a phase. This means that two numbers must be specified, and we may as well give the amplitude of the cosine and the amplitude of the sine components. This is just what the real and imaginary parts of a boldface quantity such as **p** can do. In a single symbol a complex quantity can express both the amplitude and the phase of a fluctuation.

Let us now supplement the example by assuming that

$$\mathbf{v} = 9y - i5y^2. \tag{1.26}$$

It leads to the result

$$\tilde{v} = \{9y \cos [7\alpha(x - 0.7t)] + 5y^2 \sin [7\alpha(x - 0.7t)]\}e^{1.4t}. \tag{1.27}$$

Now, the product $\tilde{p}\tilde{v}$ is often interesting because it represents the work of vertical pressure forces, per unit area and unit time. The averaged value over one wavelength can be obtained from (1.23) and (1.26) or

$$\overline{\tilde{p}\tilde{v}} = (90y^2 - 7.5y^4)e^{1.96t}. \tag{1.28}$$

The same result can be obtained directly from the original complex quantities. By using (1.22) and (1.23), substituting the exponentials of (1.20) and averaging along the x-axis, we can obtain the following convenient relation:

$$\overline{\tilde{p}\tilde{v}} = \tfrac{1}{4}(\mathbf{p}\mathbf{v}^* + \mathbf{p}^*\mathbf{v}). \tag{1.29}$$

The same method can be applied to mean square, and, for example, we have

$$\overline{\tilde{p}^2} = \tfrac{1}{2}\mathbf{p}\mathbf{p}^*. \tag{1.30}$$

It follows from (1.29) that the mean product of two quantities is zero if the two corresponding boldface quantities are orthogonal in the complex space.

2. Pressure, Vorticity, Energy

We now derive from the basic equations of motion certain special relations that will help us to understand the physical processes occurring in an oscillatory flow.

(a) THE PRESSURE EQUATION

The pressure appears in (1.2) and (1.3), and one might assume that this function could be arbitrarily selected or imposed from the outside, as a sort of potential. This is not true, because an arbitrary pressure distribution might create a velocity field that would violate (1.1); the fluid would accumulate in certain places and the density would not remain constant. In general, it takes great energy to alter the density. As long as the fluid velocity is much smaller than the speed of sound, the density remains constant and (1.1) is always satisfied. This imposes a restriction on the pressure fluctuations which can be formulated in the following sense. Consider the linearized equations (1.11), (1.12), (1.13). Differentiate (1.12) with respect to x and (1.13) with respect to y, add the two relations, and use (1.11) to simplify the results. The end product is

$$\tilde{p}_{xx} + \tilde{p}_{yy} = -2U_y\tilde{v}_x. \tag{2.1}$$

The complex form is

$$\mathbf{p}'' - \alpha^2\mathbf{p} = -2i\alpha U'\mathbf{v}. \tag{2.2}$$

This equation is similar to that of an elastic membrane loaded with some external force $\varphi(x, y)$. If the deflection of the membrane is η, the basic equation is

$$\eta_{xx} + \eta_{yy} = \kappa\varphi. \tag{2.3}$$

Thus just as the external force causes the deflection of a membrane, the product $U_y\tilde{v}_x$ is the source of pressure fluctuations. In the absence of any "source" the pressure obeys Laplace's equation and is determined solely by the boundary conditions.

(b) VORTICITY

In general the vorticity is a vector that indicates the rotation of a small mass of fluid with respect to the chosen coordinates. More precisely, let us consider a small surface element immersed in the fluid and define the circulation as the integral of the velocity along the perimeter. The circulation is equal to the flux of the vorticity through the surface; no vorticity, no circulation. In a two-dimensional flow the vorticity vector is always perpendicular to the xy-plane, and, because its orientation is fixed, the vorticity can be treated as a scalar quantity defined as

$$r = u_y - v_x. \tag{2.4}$$

A clockwise rotation gives a positive vorticity, but this is the consequence of an arbitrary definition. The vorticity of the laminar flow alone is

$$R = U_y, \tag{2.5}$$

and $\tilde{r}$, $\hat{r}$, and $\mathbf{r}$ can be defined by equations similar to (1.5) and (1.14). The vorticity fluctuations $\tilde{r}$ obey an important equation which can be obtained by differentiating (1.12) with respect to y, (1.13) with respect to x, and taking the difference. The pressure drops out and, after introducing (1.11) in several places, we obtain the result

$$\tilde{r}_t + U\tilde{r}_x + U_{yy}\tilde{v} = \nu(\tilde{r}_{xx} + \tilde{r}_{yy}). \tag{2.6}$$

The equivalent complex equation is

$$i\alpha(U - c)\mathbf{r} + U''\mathbf{v} = \nu(\mathbf{r}'' - \alpha^2\mathbf{r}). \tag{2.7}$$

This equation is analogous to that of heat conduction and diffusion in the presence of sources and sinks of heat. Indeed, in a copper sheet of constant thickness and unit specific heat, the temperature T obeys the following basic equation:

$$T_t = \kappa(T_{xx} + T_{yy}) + Q, \tag{2.8}$$

where κ is the coefficient of thermal diffusivity and $Q(x, y, t)$ is proportional to the rate of production or withdrawal of heat. If the copper is replaced by a layer of mercury moving with velocity $U(y)$, another term must be included to account for the convection of energy. The equation becomes

$$T_t + UT_x = \kappa(T_{xx} + T_{yy}) + Q. \tag{2.9}$$

In the absence of conductivity and production it has the solution $T(x - Ut, y)$, which shows that each particle of mercury retains the same temperature. In the comparison between (2.6) and (2.8) the term $U_{yy}\tilde{v}$ plays the role of Q, but it should be noted that $\tilde{v}$ is not completely independent of $\tilde{r}$, as is apparent from the definition of vorticity in (2.4).

(c) ENERGY

The kinetic energy per unit volume is $\frac{1}{2}\rho(u^2 + v^2)$. Because ρ is constant, it is convenient to define a kinetic energy E per unit mass as

$$E = \frac{1}{2}(u^2 + v^2). \tag{2.10}$$

By (1.5) and (1.6) it becomes

$$E = \frac{1}{2}U^2 + U\tilde{u} + \frac{1}{2}(\tilde{u}^2 + \tilde{v}^2). \tag{2.11}$$

The term $\frac{1}{2}U^2$ is the energy of the laminar flow in the absence of oscillations. The next term is the fluctuation of the energy caused by the oscillation and its average, when taken over one wavelength, is zero. The third term represents the energy of the oscillation, considered as a separate entity. Its mean value is given by

$$e = \tfrac{1}{2}\overline{(\tilde{u}^2 + \tilde{v}^2)} = \tfrac{1}{4}(uu^* + vv^*). \tag{2.12}$$

This quantity is a good indicator of the magnitude of the oscillation, and we now examine its behavior. By multiplying (1.12) by $\tilde{u}$, (1.13) by $\tilde{v}$, and adding and averaging, we obtain the following equation for the energy:

$$e_t + Ue_x = v(e_{xx} + e_{yy}) + \mu_1 + \mu_2 + \mu_3, \tag{2.13}$$

where the special terms are defined as

$$\mu_1 = -\overline{\tilde{u}\tilde{v}}\,U_y, \tag{2.14}$$

$$\mu_2 = -\overline{\tilde{u}\tilde{p}_x} - \overline{\tilde{v}\tilde{p}_y}, \tag{2.15}$$

$$\mu_3 = -v(\overline{\tilde{u}_x^{\,2}} + \overline{\tilde{u}_y^{\,2}} + \overline{\tilde{v}_x^{\,2}} + \overline{\tilde{v}_y^{\,2}}). \tag{2.16}$$

A positive μ_1 represents a transfer of energy from the mean flow to the oscillation, and this important effect needs some explanation. In the absence of an oscillation a particle located at a particular y has the momentum $U(y)$ per unit mass. Its neighbor, located at $y + \Delta y$, has the momentum $U + U_y\,\Delta y$. The oscillation can transport the particle from location y to location $y + \Delta y$, and if we disregard the effects of pressure gradients and viscous stresses the momentum should stay constant. In an oscillation the original momentum is $U + \tilde{u}$, where $\tilde{u}$ represents the perturbation of the momentum. In the new location at $y + \Delta y$ the mean momentum is larger by $U_y\,\Delta y$ and the perturbation is smaller by the same amount. This modification of the momentum perturbation is a simple matter of accounting. The total remains the same, but the portion attributed to the mean flow has increased. The transfer takes place in the time $\Delta t = \Delta y/\tilde{v}$, and therefore $\tilde{u}_t = -U_y\tilde{v}$. This is what is left of (1.12) if we drop the pressure forces, the viscous forces, and the transport term.

We can say that these changes of momentum are caused by fictional horizontal forces, proportional to $U_y\tilde{v}$, which represent the interaction between the mean flow and the oscillation. The power furnished by these forces is clearly $-\overline{\tilde{u}\tilde{v}}\,U_y$.

The term μ_2 corresponds to the loss of kinetic energy associated with a transfer to a region of higher pressure, or vice versa. The term μ_3 is always negative; it indicates the heat produced by viscous friction, per unit mass and unit time. This energy appears as heat and produces minute changes in temperature or density. As long as the flow is subsonic, these changes can be safely neglected.

The energy equation (2.12) is also an analog to the heat equation (or any conservation equation), but with transport by U, diffusion by v, and three different source terms.

3. Oscillations of a Uniform Stream

We now consider the case of a uniform stream with a velocity U_0 flowing between a wall at $y = 0$ and extending up to $y = \infty$. We assume that the wall moves with velocity U_0 or that it has some magic coating that makes slip possible. Any small oscillation must of course satisfy (1.17), (1.18), (1.19); but it is easier to begin our investigation with the pressure equation (2.2). With constant U it reduces to

$$\mathbf{p}'' - \alpha^2 \mathbf{p} = 0. \tag{3.1}$$

The integration is immediate, and with two arbitrary complex constants A and B the solution is

$$\mathbf{p} = Ae^{-\alpha y} + Be^{\alpha y}. \tag{3.2}$$

The term in A indicates a pressure fluctuation that dies out away from the wall, whereas the term in B diverges as $y \to \infty$. We exclude the possibility that something important occurs at very large distances from the wall and accordingly we impose the condition that $\mathbf{p} \to 0$ as $y \to \infty$. Thus B must be zero.

Let us now examine $\mathbf{u}$ from (1.18). We introduce the expression (3.2) for pressure and rearrange the terms to obtain

$$v\mathbf{u}'' - [i\alpha(U_0 - c) + v\alpha^2]\mathbf{u} = i\alpha Ae^{-\alpha y}. \tag{3.3}$$

This equation is of second order with constant coefficients and an inhomogeneous term on the right-hand side. According to classical methods, we begin by ignoring the right-hand side. The integration introduces two arbitrary complex constants C and D and leads to the homogeneous solution

$$\mathbf{u} = Ce^{-\alpha\beta y} + De^{\alpha\beta y}, \tag{3.4}$$

where the complex number β is defined as

$$\beta = \left(1 + i\frac{U_0 - c}{v\alpha}\right)^{1/2}. \tag{3.5}$$

At this stage we introduce the convention that the square root of a complex number shall always be the root that has a *positive* real part. If the real part is zero, we choose the root with positive imaginary part. With this convention, the term in C vanishes as $y \to \infty$ and the term in D diverges. Because we exclude large fluctuations far above the wall, we must take $D = 0$. If v is small, β is large, and the term in C dies out rapidly.

We must now complete our solution to solve (3.3) with the right-hand side included. This can be done by adding any particular solution to (3.4). A little guess-work shows that the following solution meets the requirements:

$$\mathbf{u} = -\frac{A}{U_0 - c} e^{-\alpha y}, \tag{3.6}$$

and the complete solution, with $D = 0$, becomes

$$\mathbf{u} = C e^{-\alpha \beta y} - \frac{A}{U_0 - c} e^{-\alpha y}. \tag{3.7}$$

We now turn our attention to $\mathbf{v}$. Because we already know $\mathbf{u}$, we can use the continuity equation (1.17). A single integration gives

$$\mathbf{v} = -i \frac{A}{U_0 - c} e^{-\alpha y} + \frac{i}{\beta} C e^{-\alpha \beta y} + F. \tag{3.8}$$

If we introduce this solution into (1.19) for $\mathbf{v}$, we find that the integration constant F must vanish.

The vorticity can be determined from the definition given in (2.4):

$$\mathbf{r} = \frac{\alpha(1 - \beta^2)}{\beta} C e^{-\alpha \beta y}. \tag{3.9}$$

Because the pressure contains only a term in A and the vorticity contains only a term in C, we can consider that the oscillation is a superposition of two independent contributions: a pressure wave and a vorticity wave. We examine these waves in some detail.

(a) The Stationary Pressure Wave

In order to look at the pressure wave alone, we take $A = p_0$, a real number, and $C = 0$. For convenience we also choose $c = 0$. From (3.2), (3.4), and (3.8) we then obtain the following expressions:

$$\tilde{p} = p_0 e^{-\alpha y} \cos (\alpha x), \tag{3.10}$$

$$\tilde{u} = -\frac{p_0}{U_0} e^{-\alpha y} \cos (\alpha x), \tag{3.11}$$

$$\tilde{v} = \frac{p_0}{U_0} e^{-\alpha y} \sin (\alpha x). \tag{3.12}$$

The corresponding flow is illustrated in Fig. 3.1, where we use the non-dimensional coordinates αx and αy. The thick arrows indicate the mean velocity U_0 and the circled positive or negative signs mark the pressure fluctuations. A large sign indicates a large pressure fluctuation. The thin horizontal arrows indicate the distribution of $\tilde{u}$ and the thin vertical arrows give that of $\tilde{v}$. In reality the amplitude of $\tilde{u}$ might be 1% of U_0 and therefore the arrows cannot be drawn to scale. The different widths are intended to convey this information. This figure is also made on the premise that p_0 is real, say $p_0 = 1$. A different value would modify the amplitude of all fluctuations. A choice such as $p_0 = e^{i\pi/5}$ would simply give the same pattern of fluctuations shifted to the right by the angle $\pi/5$.

At $y = 0$ we can imagine that $\tilde{v}$ is maintained by injection or suction of fluid through the wall. We can also imagine that the wall has a small waviness such that the slope of the wall matches that of the true streamlines. To satisfy a different condition at the wall, such as $\tilde{v} = 0$, we later consider a combination of pressure and vorticity waves.

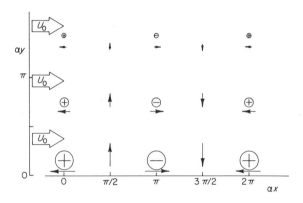

FIG. 3.1. Pressure wave in constant mean flow, with $c = 0$. The circled signs indicate the pressure fluctuations, the thin arrows indicate the velocity fluctuations.

We note in Fig. 3.1 that the low pressures occur in the regions in which $\tilde{u}$ is parallel to U_0. This is a manifestation of the theorem of Bernoulli: in an inviscid flow the sum of the pressure and the kinetic energy remain constant along a streamline. To draw the streamline it is necessary to choose a practical value of p_0 such as $0.1 U_0^2$. The qualitative results are shown in Fig. 3.2. We can also disregard the mean flow and draw the streamlines of the perturbing flow alone. These lines are shown in Fig. 3.3. They indicate a series of ascending and descending jets.

The displacement $h(x)$ of the wall from the line $y = 0$ should be given, if $c = 0$, by the relation

$$\frac{dh}{dx} = \frac{\tilde{v}(x, 0)}{U_0}.$$ (3.13)

(b) The Stationary Vorticity Wave

In order to look at the vorticity wave alone, we take $A = 0$ and $C = r_0/\alpha$. For simplicity we also choose $c = 0$. We assume that v is small enough that β from (3.5) can be approximated as $\beta = (1 + i)g$ where the factor g is

$$g = \left(\frac{U_0}{2v\alpha}\right)^{1/2}.$$ (3.14)

These expressions lead to the following flow:

$$\tilde{u} = \frac{r_0}{\alpha} e^{-\alpha\beta y} \cos\left[\alpha(x - gy)\right],$$ (3.15)

$$\tilde{v} = -\frac{r_0}{2g\alpha} e^{-\alpha\beta y} \cos\left[\alpha(x - gy) + \frac{\pi}{4}\right],$$ (3.16)

$$\tilde{r} = -gr_0 e^{-\alpha\beta y} \cos\left[\alpha(x - gy) + \frac{\pi}{4}\right].$$ (3.17)

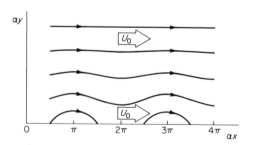

FIG. 3.2. The true streamlines of a pressure wave.

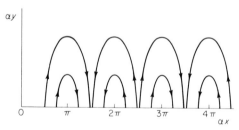

FIG. 3.3. The streamlines of the perturbation only, for a pressure wave.

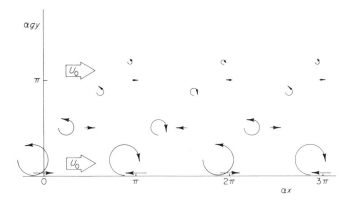

FIG. 3.4. Stationary vorticity wave. Note magnification of vertical scale.

The resulting motion is shown in Fig. 3.4, in which the curved arrows indicate the magnitude and direction of the vorticity. The scale of y is larger than the scale of x by a factor g; in the real physical space the entire flow would occur in a layer whose thickness is g times smaller than the wavelength. Note that the positive vorticity is found in streaks, inclined at an angle $dy/dx = 1/g$. Referring to the analogy between the vorticity equation and the heat equation, we can say that vorticity is injected into the flow at points such as $\alpha x = 3\pi/4$, $y = 0$, and that it diffuses away from the wall while being transported by the mean flow. There is no production of vorticity in the flow. The vorticity disappears by diffusion between the regions of opposite signs.

This injection of vorticity could be realized by using magnetic fields and electric currents. It could not be done by potential forces. A horizontal velocity fluctuation could be imposed at the wall, but it should be supplemented by a small vertical velocity to generate the pure vorticity wave.

The time to travel a half wavelength is $\pi/\alpha U_0$, and a diffusion process will spread over a distance Δy in a time of the order of $\Delta y^2/\nu$. Combining these expressions, we find that at half a wavelength from the source the spread should be roughly $\Delta y = (\nu/\alpha U_0)^{1/2}$, in accordance with our estimate of dy/dx.

Because $\tilde{v}$ is smaller (by a factor g), we did not show it in Fig. 3.4. In the expression for the vorticity we find that $\tilde{u}_y$ is larger than $\tilde{v}_x$ by a factor g. Therefore, knowing the vorticity, we can obtain $\tilde{u}$ by integrating $\tilde{r}$ along a vertical line. If the regions in which $\tilde{r}$ is found positive were vertical instead of oblique (as seen in Fig. 3.4), $\tilde{u}$ would be negative inside these regions. The oblique distribution of the vorticity introduces contributions of varying sign in the integration that gives $\tilde{u}$ when proceeding along a vertical line. Because the magnitude of the vorticity fluctuations varies exponentially with y, it is not surprising to find that $-\tilde{u}$ is controlled by the nearest amount of vorticity. This can be observed in Fig. 3.4.

(c) THE NONSTATIONARY CASES

If c is some purely real number, we find the same pressure and vorticity waves, except that $U_0 - c_r$ appears in place of U_0 and $\alpha(x - c_r t)$ in place of αx. Thus the flow patterns are not different from Figs. 3.1 and 3.4, but they move to the right with phase velocity c_r. An observer moving with this velocity c_r would observe the oscillations of a uniform stream of velocity $U_0 - c_r$.

If c is complex, we can dispose of the real part by using a frame of reference moving with velocity c_r, but the presence of c_i introduces major changes. Let us suppose that c_i is positive; then all components of the perturbation grow exponentially. Furthermore, $\tilde{u}$ will no longer be exactly in phase with $\tilde{p}$, as can be seen from (1.18) or (3.11), where U_0 must be replaced by $U_0 - ic_i$. The pressure fluctuations are no longer out of phase with $\tilde{v}$, and this means that the pressure forces must now work. This is necessary to maintain the exponential growth of the kinetic energy.

In general, the pressure wave can be viewed as the perturbation introduced in a uniform stream by waviness of the wall. If the wall amplitude increases exponentially, so does the perturbation. The extra kinetic energy is furnished by whatever mechanism is deforming the wall.

The presence of c_i affects the vorticity wave by modifying g. This affects the phase of the vorticity distribution in a complicated manner. Essentially, the perturbation is the result of the injection of vorticity disturbances at the wall as if someone agitated the fluid with a rotating device near $y = 0$. If the amplitude of these disturbances increases exponentially with the time, the vorticity found far from the wall will be weaker than in the stationary case, because it was generated long ago. Thus, if c_i is positive, the velocity will be related to the vorticity adjacent to the wall even more closely than in the stationary case.

4. Oscillations of Two Adjacent Uniform Streams

We now treat the case of two parallel flows, separated by an intermediate layer of fluid, and we neglect the friction forces. The first flow extends from $y = H$ to $y = \infty$ and proceeds with constant velocity U_1. The

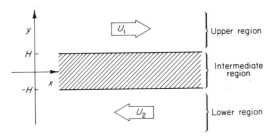

FIG. 4.1. Two uniform streams.

second flow extends from $y = -H$ to $-\infty$ with velocity $-U_2$, as seen in Fig. 4.1. We say as little as possible in this section about the flow in the intermediate region and postpone the question to Section 5. The subscript 1 is associated with the upper flow, and 2 with the lower flow.

The treatment of the upper flow is exactly the same as that of a uniform stream in Section 3. We make one minor change: instead of using the integration constant A, as in (3.2), we use the integration constant $Ae^{\alpha H}$, with the result that the pressure fluctuations are now given by

$$\mathbf{p}_1 = Ae^{-\alpha(y-H)}. \tag{4.1}$$

At $y = H$ this gives a simple expression for $\mathbf{p}_1$. Because we neglect the viscous terms, the general equations (1.18) and (1.19) lead directly to the other fluctuations, with the result

$$\mathbf{u}_1 = \frac{-A}{U_1 - c} e^{-\alpha(y-H)} \tag{4.2}$$

$$\mathbf{v}_1 = \frac{-iA}{U_1 - c} e^{-\alpha(y-H)}. \tag{4.3}$$

In the lower region the general expression for pressure fluctuations is a sum of two exponentials [see (3.2)]. Our attention is directed at oscillations involving the two streams and the intermediate region, and we do not want to consider oscillations that get very large far below the intermediate layer. Thus we require $\mathbf{p}_2$ and the other fluctuations $\mathbf{u}_2$ and $\mathbf{v}_2$ to vanish as $y \to -\infty$. In place of B we substitute $Be^{\alpha H}$ and finally the oscillations of the lower stream are given by

$$\mathbf{p}_2 = Be^{\alpha(y+H)}, \tag{4.4}$$

$$\mathbf{u}_2 = \frac{B}{U_2 + c} e^{\alpha(y+H)}, \tag{4.5}$$

$$\mathbf{v}_2 = \frac{-iB}{U_2 + c} e^{\alpha(y+H)}. \tag{4.6}$$

We now consider two possible relations between the upper and lower solutions.

(a) A NEUTRAL OSCILLATION

Let us search for a neutral solution; that is, let us write $c_i = 0$. For simplicity we also assume that the streams have equal but opposite mean velocities; thus $U_1 = U_2$. If the intermediate layer is thin, it cannot support a pressure difference and we can write with reasonable certainty that

$$\mathbf{p}_1(H) = \mathbf{p}_2(-H). \tag{4.7}$$

This leads to the relation $A = B$ from (4.1) and (4.4) with the consequence that $\mathbf{v}_1(H)$ and $\mathbf{v}_2(-H)$ are also equal, which is encouraging, for it means that the fluid simply traverses the intermediate layer. The matching of the pressure means also that $\mathbf{u}_1(H) = -\mathbf{u}_2(-H)$, which indicates that the intermediate layer must provide the transition between these fluctuations. A sketch of the oscillation is shown in Fig. 4.2 for a real value of A.

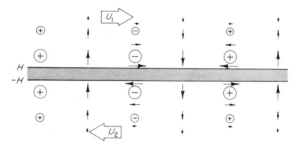

FIG. 4.2. A neutral oscillation, with an intermediate region.

Let us now discuss the streamlines near the boundary. We could inject a filament of ink or smoke near $y = H$, and, as the flow oscillates, it will form a line at some height $\tilde{h}(x, t)$. At first, we might be tempted to say that the variations of $\tilde{h}$ are caused by the $\tilde{v}$ component and write $\partial \tilde{h}/\partial t = \tilde{v}$. However, in Fig. 4.2, with a stationary flow, this relation would lead to $\tilde{h} = At \sin(\alpha x)$: for infinite t it gives an infinite $\tilde{h}$! The fault is that we have not yet accounted for the fact that the ink moves with the fluid velocity. In the linearized approximation this leads to the equation

$$\tilde{h}_t + U\tilde{h}_x = \tilde{v}. \tag{4.8}$$

With this equation we find that in a neutral oscillation the maxima of $\tilde{h}$ occur after the particles have traversed a region in which $\tilde{v}$ is ascending.

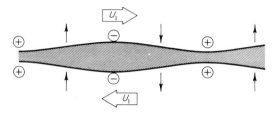

FIG. 4.3. Displacements for a neutral flow, with matched pressure fluctuations.

The locations of two filets mark the edges of the intermediate layer shown in Fig. 4.3; although the $\tilde{v}$ are matched through the layer, the difference in mean flow produces a periodic thickening of the intermediate layer. Clearly this solution is inacceptable in the limit in which $H \to 0$, and we can expect that finite intermediate layers will have some structure.

(b) THE THIN INTERMEDIATE REGION

Because the displacements of the boundaries of the intermediate region can cause difficulties, let us examine the consequences of the following pair of assumptions: (a) the pressures are matched through the intermediate layer and (b) the displacements are also matched. As before, the pressure condition gives $A = B$, and from (4.8) we now have

$$\tilde{h}_{1t} + U_1 \tilde{h}_{1x} = \tilde{v}_1 \quad \text{at} \quad y = H, \tag{4.9}$$

$$\tilde{h}_{2t} + U_2 \tilde{h}_{2x} = \tilde{v}_2 \quad \text{at} \quad y = -H. \tag{4.10}$$

After some manipulations the second condition leads to

$$\frac{\tilde{h}_1}{\tilde{h}_2} = -\left(\frac{U_2 + c}{U_1 - c}\right)^2 = 1. \tag{4.11}$$

This means that c is no longer arbitrary and the above relation yields

$$c = \frac{U_1 - U_2}{2} \pm i \frac{U_1 + U_2}{2}. \tag{4.12}$$

Thus there are two solutions, one exponentially growing with t and the other decaying. A discussion of initial conditions reveals that, except for extraordinary cases, the growing solution becomes dominant and the flow is unstable. The values of $\tilde{v}(H)$ can also be determined, and we find

$$\frac{v_1(H)}{v_2(H)} = \pm i. \tag{4.13}$$

There is, therefore, a phase difference of $\pi/2$ between the $\tilde{v}$ fluctuations across the intermediate layer, but it is perfectly acceptable. The flow is

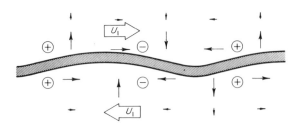

FIG. 4.4. The Helmholtz instability.

sketched in Fig. 4.4 for $A = 1$ and $U_1 = U_2$. This motion has been well known since its first description by Helmholtz in 1868.

(c) REMARKS ON VISCOUS EFFECTS

If we examine the effects of viscosity, we find that vorticity waves could exist in the upper and the lower flows, vanishing far away from the intermediate region. This would allow a more sophisticated matching of the flows. In exponentially growing oscillations [sign plus in (4.12)] a small viscosity would add only a small amount of damping. In damped oscillations, however, one finds that a small viscosity has a different effect. Let us suppose that the initial conditions are such that the motion will decay exponentially [sign minus in (4.12)]. The vorticity is zero except in the thin intermediate region. By viscous diffusion some vorticity will spread outside the intermediate region, where it can linger long after the main oscillation has died out. Thus a small viscosity may have only a small effect on a flow. If the flow is unstable, this effect remains unimportant. However, if the inviscid solution decays, there may be a time when everything will have vanished except for the consequences of viscosity. This means that the inviscid decaying solution can be observed only for a limited time and after some very special initial conditions. In the long run, it is masked by viscous effects.

5. The Inviscid Shear Layer

We now penetrate a little deeper into the real difficulties of hydrodynamics by studying the stability of a shear layer between two parallel streams. For this purpose it is convenient to specify a particular mean velocity profile, and we select the function

$$U = U_0 \tanh \frac{y}{H}. \tag{5.1}$$

The free-stream velocities are $\pm U_0$ and H is a measure of the thickness of the layer. At this point we could define nondimensional velocities and lengths, but it is simpler to imagine that we are using a system of physical units in which H happens to be the unit of length and U_0, the unit of velocity. Thus the unit of time would be H/U_0. This leaves no further adjustable quantities as long as masses are not involved. If we find $\alpha = 0.8$ or $c_i = 0.1$, it will simply mean that in any other systems of units we have the nondimensional relations $\alpha H = 0.8$, $c_i = 0.1 U_0$.

For simplicity we neglect viscous effects in this section, and, with $v = 0$, the basic equations reduce to the following second-order system:

$$\mathbf{v}' = -i\alpha\mathbf{u}, \tag{5.2}$$

$$\mathbf{p}' = -i\alpha(U - c)\mathbf{v}, \tag{5.3}$$

$$i\alpha(U - c)\mathbf{u} + U'\mathbf{v} + i\alpha\mathbf{p} = 0. \tag{5.4}$$

It should be noted that the inviscid assumption also permits us to omit any concern of the mean flow not being parallel, that is U can be taken as a function of y only.

Because the pressure equation (2.2) contains a term on the right-hand side, we can no longer start from a simple Laplace equation.

The pressure can be eliminated by differentiating (5.4) and substituting it into (5.3). The result is the following system:

$$\mathbf{v}' = -i\alpha\mathbf{u} \tag{5.5}$$

$$\mathbf{u}' = \frac{i}{\alpha}\left(\frac{U''}{U - c} + \alpha^2\right)\mathbf{v}. \tag{5.6}$$

By differentiating (5.5) and eliminating $\mathbf{u}$ we can obtain a more compact equation:

$$\mathbf{v}'' = \left(\frac{U''}{U - c} + \alpha^2\right)\mathbf{v}. \tag{5.7}$$

This important equation governs the stability of parallel inviscid flows and was first obtained by Rayleigh (1880). From the solution of this equation **u** and **p** can be derived according to (5.5) and (5.4). Because these equations contain coefficients that vary with y, the solutions can no longer be spelled out as a sum of exponentials, and it becomes advantageous to use a digital computer. We next discuss some of the general computing practices with applications to the inviscid shear layer.

(a) THE PROGRAM

Essentially, a digital computer can store numbers, perform the four operations of arithmetic, and display information. By using power series, special routines, and procedures, the computer can also extract roots, decide which number is the largest, and perform logical tasks. A program of operations must be prepared by a specialist, but certain simplified techniques such as FORTRAN can be learned in a matter of days. The programmer makes abundant use of certain subroutines, which are little programs of limited scope but frequently called upon: interpolation between entries of a table, integration, and printing or even plotting data.

In general, the digital computer deals strictly with real numbers, and it is necessary to separate each equation into its real and imaginary parts. (Certain subroutines which handle complex numbers are available; in effect, they manipulate pairs of numbers according to the appropriate rules.)

The computer uses all numbers between extremes, such as $\pm 10^{38}$ and $\pm 10^{-38}$. If the upper limit is crossed, the computer stops and announces OVERFLOW. In the typical computer each number is known to eight decimal places, but various sources of error can gradually contaminate the results, so that there are only five or six significant figures. Roughly, each operation performed on a number introduces an error of the order of 10^{-8}. If a number is processed 10^6 times (which might take 10 seconds of machine time) and the errors have random signs, they will combine quadratically and the accuracy will drop to five significant figures.

A pair of eight-digit numbers can be used to form a single number of sixteen digits. Such pairs can be processed to form the sums or products of the corresponding single numbers. These operations are collectively known as the method of double precision. The basic error is reduced to 10^{-16} but the computing time is more than doubled.

A theoretical physicist will consider (5.7) as the most compact statement of the problem and ask for a program. The person in charge of writing the program will recognize the need for dealing with $\mathbf{v}'$ as well as with $\mathbf{v}$ and $\mathbf{v}''$. Because the computer performs one integration at a time, the programmer will write the equations

$$\mathbf{v} = \int \mathbf{v}'\, dy, \tag{5.8}$$

$$\mathbf{v}' = \int \mathbf{v}''\, dy = \int \left(\frac{U''}{U - c} + \alpha^2\right)\mathbf{v}\, dy. \tag{5.9}$$

This is equivalent to the integrals of (5.5) and (5.6). In a simple problem the difference is trivial, but in more complicated oscillations there is no need for compact high-order equations because the programmer will always regress and cast a system of first-order equations. Thus the system of equations (5.5) and (5.6) is ready for numerical integration, whereas (5.5) is perhaps more convenient for the human memory.

The numerical integration must, of course, start at some value of y with some specified values of $\mathbf{v}$ and $\mathbf{v}'$ [which by (5.5) is equivalent to a specification of $\mathbf{u}$]. The computer proceeds in steps Δy on the assumption that every function can be approximated between y and $y + \Delta y$ by a polynomial of degree n. For $n = 1$ we have Newton's integration method. For $n = 2$ we have the method introduced by Euler. With modern computers it is common practice to choose $n = 4$ and to use a method due to Runge and Kutta.

The step size can either be prescribed and kept constant or it can be selected by the computer itself to maintain some error-barometer within a prescribed safety zone. Because the computer operates in the binary system, some accuracy can be gained by choosing $\Delta y = 2^{-M}$, where M is an arbitrary integer. This trick eliminates the errors in the decimal to binary and binary to decimal conversions.

(b) Testing the Program

Let us suppose that a program has been prepared to solve (5.7). The probability that the computer will fail to follow the instructions exactly and faithfully is quite small, but it is quite probable that the first version of the program will contain at least one error of human origin. It is strongly advisable to perform a systematic verification before engaging in actual work. For instance, we can first take $U'' = 0$ and verify the α^2 term in (5.7). Then we can take a very small α, a constant U, and a constant U''; or, with $U'' = 2$, $c = 0$, $U = y^2$, we could verify that $\mathbf{v} = 1/y$ is a solution. Note that for the purpose of verification U'' is considered as an arbitrary function, not necessarily related to U. This requires a certain flexibility of the program, such as a table of inputs for U and U'' or some options. As a test of the inter-action between real and imaginary parts, we could take $U'' = 2$, $c = i$, and $U = 0$, with a very small α. One of the solutions should be $e^{(1+i)y}$. The reason for taking a very small α, rather than 0, is that if the program includes a calculation of $\mathbf{u}$ by (5.5) the result would be OVERFLOW.

Sometimes a particular analytic solution is known and the accuracy of the program can be evaluated. Thus, with $\alpha = 1$ and $c = 0$, (5.7) admits the solution $\mathbf{v} = \cosh 3y/\cosh y$. Starting at $y = -3$, with $\mathbf{v} = \mathbf{v}' = 1$, the computer advances by steps $\Delta y = 0.01$ and finds $\mathbf{v}(+3) = 1.005$ in place of 1, together with $\mathbf{v}'(+3) = -0.995$ in place of -1. These errors are probably due to the fact that we used a table for U and U'' with 18 entries between $y = 0$ and $y = 3$ in four significant figures.

This result is encouraging, but it does not exclude the possibility of an error involving the imaginary part of the solution! Every wheel of the machine must be tested once.

For complicated programs it is helpful to insert special coefficients, all equal to unity in the actual performance of the calculations. By setting these coefficients to zero it becomes possible to perform numerous veri-fications. Thus every wheel of the machine can be tested independently and at least once.

(c) An Example of Computations

Let us now examine the problem. Outside the range $-3 < y < 3$ the mean velocity is nearly constant and the pressure is given by

$$\mathbf{p} = Ae^{-\alpha y} + Be^{\alpha y}. \tag{5.10}$$

This corresponds to

$$\mathbf{v} = \frac{-iA}{U - c} e^{-\alpha y} + \frac{iB}{U - c} e^{\alpha y}. \tag{5.11}$$

We are not interested in perturbations that become large far away from the layer, and therefore we specify that B must vanish above the layer and A must vanish below. This takes care of the solution outside the shear region in which the coefficients vary.

Starting from below, we choose $B_2 = -i(1 - c)e^{-3\alpha}$ because it leads to the simple initial values $\mathbf{v}(-3) = 1$ and $\mathbf{v}'(-3) = \alpha$. Another choice of B_2 would alter the amplitude and phase of the entire solution by constant

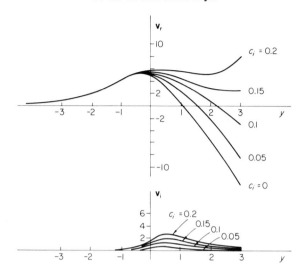

FIG. 5.1. Tentative numerical solutions for the inviscid shear layer: $\alpha = 0.8$, $c_r = 0$, $U = \tanh y$.

factors. It is now necessary to choose some numerical values, and we shall try our chances with $\alpha = 0.8$ and $c = 0$.

The computer starts from $y = -3$ and proceeds to $y = +3$. The resulting functions $\mathbf{v}_r$ and $\mathbf{v}_i$ are shown in Fig. 5.1. Clearly, $\mathbf{v}_r$ does not decay exponentially in the region $y > 3$ and something is wrong. As we try other values of c_i, such as 0.05, 0.1, 0.15, and 0.2, we notice that the asymptotic behavior of $\mathbf{v}_r$ changes; there is some hope of matching the conditions at $y = 3$ if c_i is in the vicinity of 0.15.

The corresponding curves for $\mathbf{v}_i$ are also shown and they offer the same hope. It now remains to find the correct value of c_i. In general, the value of c, which complies with boundary conditions, is called an eigenvalue.

If $y > 3$ and $B = 0$, then $\mathbf{v}$ must fall exponentially. We can always determine the complex constant A from the value of $\mathbf{v}$ given by the computer at $y = 3$, but this will not guarantee the continuity of the derivative. In order to determine how closely the computed solution matches the exponential solution, we instruct the computer to print out the ratio F, defined as

$$F = \frac{-\mathbf{v}'(3)}{\alpha \mathbf{v}(3)}. \tag{5.12}$$

If $\mathbf{v}$ matches the parallel stream solution, we must have $F = 1$. If $\mathbf{v}$ matches the diverging solution, F should be -1. The values of F are shown

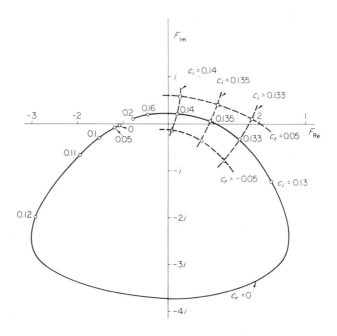

FIG. 5.2. Search for an eigenvalue giving $F(c) = 1$, $\alpha = 0.8$,
$U = \tanh y$, $F = -\mathbf{v}'(3)/\alpha\mathbf{v}(3)$.

in Fig. 5.2 for $\alpha = 0.8$ and a variety of values of c. The round dots correspond to $c_r = 0$, and we find the eigenvalue $c_r = 0$, $c_i = 0.1345$. If we give c_r some small values, we obtain the squares or the triangles and the target is missed.

The eigenvalue for c corresponds to the eigenfunction $\mathbf{v}$ shown in Fig. 5.3. In practice, each step of integration requires about 100 operations; there are four functions, $\mathbf{v}_r$, $\mathbf{v}_i$, $\mathbf{v}_r'$, $\mathbf{v}_i'$, and 600 steps. At the rate of 100 μsec per operation the computer takes about 2.5 seconds and the cost is about that of a pack of cigarettes.

Two remarks are now appropriate. If we take the complex conjugate of c, the new solution should be the complex conjugate of the function shown in Fig. 5.3. Indeed, U, U'', and α are real, so that $\mathbf{v}$ and c are the only quantities affected. However, this new solution is damped, and there may be some effects of the lingering vorticity, even in the limit of vanishing viscosity. The seemingly anomalous result obtained by taking the complex conjugate is sometimes termed a paradox in stability theory and we will have occasion to comment on it often. Strictly speaking, the answer is found when we seek to ascertain whether or not the inviscid solutions of the Rayleigh equation (5.7) are solutions of the complete viscous problem in the limit as the viscosity tends to zero. An analysis of these features has been best put forward by the work of Morawetz (1952).

We also remark that in Fig. 5.2 the curves giving F for some constant c_i seem to be orthogonal to the curves taken at constant c_r. This should not be a surprise, for F is a complex function of the complex variable c. Verify that near the eigenvalue a rough evaluation produces $dF/dc = 4 + i8$ for a real and/or an imaginary Δc.

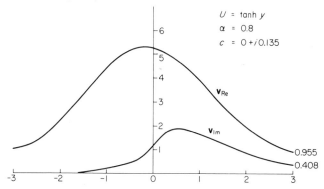

FIG. 5.3. Example of eigenfunction: $U = \tanh y$, $\alpha = 0.8$, $c = 0 + i(0.135)$.

(d) COMPLETE CALCULATIONS

Calculations for other values of α between 0 and 1 follow the same procedure. All eigenvalues are found with $c_r = 0$, and the values of c_i are shown in Fig. 5.4. In the same figure we indicate the product αc_i which

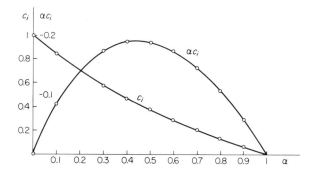

FIG. 5.4. Eigenvalues and amplifications for a shear layer: $U = \tanh y$.

corresponds to the rate of growth of the oscillation. A maximum rate of growth is found near $\alpha = 0.45$. Thus the inviscid shear layer is unstable in the range $0 < \alpha < 1$. If we examine $\alpha = 1.05$ for any small value of c_i, we find that $\mathbf{v}_r$ stays positive and will eventually go to $+\infty$. This veering away from the horizontal axis becomes more pronounced if we increase either c_r or c_i. The conclusion is simply that there are no eigenvalues for $\alpha > 1$ and that somehow the viscosity should not be ignored. The eigenfunctions for various values of α are plotted in Fig. 5.5 in the complex plane. The curve identified as $\alpha = 0.6$ and $\mathbf{v}(-3) = 1$ corresponds to the results of Fig. 5.3. Because it has an obvious symmetry about the origin, we rotate $\mathbf{v}$ in such a way that $\mathbf{v}_i$ becomes maximum at the origin. This gives the other curve labeled with $\alpha = 0.6$ and shows that, with the proper phase, $\mathbf{v}_r$ can be asymmetric and $\mathbf{v}_i$ can be symmetric. Rotation by another angle $\pi/2$ can reverse this result. Other rotated eigenfunctions are shown in Fig. 5.5, which all have the property that the magnitude of $\mathbf{v}$ is unity at $y = -3$.

This symmetry of the solution could have been inferred from the general nature of (5.1) and (5.7). We could also have predicted that c_r must vanish (Tatsumi and Gotoh, 1960), but we preferred to describe a general approach to inviscid stability problems, which can be applied to a broad variety of velocity profiles.

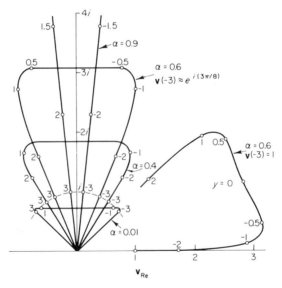

FIG. 5.5. A family of eigenfunctions: $U = \tanh y$, $c_r = 0$.

(e) The Vorticity Equation

The vorticity fluctuations can be obtained as a by-product of the calculations. For $\alpha = 1$ the vorticity fluctuations are maximum at $y = 0$ and exactly out of phase with the pressure. If we refer to the vorticity equation (2.7), we see that, with $c = 0$ and $v = 0$, the vorticity is directly proportional to $\mathbf{v}$, with a factor U''/U. This factor is negative through the entire shear layer, even at $y = 0$, where L'Hôpital's rule must be applied. For $\alpha < 1$ the situation is different because c_i is not zero. The vorticity is now given by

$$\mathbf{r} = \frac{U''(-c_i + iU)}{U^2 + c_i^2}\,\mathbf{v}. \tag{5.13}$$

Because $\mathbf{v}$ is almost constant near $y = 0$, the vorticity is zero at $y = 0$ and has peaks on each side near $y = \pm\,|c_i|$. Thus the unstable solution has two regions of concentrated vorticity in every half wavelength, which merge into one when c_i becomes zero. This merging is associated with steep vorticity gradients (Michalke, 1964), and a small amount of viscosity is likely to produce serious changes. Thus the viscous problem may have eigenvalues if $\alpha > 1$ with $c_i < 0$. [See Section 13(d).]

(f) Algebraic Methods

A method introduced by Rayleigh (1880) leads to interesting results. He approximated the shear profile by the following polygonal contour:

$$U = \begin{cases} 1 & y > 1, \\ y & -1 < y < 1, \\ -1 & y < -1. \end{cases} \tag{5.14}$$

Between the edges located at $y = \pm 1$, U'' vanishes and the solution of (5.7) can be given as a sum of two exponentials. However, at the edges U'' becomes infinite and precautions are necessary. We can assume that U'' is large but finite within the small layer $y = 1 \pm \varepsilon$ and that $\mathbf{v}$ will remain almost constant within this layer. Integration of (5.7) shows that the value of $(U - c)v' - U'v$ at $y = 1 - \varepsilon$ must be equal to the value at $y = 1 + \varepsilon$. This leads to the simple expression $c^2 = (1 - 1/2\alpha)^2 - (1/4\alpha^2)e^{-4\alpha}$, which agrees qualitatively with Fig. 5.4. The unstable waves occur in the range $0 < \alpha < 0.64$. For larger values of α this method gives real values of c, tending to ± 1 as $\alpha \to \infty$. These real eigenvalues are encountered in many other applications of the algebraic method, when the curvature of the mean

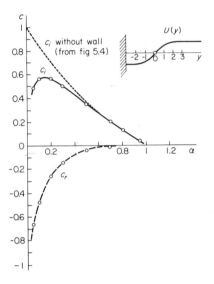

Fig. 5.6. Eigenvalues for a shear layer in the vicinity of a wall. The viscosity is disregarded.

profile is concentrated at a discrete set of points. Indeed, for large α each angle of a profile has its local oscillation. The special profile, which has a single corner and two constant slopes U_1' and U_2', leads to the eigenvalue

$$c = U_0 - \frac{U_1' - U_2'}{2\alpha}. \tag{5.15}$$

Thus, as $\alpha \to \infty$, a solution exists which is traveling with the corner speed. These eigenvalues do not appear when continuous mean profiles are employed.

Note that the production of vorticity is concentrated near the edges of the flow, whereas $U = \tanh y$ has maximum vorticity production near the center line. This case, and that of more refined polygonal contours, has been discussed by Michalke and Schade (1963).

(g) THE EFFECT OF A WALL

We now examine the effect of a wall placed at $y = -3$. Thus the solution of (5.7) is still proportional to $e^{-\alpha y}$ in the free stream located above the shear layer for $y > 3$. However, at $y = -3$ we impose the boundary condition $\mathbf{v} = 0$, which indicates that the fluid cannot pass through the wall. Starting with a set of tentative values for c and a specific value of α, the computer proceeds from $y = 3$ to $y = -3$ and indicates $\mathbf{v}(-3)$. A map of $\mathbf{v}(-3)$, similar to the map of F (see Fig. 5.2), soon indicates the appropriate value of c. These eigenvalues, which satisfy the condition at the wall, have been collected in Fig. 5.6. For $\alpha > 0.5$ the results are not very different from those obtained without a wall. For lower values of α the wall reduces the values of c_i and the phase velocity takes negative values. The results for $\alpha > 0.5$ are not too surprising, for we have seen that the eigenfunctions for such wavenumbers have only a small amplitude into the free stream on either side. For low wavenumbers, however, the free-stream oscillation plays an essential role, mainly in the build up of pressure fluctuations.

In a plot of αc_i the presence of the wall is not so apparent as it is in a plot of c_i, the maximum occurring for a slightly larger value of α, as shown in Fig. 5.4.

6. Inviscid Jets and Wakes

In this section we consider the stability of more complicated flows such as jets and wakes. We rely primarily on the computer, but the algebraic method mentioned in Section 5 is also used. Moreover, a comparison will show that the eigenvalues of the algebraic method are surprisingly good. This agreement does not extend to the eigenfunctions.

(a) SYMMETRIC JET

As a simple example of a symmetric jet, let us consider

$$U = \mathrm{sech}^2 \, y = \frac{1}{\cosh^2 y}, \tag{6.1}$$

where the maximum velocity is unity and U is almost zero if $|y| > 3$. The equation governing the oscillations (Section 5) is, when considering the inviscid problem:

$$\mathbf{v}'' = \left(\frac{U''}{U - c} + \alpha^2 \right) \mathbf{v}. \tag{6.2}$$

The method of integration is the same as in Section 5.

This profile has the advantage that three particular solutions of (6.2) have been found. Let us examine the possibility of a solution in the form of

$$\mathbf{v} = (\cosh \beta y)^{-m}, \qquad m \geq 0. \tag{6.3}$$

Some calculations lead to the equation

$$m^2\beta^2 - m(m+1)\beta^2(\cosh \beta y)^{-2} = -6\cosh^{-2}y\left(\frac{1 - \frac{2}{3}\cosh^2 y}{1 - c\cosh^2 y}\right) + \alpha^2. \quad (6.4)$$

This leads to a solution with

$$c = \tfrac{2}{3},$$

$$\beta = 1,$$

$$m = \alpha = 2,$$

which is compatible with the boundary conditions. Another possibility exists for this choice when $c = 0$, $\beta = 1$, $m = 2$, and $\alpha = 0$.

A second choice could be made by writing

$$\mathbf{v} = \frac{\sinh \beta y}{(\cosh \beta y)^2}. \quad (6.5)$$

Substitution and further calculations lead to

$$\beta^2 - 6\beta^2 \cosh^{-2}\beta y = -6\cosh^{-2}y\left(\frac{1 - \frac{2}{3}\cosh^2 y}{1 - c\cosh^2 y}\right) + \alpha^2. \quad (6.6)$$

A solution exists if

$$c = \tfrac{2}{3},$$

$$\beta = \alpha = 1.$$

Note that these two particular solutions are neutral and that the wave velocity is equal to the velocity at the points of inflection of the profile. Indeed, at the height $h = 0.662$ we have an inflection point at which $U''(h) = 0$ and $U(h) = \tfrac{2}{3}$. For a neutral solution this is not surprising, for $U - c$ may vanish for some value of y, in which case $\mathbf{v}''$ becomes infinite unless U'' or $\mathbf{v}$ also vanishes.

The computer can now be started from values of α and c close to any one of the three solutions given, such as $\alpha = 1.9$, $c_i = 0.02$, and $c_r = 0.66$, and the search in c leads to new eigenvalues. As long as c_i is finite and the step size Δy is small enough, no difficulty is encountered. The results are shown in the traditional manner in Fig. 6.1. Another way of plotting the same points is given in Fig. 6.2.

The eigenfunctions $\mathbf{v}$ are symmetric for curve I and antisymmetric for curve II. It is worth noticing that if $0 < \alpha < 1$ the jet can oscillate in two distinctly different ways. We say that curves I and II correspond to two

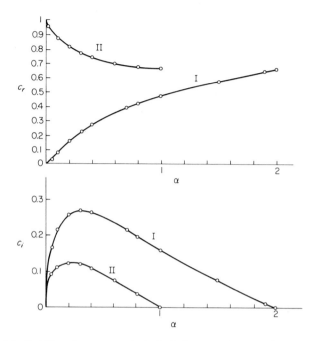

FIG. 6.1. Stability of a symmetric inviscid jet: $U = (\cosh y)^{-2}$ on curve I, $\mathbf{v}$ is symmetric; in curve II, $\mathbf{v}$ is antisymmetric.

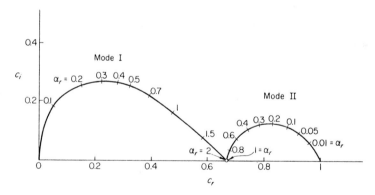

FIG. 6.2. Eigenvalues of a symmetric jet. The function α in the complex c plane.

different modes of the flow. Curve I was computed by Sato and Kuriki (1961) and by Kaplan (1964). Incidentally, some authors refer to $\mathbf{u}$ and speak of mode I as being asymmetric. Some analytical work related to such

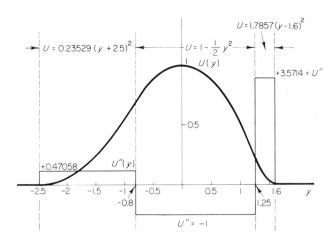

FIG. 6.3. An example of asymmetric jet.

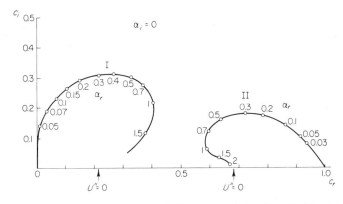

FIG. 6.4. Neutral stability curve of the asymmetric jet shown in Fig. 6.3.

flows has been done or reviewed by Drazin and Howard (1962), who made use of power series techniques.

As discussed in Section 5, another set of eigenvalues can be found by taking the complex conjugate of the present set of eigenvalues and eigenfunctions. However, the presence of lingering vorticity will invalidate these solutions, even in the case of extremely small viscosity.

As long as α is real and ω complex, a constant mean velocity can be superposed on a jet flow without modification of c_i. Indeed, this problem of parallel flow is invariant to a Galilean transformation. Thus the results obtained for jets are immediately applicable to wakes.

(b) Asymmetric Jet

In the symmetric jet a major difficulty has been avoided. For a neutral solution the choice of $c_r = 0.66$ guarantees that $U''/(U - c)$ will remain finite on both sides of the jet. The phase velocity is the same as the inflection velocity. In order to look further into this matter, we have chosen the asymmetric jet shown in Fig. 6.3 and defined as

$$
\begin{aligned}
U &= 0, & -\infty &< y < -2.5, \\
U &= 0.23529(y + 2.5)^2, & -2.5 &< y < -0.8 \\
U &= 1 - 0.5y^2, & -0.8 &< y < 1.25, & (6.7) \\
U &= 1.7857(y - 1.6)^2, & 1.25 &< y < 1.6, \\
U &= 0, & 1.6 &< y < \infty.
\end{aligned}
$$

The stability results for this jet profile are given in Fig. 6.4. Note that these curves can be compared with those of the symmetric jet, as given in Fig. 6.2. There are two modes, but they do not simply come from a symmetric or an asymmetric eigenfunction. In the complex c-plane the results form two separate curves. It is clear why there is a gap between these curves: the two points of inflection are no longer symmetric with respect to $y = 0$, and the phase velocities tend to different values as $c_i \to 0$.

Of further interest are the eigenfunctions of the asymmetric jet. Two examples are shown in Fig. 6.5, each corresponding to a different mode of

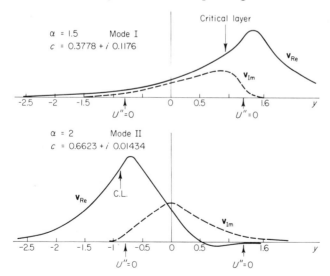

Fig. 6.5. Some eigenfunctions of the asymmetric jet of Fig. 6.3.

the jet. These functions bear some analogy to those of the shear layer, as shown in Fig. 5.3. Thus it appears that each inflection introduces a separate mode of instability. A general theorem due to Howard (1964) shows that this applies to any number of inflection points.

Essentially, the computations leading to the results for the asymmetric jet are straightforward, except for the usual difficulties (cf. Section 5), if $c_i \to 0$.

In a jet defined by a symmetric trapezoid Michalke and Schade obtained the two modes in Fig. 6.1. As the sides of the trapezoid become more vertical, the neutral lines move toward one another. They coalesce for a jet

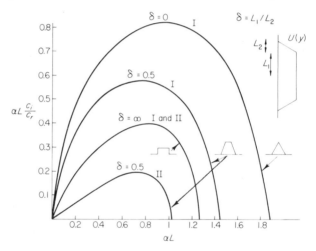

FIG. 6.6. Amplification in a family of symmetric jets. For Mode I, $\mathbf{v}$ is symmetric. (From Michalke and Schade, 1963.)

with a rectangular profile. These results are shown in Fig. 6.6. They also consider wakes and axisymmetric jets and wakes in detail. Finally, they study shear layers, which present three corners, and obtain interesting results. Essentially, they find that when two corners are sufficiently distinct the eigenvalues tend to form two separate groups.

(c) SOME DECEPTIVE SIMPLIFICATIONS

The general equation (5.7), used in the study of shear layers, jets, and wakes, is of second order. In addition, this equation is linear and homogeneous, and it is tempting to try a transformation of the sort

$$\mathbf{v}(y) = e^{\int g(y)\,dy}. \tag{6.8}$$

Having done this, the original equation is reduced to a first-order, but nonlinear, equation, namely,

$$g' + g^2 = \frac{U''}{U - c} + \alpha^2. \tag{6.9}$$

This is known as a Riccati equation in most references.

In a jet we can begin to integrate (6.9) on one side (say, at $y = -3$); this means that we start with $g = \alpha$ at this point. With the correct value of c, we should emerge on the other side (at $y = +3$) with $g = -\alpha$. In a symmetric jet we found that the solution corresponding to a tentative value of c behaved in a tortuous fashion in the complex plane. It is only when the choice of c is very close to the eigenvalue that the curve ends in the vicinity of the desired point. Thus the search for c must proceed in small steps and with caution. This difficulty in finding the eigenvalues more than cancels the advantage of having only a first-order equation to integrate.

In the same vein let us suppose that we are studying stability of a jet between two walls located $y = +a$, $y = -b$, where a and b are positive. We know that the general solution of the governing equation will contain two constants A and B and will be in the form of

$$\mathbf{v} = A f(y) + B g(y). \tag{6.10}$$

The functions f and g are different solutions of (6.2). The computer can easily obtain two such functions by starting anywhere from some initial conditions.

The boundary conditions for (6.10) give

$$A f(a) + B g(a) = 0, \tag{6.11}$$

$$A f(-b) + B g(-b) = 0. \tag{6.12}$$

The combination of (6.11) and (6.12) indicate that the determinant of the coefficients A and B must vanish for a nontrivial solution. Writing this out, we have

$$\Delta = f(a) g(-b) - f(-b) g(a). \tag{6.13}$$

Using the computer, we make a search for values of c such that $\Delta = 0$. Practically speaking, it is found that Δ varies with c in such a complicated manner that the search is delicate indeed. It is advisable instead to consider (6.11) and (6.12) as expressions that give $\mathbf{v}(a)$ and $\mathbf{v}(-b)$. We then specify that $\mathbf{v}(a) = 0$ and obtain the new expression

$$\mathbf{v}(-b) = A \left[f(-b) - \frac{f(a)}{g(a)} g(-b) \right]. \tag{6.14}$$

The computer then searches for the zeros of $\mathbf{v}(-b)$, and it is found that their variations are more predictable than those of Δ.

These two examples of search in g or Δ illustrate a general lesson learned from experience: the zeros of a function are easier to locate when the function represents a simple physical quantity.

Chapter II

Simplified Analysis of the Boundary Layer

7. Generalities and Boundary Conditions

We now examine the stability of a laminar boundary layer. In the preceding sections we established the general equations and outlined certain numerical techniques. We are now equipped to make further progress. Viscous effects are now retained. In this chapter we shall be satisfied with approximate results and accept our goal as that of grasping the essential physical mechanisms. The more reliable mathematical techniques will be introduced later.

As we examine the case of a boundary layer, we meet new difficulties, and special tactics are required for its formulation.

In a laminar boundary layer the mean velocity $U(y)$ starts from zero along a wall located at $y = 0$ and increases until it becomes constant in the free stream. Thus we assume that $U = U_0$ if $y > \delta$, and, as in the shear layer, we select units of length and time such that $\delta = 1$ and $U_0 = 1$. We focus our attention on neutral oscillations and often examine the flow from the point of view of an observer moving with the phase velocity c_r. In this frame the observer sees the wall move to the left with a velocity $-c_r$; the free stream moves to the right with $U = 1 - c_r$. In general, $0 < c_r < 1$, and the moving observer will notice that a layer of fluid remains at rest. This layer is the seat of unusual effects and has been called the "critical" layer. It is located at some height h such that $U(h) = c_r$.

For added facility this study consists of four sections:

1. The inviscid approximation (Section 8).
2. The correction due to wall friction (Section 9).
3. The correction to critical layer effects (Section 10).
4. Final results for boundary layers (Section 11).

The inviscid approximation is given by the complex functions $\mathbf{u}_1$, $\mathbf{v}_1$, $\mathbf{p}_1$, and $\mathbf{r}_1$. The wall correction adds the terms $\mathbf{u}_2$, $\mathbf{v}_2$, $\mathbf{p}_2$, and $\mathbf{r}_2$, and the critical layer supplements it with terms denoted as $\mathbf{u}_3$, $\mathbf{v}_3$, $\mathbf{p}_3$, and $\mathbf{r}_3$. The final solution then becomes the sum of these three contributions. Far above the layer we require that the fluctuations vanish along with conditions applied at the wall, which is considered rigid. Thus at $y = 0$ we have

$$\mathbf{u}_1 + \mathbf{u}_2 + \mathbf{u}_3 = 0, \tag{7.1}$$

$$\mathbf{v}_1 + \mathbf{v}_2 + \mathbf{v}_3 = 0. \tag{7.2}$$

If we refer to (1.18), we shall see that if $\mathbf{u}$ and $\mathbf{v}$ vanish at the wall the pressure forces must be balanced by the friction forces. Therefore at $y = 0$ we have a special force relation which, after some use of (2.4) and (1.17), becomes

$$i\alpha(\mathbf{p}_1 + \mathbf{p}_2 + \mathbf{p}_3) = v(\mathbf{r}_1' + \mathbf{r}_2' + \mathbf{r}_3'). \tag{7.3}$$

Because the viscous processes generate small pressure differences, we neglect $\mathbf{p}_2$ and $\mathbf{p}_3$. Furthermore, the term $\mathbf{r}_2'$ is the result of effects of friction at the wall, and we assume that it is larger than $\mathbf{r}_1'(0)$ and $\mathbf{r}_3'(0)$. These assumptions can be verified after all results have been collected. Thus (7.3) reduces to

$$i\alpha\, \mathbf{p}_1(0) = v\, \mathbf{r}_2'(0). \tag{7.4}$$

8. The Inviscid Solution

If the viscosity is neglected, the integration of the basic equations (1.17), (1.18), and (1.19) can be reduced to that of the Rayleigh equation (5.7), or

$$\mathbf{v}'' = \left(\frac{U''}{U - c} + \alpha^2\right)\mathbf{v}, \tag{8.1}$$

where we now take for U and U'' the functions given by Blasius (1908). These functions correspond to flow past a flat plate without a pressure gradient and are shown in Fig. 8.1. Starting from a pressure wave in the free stream, we take $\mathbf{v}_1 = e^{-\alpha(y-1)}$ for $y > 1$ and integrate downward. The result with $\alpha = 1$ and $c = 0$ is shown in Fig. 8.2. The term U'' tends to bend $\mathbf{v}_1$ toward the origin but not enough to approach the boundary condition $\mathbf{v}_1(0) = 0$. If we give c_r some positive value, we run into difficulties at $y = h$, where $\mathbf{v}_1''$ becomes infinite. As long as the computer does not take a value of y so close to h that $(U - c)^{-1}$ becomes of the order of 10^{38}, the integration will proceed normally. The results will merely show a sharp nick near $y = h$. We can specify that the computer must not let $\mathbf{v}'_1$ exceed some arbitrary limit, say $|\mathbf{v}'_1| < 10^3$, or we could maintain c_i at some small but nonzero value. In any case, it is essential to note that something becomes infinite and that a re-examination of the situation is necessary from a mathematical as well as a physical point of view. This will be the object of Section 10.

With a limit on $\mathbf{v}'_1$, we obtained the curves shown in Fig. 8.2, and it appears that the boundary condition on $\mathbf{v}_1$ is satisfied if c_r is nearly 0.4. Minor variations of c_r will later allow for the presence of corrective terms $\mathbf{v}_2(0)$ and $\mathbf{v}_3(0)$.

Similar calculations for other values of α lead to the curves shown in

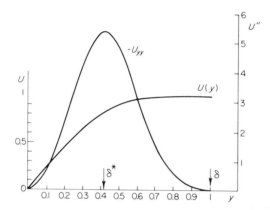

FIG. 8.1. The velocity profile of Blasius.

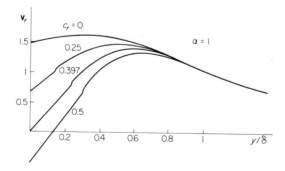

FIG. 8.2. Inviscid solutions for various c_r.

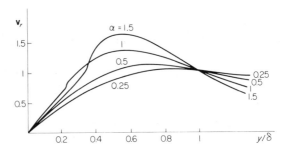

Fɪɢ. 8.3. Inviscid solutions for various α.

Fig. 8.3. We have verified that the ceiling imposed on $\mathbf{v}'(h)$ has only minor effects on the results.

This behavior of $\mathbf{v}_1$ is due to the appearance of vorticity fluctuations, and we shall now look into this subject. The Rayleigh equation (8.1) is formally equivalent to the inviscid vorticity equation. Indeed, the vorticity fluctuations obey the equation [see (2.6)]

$$\tilde{r}_t + U\tilde{r}_x + U''\tilde{v} = 0. \tag{8.2}$$

The complex form for the definition of $\tilde{r}$ [see (2.4)] is

$$\mathbf{r} = \mathbf{u}' - i\alpha\mathbf{v}. \tag{8.3}$$

By the condition of incompressibility $\mathbf{u}$ can be eliminated to give

$$\mathbf{r} = -\frac{i}{\alpha}(\mathbf{v}'' - \alpha^2\mathbf{v}). \tag{8.4}$$

It can now be easily verified that the vorticity equation, expressed in terms of $\mathbf{r}$, is identical to the Rayleigh equation.

In the free stream U'' vanishes and there are no vorticity fluctuations. In the layer the term $U''\tilde{v}$ acts as a source of vorticity fluctuations which modify the simple pressure wave. The vorticity equation (8.2) is simply a form of the Helmholtz theorem which states that in the absence of viscosity or other forces capable of producing a torque a fluid particle retains its vorticity during the motion. This is the equivalence of the conservation of angular momentum in rigid body mechanics. Thus a particle moving upward with velocity v will retain its vorticity r. It will move from a region in which the mean vorticity is U' to a region in which it is $U' + U'' \Delta y$. The vorticity fluctuation has been defined as $\tilde{r} = r - U'$ and therefore $\tilde{r}$ will change by the amount $-U'' \Delta y$. The rate of change will be $-U''\tilde{v}$. Similar perturbations caused by an invariable corpus moving to various environments have been contemplated by Swift (1726).

The production of vorticity fluctuations is in phase with $\tilde{v}_1$, but because the production is balanced by the transport term the largest vorticity fluctuations along a horizontal line occur in the regions in which $\tilde{v}_1$ is zero. This vorticity induces a change in the horizontal velocity fluctuation. Let us now examine the velocity fluctuations corresponding to the

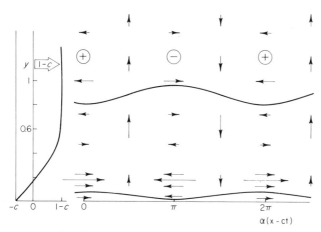

FIG. 8.4. Inviscid boundary layer oscillations.

numerical solution when $v(0) = 0$. The flow is sketched in Fig. 8.4 in the moving frame of reference. Note that u_1 is proportional to v'_1 and therefore changes sign near the point $y = 0.6$.

The pressure fluctuations build up exponentially in the free stream and remain almost constant through the boundary layer. If we indicate by p_e the pressure fluctuations at the edge of the layer ($y = 1$), the relations (3.11) and (3.12) for a pressure wave lead to the following value for the velocity fluctuations at the edge of the layer:

$$u_e = -\frac{p_e}{1 - c}, \tag{8.5}$$

$$v_e = -\frac{ip_e}{1 - c}. \tag{8.6}$$

We now give an approximate expression for the solution to the equation (8.1) within the layer. The results shown in Fig. 8.3 suggest that if α is small v_1 could be approximated by a parabola starting from the origin and meeting the exponential tail at $y = 1$. This approximation simply ignores the singularity at the critical layer. It is easy to show that the following parabolic function matches v_1 and v_1' at $y = 1$:

$$\mathbf{v}_1 = -i[(2 + \alpha)y - (1 + \alpha)y^2] \frac{\mathbf{p}_e}{1 - c}. \tag{8.7}$$

The subscript 1 indicates that we are dealing with the inviscid solution of the problem. From (8.7) we can obtain the corresponding equation

$$\mathbf{u}_1 = \left(\frac{2 + \alpha}{\alpha} - \frac{2 + 2\alpha}{\alpha} y\right) \frac{\mathbf{p}_e}{1 - c}. \tag{8.8}$$

For larger values of α the computed curves indicate that $\mathbf{u}_1(0)$ is more accurate if the factor $2 + \alpha$ is replaced by a factor 2 or even less. (Cf. Betchov, 1960b).

For the vorticity we have

$$\mathbf{r}_1 = -2 \frac{1 + \alpha}{\alpha} \frac{\mathbf{p}_e}{1 - c}. \tag{8.9}$$

The pressure is assumed constant; hence

$$\mathbf{p}_1 = \mathbf{p}_e. \tag{8.10}$$

To summarize this section we can say that the phase velocity controls the magnitude of the vorticity fluctuations and, for a well chosen c_r, the motion induced by $\tilde{r}$ just cancels $\tilde{v}$ along the wall. We must also note that $\tilde{v}'$ tends to become infinite at the critical layer and that a re-examination of this part of the flow is necessary. A parabolic approximation can be used if α is not too large; say, $\alpha < 1$.

9. Friction at the Wall

The inviscid solution produces an oscillating slip flow along the wall of magnitude $\mathbf{u}_1(0)$. This is not realistic, and a viscous correction must be introduced. Physically speaking, the viscosity will create stresses that will reduce $\mathbf{u}$ to zero in some thin layer. This is described by a correction $\mathbf{u}_2$ which is to be added to $\mathbf{u}_1$. Clearly, $\mathbf{u}_2$ must vanish far from the wall. The effects of friction can also be described in terms of vorticity. As the oscillation described by the inviscid solution develops, the fluid in contact with the wall acquires vorticity just as the wheels of a landing aircraft gain angular momentum at touch down. This vorticity then diffuses into the boundary layer and induces a modification $\mathbf{u}_2$ of the velocity, such that

$\mathbf{u}_1 + \mathbf{u}_2$ almost vanishes at the wall. As we shall see later, another contribution $\mathbf{u}_3$ will emerge.

The vorticity $\mathbf{r}_2$, seeping out of the wall, must satisfy (2.7) which reduces to

$$(U - c)\mathbf{r}_2 = \frac{-i\nu}{\alpha}(\mathbf{r}_2'' - \alpha^2 \mathbf{r}_2). \tag{9.1}$$

Because we are dealing with a viscous correction, the term $\alpha^2 \mathbf{r}_2$ can be neglected, for it is much smaller than $\mathbf{r}_2''$. There is no production term, for we assume that the vertical motion produced by viscous effects is negligible, and, furthermore, U'' often vanishes near the wall. Because (9.1) is of second order, there are two arbitrary constants. The condition $\mathbf{r}_2(\infty) = 0$ determines one constant and the other is specified by the special relation existing at the wall between pressure and viscous stresses. The solution to (9.1) will yield an induced velocity given by the relation

$$\int_y^\infty \mathbf{r}_2 \, dy = \mathbf{u}_2(\infty) - \mathbf{u}_2(y) = -\mathbf{u}_2(y), \tag{9.2}$$

where we have neglected the term in $\mathbf{v}_2$.

Let us now integrate (9.1). If we assume that U is negligible in the region of interest, the integration is immediate but the results are somewhat disappointing. If we assume that U varies linearly with y, (9.1) becomes a Bessel equation and the vorticity is given by a Hankel function of fractional order and complex argument. Although numerical tables are available (Harvard Computation Laboratory, 1945; also Singh *et al.*, 1963; Smirnov, 1960; and Nosova and Tumarkin, 1965), we prefer a shortcut. The essential property of U is that it makes the coefficient $(U - c)$ negative below the critical layer and positive above. This point directs us to assumptions that preserve this behavior.

Below the critical layer we can use the approximation $U = 0$ which suggests the solution

$$\mathbf{r}_{2(\text{below})} = M \exp\left[-(1 - i)\left(\frac{\alpha c}{2\nu}\right)^{1/2} y\right], \tag{9.3}$$

where M is an arbitrary complex constant. The corresponding distribution is shown in Fig. 9.1 between $y = 0$ and $y = h$.

The lines of zero vorticity fluctuation are tilted to the left, and, in between, the magnitude of the vorticity fluctuations decays exponentially with the distance from the wall.

Above the critical layer we use the approximation $U = 2c$ to gain some qualitative picture. This leads to

$$\mathbf{r}_{2(\text{above})} = Q \exp\left[-(1 + i)\left(\frac{\alpha c}{2\nu}\right)^{1/2} y \right]. \tag{9.4}$$

The change in the sign of $U - c$ causes the line of zero vorticity to tilt to the right (see Fig. 9.1). Although we could give a more accurate picture of the distribution of the vorticity introduced by the presence of the wall, the essential elements are already before us.

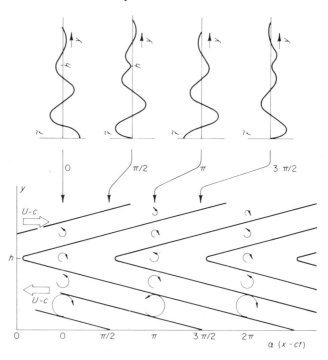

FIG. 9.1. Distribution of vorticity introduced at the wall in the x-y plane and at specific positions. Note the double beat at $\pi/2$.

Let us now consider the velocity fluctuations induced by the vorticity distribution shown in Fig. 9.1. If (9.3) is applied from $y = 0$ to $y = \infty$, as if U did not change, the velocity induced at the wall [see (9.2)] becomes

$$\mathbf{u}_{2(\text{under})} = -(1 + i)M\left(\frac{\nu}{2\alpha c}\right)^{1/2}, \tag{9.5}$$

where the subscript "under" indicates that this velocity would be observed if the viscous effects were confined below the critical layer. This is also applicable in the limit as $v \to 0$. The constant M can be determined by application of the special relation of (7.4), which by (9.3) becomes

$$i\alpha \mathbf{p}_e = (-1 + i)M\left(\frac{\alpha c v}{2}\right)^{1/2}. \tag{9.6}$$

Elimination of M between (9.5) and (9.6) gives

$$\mathbf{u}_{2(\text{under})} = -\frac{\mathbf{p}_e}{c}. \tag{9.7}$$

In fact $\mathbf{u}_2$ differs from $\mathbf{u}_{2(\text{under})}$ because of the changes occurring above the critical layer and this departure is very important for finite viscosity.

The repartitions of the vorticity fluctuation along some selected vertical lines are shown in the upper part of Fig. 9.1. If $\alpha(x - ct) = 0$ or π, the distributions are the same as if $U - c$ had not changed sign. The vorticity coming from the left simply takes the place that vorticity from the right would have occupied. At such angles as $\alpha(x - ct) = \pi/2$ or $3\pi/2$ the distribution of vorticity shows a double beat due to the folding of a particular streak of vorticity. In calculating $\mathbf{u}_2$ by integration of $\mathbf{r}_2$ from $y = \infty$ downward, we therefore encounter contributions of alternate sign, except through the critical layer, where the double beat adds a special amount. Hence the change of sign of the coefficient $U - c$ adds a special contribution $\Delta \mathbf{r}_2$ to the vorticity already described by (9.3).

The magnitude of $\Delta \mathbf{r}_2$ is that of $\mathbf{r}_{2(\text{below})}$ at $y = h$, and the phase of $\Delta \mathbf{r}_2$ is shifted to the right by an angle θ which appears to be near $\pi/2$. Therefore, by (9.4), we have at $y = h$

$$\Delta \mathbf{r}_2 = M \exp\left[-(1 + i)\left(\frac{\alpha c}{2v}\right)^{1/2} h - i\theta\right]. \tag{9.8}$$

We consider that this special vorticity appears in a region defined by $y = h \pm l$, where l is perhaps one quarter of the wavelength in the y-direction. To allow for later adjustments we introduce a corrective factor g and define

$$l = g\left(\frac{2v}{\alpha c}\right)^{1/2}. \tag{9.9}$$

Thus we have defined a correction due to bends in the vorticity distribution. This simply amounts to a special row of vortices at the critical layer.

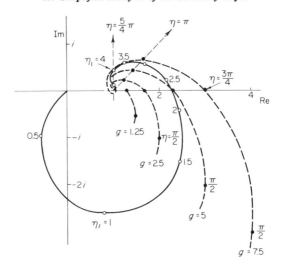

FIG. 9.2. The function $\mathbf{u}_2(0)/(-\mathbf{p}_e/c)$ in the complex plane:
——— if $U' = $ constant, by Hankel functions.
- - - - - by Eq. (9.13), with $\theta = 1.6$, $g = 1.25, 2.5, 5, 7.5$.

The velocity fluctutations induced by $\Delta\mathbf{r}_2$ must be zero when y is large. If $y < h - l$, we find by (9.2)

$$\Delta\mathbf{u}_2 = -2gM\left(\frac{2v}{\alpha c}\right)^{1/2}e^{-\eta}e^{i(\eta-\theta)} \tag{9.10}$$

with

$$\eta = \left(\frac{\alpha ch^2}{2v}\right)^{1/2}. \tag{9.11}$$

Elimination of M by (9.6) gives

$$\Delta\mathbf{u}_2 = -\frac{\mathbf{p}_e}{c}2ge^{-\theta}Z(\eta), \tag{9.12}$$

where $Z(\eta)$ is a simple spiral in the complex plane defined as

$$Z(\eta) = (1-i)e^{-(1-i)(\eta-\theta)}. \tag{9.13}$$

The velocity induced at the wall is $\mathbf{u}_2 = \mathbf{u}_{2(\text{under})} + \Delta\mathbf{u}_2$, and we find

$$\mathbf{u}_2(0) = -\frac{\mathbf{p}_e}{c}[1 + 2ge^{-\theta}Z(\eta)]. \tag{9.14}$$

This function is shown in Fig. 9.2 for various values of g and for $\theta = \pi/2$.

In the same figure we show the velocity $\mathbf{u}_2$ obtained by the exact integration of (9.1) after assuming that U is proportional to y and neglecting the term in $\alpha^2 \mathbf{r}$. This can be done with Hankel functions or with the help of a special function introduced by Tietjens (1925) and discussed by Holstein (1950) and Lin (1955). As we can see, a particular choice of g gives a crude approximation. The fact that the better curve does not spiral about the terminal point is not very interesting and it rules out certain modes of oscillation that are of little physical interest.

In the integration with constant U' the Hankel functions contain a parameter η_1 defined as

$$\eta_1 = \left(\frac{\alpha U' h^3}{\nu}\right)^{1/3} = 2^{1/3} \eta^{2/3}. \tag{9.15}$$

The different power is due to the variation of the transporting velocity with the distance from the wall.

Exact integration with the assumption $U = 0$ below h and $2c$ above h with precautions to keep $\mathbf{r}$ and $\mathbf{r}'$ continuous leads to another curve that coincides almost exactly with $g = 1.25$. A realistic profile has not been integrated explicitly.

The interesting portion of the curve has the positive imaginary part. It is worth noticing that the inviscid velocity fluctuations $\mathbf{u}_1(1)$ at the edge of the layer amount precisely to $-\mathbf{p}_e/c$. Thus, if $\mathbf{v}_1(y)$ had the same slope at the wall as at the outer edge, $\mathbf{u}_2(0)$ should be equal to $-\mathbf{p}_e/c$ to cancel $\mathbf{u}_1(0)$. This corresponds to $\eta = \infty$; but we have seen that $\mathbf{v}_1'$ is steeper at the wall than at the edge and therefore the real part of $-\mathbf{u}_2 c/\mathbf{p}_e$ should be larger than one.

In reality the gradual change in velocity makes all parts of the layer contribute to $\Delta \mathbf{u}_2$, but the folding near the critical layer is the dominant effect. For a particular η any $\mathbf{u}_2(0)$ can be "explained" by fitting g and θ. Thus we can define a generalized $\mathscr{F}(\eta)$ such that

$$\mathbf{u}_2 = -\frac{\mathbf{p}_e}{c}(1 + \mathscr{F}(\eta)). \tag{9.16}$$

Equation (9.16) relates the viscous stresses at the wall to the velocity induced by the vorticity injected at the wall.

10. The Critical Layer

In the treatment of the inviscid solution we encountered difficulty at $y = h$ which we have heretofore shirked. We must now face up to the fact that if the inviscid solution leads to infinities some of the viscous terms will become large enough to contribute to the true solution and any infinities are prevented. Therefore we must reconsider the solution in the vicinity of $y = h$. Because we are concerned only with a region immediately above or below h, we can consider U'' and $\mathbf{v}_1$ as near constants. On the other hand, we should expect some modification of the corresponding $\mathbf{u}_1$, but by comparison it would be surprising to find a substantial change in $\mathbf{v}_1$. Temporarily, we omit the subscripts 1 and 3 in determining the viscous solution.

(a) THE VISCOUS SOLUTION

We begin by writing the viscous vorticity equation by neglecting the term giving the diffusion in the x-direction and, temporarily omitting the subscripts, we have:

$$(U - c)\mathbf{r} - \frac{iU''\mathbf{v}}{\alpha} = -\frac{iv}{\alpha}\mathbf{r}''. \qquad (10.1)$$

Because $U - c$ changes sign across the critical layer, we use an approximation that retains this property and suggests

$$U(y) = c + U'(h)(y - h). \qquad (10.2)$$

The inviscid solution is

$$\mathbf{r} = \left(\frac{iU''\mathbf{v}}{\alpha U'}\right)_{y=h} \frac{1}{y - h}. \qquad (10.3)$$

This solution is acceptable as long as the magnitude of the viscous terms remains inferior to that of the transport term. The difficulties begin when diffusion is comparable with transport, and this occurs at locations such that the right-hand side of (10.1) becomes equal to one of the terms of the left-hand side. This leads to

$$U'(h)(y - h)\left(\frac{U''\mathbf{v}}{\alpha U'}\right)_{y=h} \frac{1}{(y - h)} = \pm \frac{v}{\alpha}\left(\frac{U''\mathbf{v}}{\alpha U'}\right)_{y=h} \frac{1}{(y - h)^3}. \qquad (10.4)$$

This relation defines a zone between $y = h + l$ and $y = h - l$, where l is obtained from (10.4) and amounts to

$$l = \left(\frac{v}{\alpha U'}\right)^{1/3}. \qquad (10.5)$$

Let us accept the inviscid solution outside this zone and try another drastic assumption to obtain a solution inside the zone. To do this we

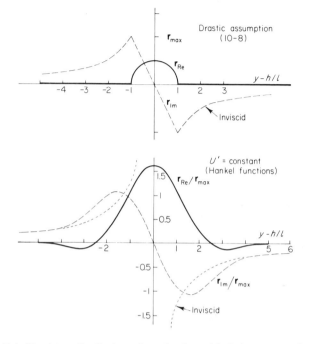

Fig. 10.1. Vorticity distribution through the critical layer according to two approximations.

simply drop the transport term inside the zone. Then the solution of (10.1) becomes a parabola, and we write

$$\mathbf{r} = a + b(y - h) + \left(\frac{U''\mathbf{v}}{v}\right)_{y=h} \frac{(y - h)^2}{2}. \tag{10.6}$$

The two constants, a and b, can be determined in such a way that the viscous solution (10.6) matches the inviscid solution at $y = h \pm l$. With the convenient definition

$$\mathbf{r}_{\text{max}} = -\left(\frac{U''\mathbf{v}}{l\alpha U'}\right)_{y=h}, \tag{10.7}$$

this leads to the following approximate solution:

$$\mathbf{r} = \mathbf{r}_{\text{max}}\left[\frac{1}{2} - \frac{(y-h)^2}{2l^2} - i\frac{(y-h)}{l}\right]. \tag{10.8}$$

The real and imaginary parts of this solution are shown in Fig. 10.1.

We could integrate (10.1) through the zone with the more realistic (10.2) and the condition that the solution must be asymptotic to the inviscid

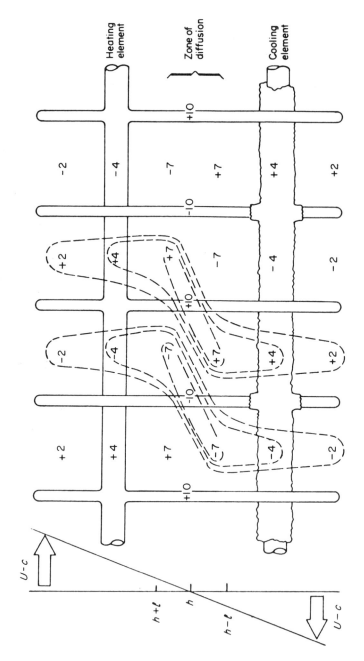

FIG. 10.2. Analogy with vorticity distribution through the critical layer. A shear flow over a system of heating and cooling elements would result in the indicated temperature distribution. The vertical scale is greatly expanded.

solution (10.3) above and below the layer. This leads again to the Hankel functions and their integrals. The results are shown in Fig. 10.1; the comparison reveals that our drastic assumption contained all the essential ingredients. However it gives a low value of r_{Re}.

We must now analyze these results and understand the physical reality concealed by these equations. This becomes easy if we use the analogy between the vorticity and heat equations. Let us imagine that heat is forced into the fluid at some fixed location and out of the fluid at others and that this heat is convected and diffused. The general equation is

$$\frac{\partial T}{\partial t} + U(y)\frac{\partial T}{\partial x} = k\left(\frac{\partial^2 T}{\partial x^2} + \frac{\partial^2 T}{\partial y^2}\right) + Q(x, y, t), \tag{10.9}$$

where T is the temperature, k, the diffusivity, and $Q(x, y, t)$, the forcing function.

In the analogy with the critical layer flow we take Q as independent of y and periodic with x and t (we restrict ourselves to the neutral case). The observer moving with the wave velocity will see a stationary situation and, taking, for example, $Q = Q_0 \cos[\alpha(x - ct)]$, he will write

$$U'(h)(y - h)T_x = kT_{yy} + Q_0 \cos \alpha x. \tag{10.10}$$

This forcing function could be produced by a periodic array of hot and cold pipes parallel to the y-axis, shown in Fig. 10.2. Away from the critical layer we have the "inviscid" solution

$$T = \frac{Q_0}{\alpha U'} \frac{\sin \alpha x}{(y - h)}. \tag{10.11}$$

It simply tells us that at a given altitude y the highest temperatures are found downstream of the heating elements. Indeed, the fluid accumulates heat while moving over half a wavelength. As the altitude above h increases, the fluid takes less time to pass over a heater, and therefore it accumulates less heat. Thus T drops with $1/y$. Typical temperatures are indicated in Fig. 10.2. In drawing this figure, it was necessary to compress the horizontal dimensions and to expand the vertical scales. Thus, in reality, the spacing between pipes is much larger than indicated. As a result, the greatest temperature gradients are not along horizontal lines but along vertical lines, across the critical zone. The velocity reversal causes a hot region above the zone to be adjacent to a cold region below the zone. In the absence of any diffusion the temperature would become so high that the forcing function would be modified and the analogy would fail.

With diffusion, we find that a special process occurs.

In the vicinity of $h = 0$ let us imagine that a special amount of heat is released at some initial time from a heating element. Diffusion will cause the heat to spread and, after a time Δt, it will occupy a region of the order of Δy with $\Delta y^2 = k \, \Delta t$.

At the edges of this region the fluid moves with velocity $U' \, \Delta y$, and the horizontal displacement is therefore $\Delta x = U' \, \Delta y \, \Delta t$. The limits of the diffusion zone are found where the horizontal displacement associated with the diffusion time corresponds roughly to one quarter of a wavelength. Thus we have

$$\Delta x = \frac{U' \Delta y^3}{k} = \frac{\pi}{2\alpha}. \tag{10.12}$$

This gives an expression for Δy similar to (10.5), which defined l. If we neglect the transport term inside the diffusion zone and match the solution with (10.11), we obtain

$$T = \frac{Q_0 l^2}{2\nu} \left[\left(1 - \frac{(y - h)^2}{l^2} \right) \cos \alpha x + 2 \frac{y - h}{l} \sin \alpha x \right]. \tag{10.13}$$

The analogy with the vorticity is now obvious. The term Q in (10.9) plays the role of $U''v$ and k corresponds to ν. It is also clear that the maxima inside the zone must be 90° out of phase with the maxima outside the zone. In Fig. 10.2 we have outlined the isotherms and indicated some typical values of T. You can easily convince yourself that the information given in Fig. 10.1 regarding the vorticity leads to similar curves of constant vorticity.

(b) Effect of the Critical Layer on the Velocity Field

The viscous vorticity distribution differs from the inviscid vorticity in two respects. First, let us examine the asymmetric part of $\mathbf{r}$ which corresponds to $\mathbf{r}_{Im}$ in Fig. 10.1, where we assumed that $\mathbf{v}(h)$ was real. Instead of jumping from one infinity to the other this function now passes through zero. Second, a symmetric distribution, which forms lumps of vorticity in phase with $\mathbf{v}$, has appeared.

Let us examine $\mathbf{u}$. Because the gradients in y are much larger than the gradients in x, we have $\mathbf{u}_y \gg \alpha \mathbf{v}$, and the equation defining the vorticity [see (2.4)] can be integrated as follows:

$$\mathbf{u}(y) = \mathbf{u}(h + l) + \int_{h+l}^{y} \mathbf{r} \, dy. \tag{10.14}$$

If $y > h + l$, we are above the critical zone and $\mathbf{u}$ is identical to the inviscid solution $\mathbf{u}_1$. At $y = h - l$ we find that the integral of the asymmetric part of $\mathbf{r}$ amounts to zero, whereas the integral of the symmetric part amounts to the quantity $\mathbf{u}_3$ defined as

$$\mathbf{u}_3 = -\int_{h-l}^{h+l} \mathbf{r}\, dy = \frac{2}{3}\left(\frac{U''\mathbf{v}}{\alpha U'}\right)_{y=h}. \tag{10.15}$$

Thus the vorticity located within the critical zone adds the special contribution $\mathbf{u}_3$ to the fluctuation $\mathbf{u}_1$. This is merely the velocity induced by the lumps of vorticity undergoing a diffusion process and located in phase with $\mathbf{v}_1$.

Similar calculations can be made with a polynomial of higher order (Betchov, 1960b), with Hankel functions (Holstein, 1950), or by using a detour in the complex plane (Lin, 1955). The exact calculations lead to similar results (see Fig. 10.1), but the approximate factor $\frac{2}{3}$ in (10.15) must be replaced by the factor π. Below the layer the vorticity is simply that of the inviscid solution, and we have

$$\mathbf{u}(y) = \mathbf{u}_1(y) + \mathbf{u}_3. \tag{10.16}$$

Essentially, the critical layer acts as a row of alternating vortices, extremely elongated in the x-direction. Because $\mathbf{u}_3$ is in phase with $\mathbf{v}_1$, the product $\bar{u}\bar{v}$ varies from zero above the critical zone to some slowly varying quantity below the zone.

Because $\mathbf{v}_1(y)$ is related to $\mathbf{p}_e$ by (8.10), we can pass from (10.15) to the following relation (with π in place of $\frac{2}{3}$):

$$\mathbf{u}_3 = i\pi \frac{-hU''(h)}{U'(h)} \frac{2+\alpha}{\alpha} \frac{\mathbf{p}_e}{1-c} \tag{10.17}$$

or, by referring to (8.8),

$$\mathbf{u}_3 = i\pi \frac{-hU''(h)}{U'(h)} \mathbf{u}_1(0). \tag{10.18}$$

Some remarks on the role of v are now appropriate. If $v \to 0$, it follows from (10.5) that l decreases as $v^{1/3}$ and, from (10.7), that $\mathbf{r}_{\max}$ increases as $v^{-1/3}$. Thus $\mathbf{u}_3$ remains constant. The strength of the diffusing vortices is independent of the viscosity in the limit $v \to 0$. Furthermore, if we return to the vorticity equation (10.1) and replace v by some small *negative* quantity, we can easily repeat the procedure of integration. The only final difference will be a reversal in the signs of $\mathbf{r}_{\max}$ and $\mathbf{u}_3$. Thus the sign of v is more important than its magnitude. No matter how small the viscosity, it has a profound influence on the flow.

The critical layer also has an effect on **v**. From the continuity equation (1.8) we find that $\mathbf{u}_3$ adds a new contribution $\mathbf{v}_3$ such that

$$\mathbf{v}_3 = 0, \qquad\qquad y > h + l,$$
$$\mathbf{v}_3 = -i\alpha\mathbf{u}_3(y - h), \qquad y < h - l. \tag{10.19}$$

11. Final Results for Boundary Layers

We are now ready to collect and assimilate the results of the preceding sections. As a guideline we use the pressure $\mathbf{p}_e$ which permeates the entire oscillating layer. If we make the convenient (and immaterial) choice that $\mathbf{p}_e$ is real and refer to (8.5) for $\mathbf{u}_1(0)$, to Fig. 9.2 for $\mathbf{u}_2(0)$, and to (10.17) for $\mathbf{u}_3(0)$, we find that the phases of the various contributions are related to one another, as shown in Fig. 11.1. To satisfy the boundary condition on **u** the sum $\mathbf{u}_1 + \mathbf{u}_3$ must correspond to the value of $-\mathbf{u}_2$, for some choice of η.

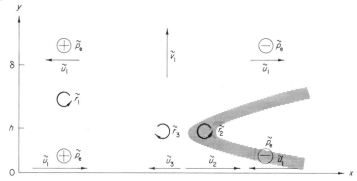

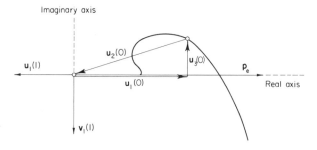

FIG. 11.1. Phase diagram for various components of a boundary layer oscillation.

After collecting results from (8.8), (9.16), and (10.17) and rearranging terms, we arrive at the relation

$$\frac{2+\alpha}{\alpha}\frac{c}{1-c}\left\{1+i\pi\left[\frac{-h\,U''(h)}{U'(h)}\right]\right\} = 1 + \mathscr{F}(\eta), \qquad (11.1)$$

where η is given by (9.15). Because c is approximately equal to $hU'(0)$, this expression can be simplified further to read

$$\frac{2+\alpha}{\alpha}\frac{U'(0)h}{1-U'(0)h}\left\{1+i\pi\left[\frac{-h\,U''(h)}{U'(0)}\right]\right\} = 1 + \mathscr{F}(\eta). \qquad (11.2)$$

The left-hand side is a function of α multiplying a function of h, and the right-hand side, for given α and h, is a function of the viscosity. Thus the solutions of (10.2) can be found graphically by the construction shown in Fig. 11.2. The left-hand side is plotted for various values of α, in the case of Blasius velocity profile, and the lines of constant h are shown radiating from the origin. We use a displacement thickness $\delta^* = 0.42$, corresponding to $U'(0) = 2.35$ and $U'' = -64y^2$ near the wall. Because the parabolic approximation is not very good when α exceeds 1, we have used the factor 2 in place of the factor $2 + \alpha$, as discussed in (8.8). The right-hand side corresponds to constant U' and has been computed from Hankel functions (see Fig. 9.2).

For a particular α there are two intercepts. For each intercept we have a particular value of h and η. From h we can obtain c, and because η is a function of α, h, and v we can also determine v and the Reynolds number $\mathscr{R} = U_0\,\delta/v$. The neutrally stable oscillations correspond to the two branches of the curve shown in Fig. 11.3a, the lower branch for low values of η_1 and the upper branch for high values of η_1. Merging occurs near $\alpha = 1.3$ for $\eta_1 = 3.8$.

If we refer to the simple model of the fold in the vorticity distribution shown in Fig. 9.1 and the broken curves of Fig. 9.2, we notice that the interesting part of $\mathscr{F}$ corresponds to an increase of η_1 of about $\pi/2$. This means that as we follow the neutral curve in the range of $\alpha > 0.1$ the phase difference between the pressure wave and the elbows of the vorticity distribution (see Fig. 9.2) varies by about 90°. If η_1 is about 3.6, the vorticity $\Delta\tilde{r}_2$ is maximum at a position about $\pi/4$ to the left of the region of maximum $\tilde{p}$. From this position it can cancel the flow induced at the wall by the vorticity produced by $U''\tilde{v}$ and stagnating at the critical layer. It also adds its effect to that of $\tilde{u}_{2(\text{below})}$ when compensating for $\tilde{u}_1(0)$. The upper branch of the neutral curve corresponds to locations of $\Delta\mathbf{r}_2$ farther to the left,

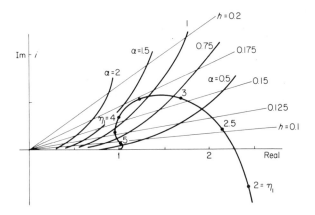

FIG. 11.2. Stability of Blasius boundary layer. Graphical solution of

$$\frac{2}{\alpha} \frac{h}{0.425\text{-}h} (1 + i85h^3) = 1 + \mathcal{F}(\eta_1).$$

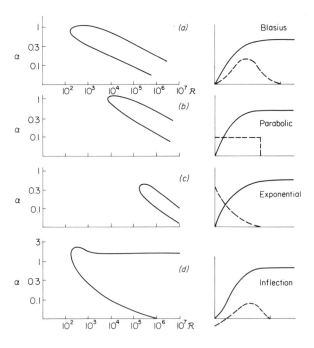

FIG. 11.3. Stability of various profiles: (*a*) Blasius, (*b*) parabolic, (*c*) exponential, and (*d*) inflection.

and η_1 is the phase angle that controls the oscillation. With other mean velocity profiles the function $\mathscr{F}$ may look different, but in order to find neutral solutions it will always be essential to have positive imaginary parts of $\mathscr{F}$ as long as U'' is negative.

Similar graphical constructions, in which the same $\mathscr{F}$ and $U'(0)$ as in Fig. 11.2 are kept, but different functions $U''(h)$ are considered, have produced the results shown in Fig. 11.3 *b* and *c*. The exponential mean velocity profile is encountered in the flow over a porous plate, when there is continuous suction of the fluid through the plate. Because the curvature of the mean profile is concentrated near the wall, we see that the oscillations occur at higher Reynolds numbers and therefore suction has a stabilizing effect. (For Fig. 11.3*c* see Chiarulli and Freeman 1948 for further information.)

Figure 11.3*d* deals with the profile encountered when the mean flow tends to separate from the wall, as in the presence of an adverse mean pressure gradient. (Cf. Pretsch, 1952.) Close to the wall U'' is positive and the left-hand side of (11.2) has a negative imaginary part (see Fig. 11.4). Thus no solutions are possible unless h is above the inflection point. If h coincides with the inflection point, the two branches correspond to $\eta_1 = 2.3$ and $\eta_1 = \infty$. This means that the upper branch of the neutral curve extends to infinite Reynolds numbers. This is a general property of profiles presenting an inflection.

Of course, as we meet lower Reynolds numbers and use profiles that differ greatly from the simple one, the accuracy of the graphical method collapses. Because the results shown in Fig. 11.3 are in qualitative agreement with more accurate calculations obtained when the method of Lin or others is followed, we can conclude that the present method permits us to evaluate the different factors and to understand the physical mechanisms.

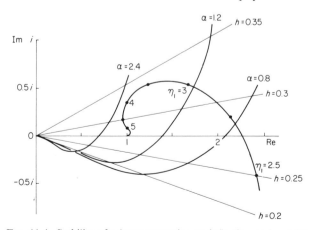

FIG. 11.4. Stability of a layer presenting an inflection at $h = 0.28$.

12. The Unstable Boundary Layer

In this section we examine the problem of the oscillating boundary layer when the solution is not neutrally stable. We start from the same basic equations as those of the neutral case. However, we must note that c is complex. For an unstable solution the imaginary part c_i is positive. Then we find that the analysis given for neutral waves can be adapted without too much difficulty. In damped solutions c_i is negative, and we encounter a very peculiar effect at the critical layer which again stresses the importance of viscosity, even though very small.

(a) Unstable Solutions

The equations derived for inviscid free-stream solutions are exactly applicable if c is complex. The function $\mathbf{p}$ must still decay as $e^{-\alpha y}$ to satisfy the boundary condition at infinity; the same is true for $\mathbf{u}(y)$ and $\mathbf{v}(y)$. Furthermore, we take $\mathbf{p} = Ae^{-\alpha y}$, where A is real and positive. This convenient choice fixes the phase of the oscillation to that of the free-stream pressure fluctuation. The continuity equation indicates that $\mathbf{u}$ and $\mathbf{v}$ will still be out of phase by $90°$, a condition that exists only if they both decay as $e^{-\alpha y}$ or both grow as $e^{\alpha y}$. The phase relation between $\mathbf{p}$ and $\mathbf{v}$, however, is modified. We find that

$$\frac{\mathbf{p}\mathbf{v}^* + \mathbf{p}^*\mathbf{v}}{4} = \frac{1}{2}c_i\frac{\mathbf{p}\mathbf{p}^*}{(1-c_r)^2 + c_i{}^2}. \tag{12.1}$$

This means that if we draw a plane parallel to the wall, somewhere in the free stream, the pressure forces will transfer energy across this plane from below and in proportion to c_i. This is the energy required to increase the kinetic energy of the flow. Indeed, the amplitude of the oscillation is growing exponentially in time.

Let us now direct our attention to the boundary layer proper. Neglecting the viscous terms, we have the Rayleigh equation which, by separating the real and the imaginary contributions, can be written as follows:

$$\mathbf{v}'' = \frac{U''(U - c_r)}{(U - c_r)^2 + c_i{}^2}\mathbf{v} + i\frac{U''c_i}{(U - c_r)^2 + c_i{}^2}\mathbf{v}. \tag{12.2}$$

The first term on the right-hand side is similar to the term used in the preceding inviscid analysis but has built-in protection against divergence at the critical layer. Thus we can assume that $\mathbf{v}(y)$ will be related to the fluctuation $\mathbf{v}(1)$ at the edge of the layer, as in the inviscid treatment. In addition, we have a contribution from the second term of the right-hand side of (12.2). In the limit, as $c_i \to 0$, this second term can be integrated across the critical layer:

$$i \int_{h-\varepsilon}^{h+\varepsilon} \frac{U''c_i\mathbf{v}}{(U-c_r)^2 + c_i^2} \, dy = i \frac{U''(h)\,\mathbf{v}(h)}{U'(h)} \int_{-\infty}^{+\infty} \frac{d\eta}{1+\eta^2}$$

$$= i\pi \frac{U''(h)\,\mathbf{v}(h)}{U'(h)}. \tag{12.3}$$

The same results can be obtained by taking $c_i = 0$ and integrating around the singularity located at $y = h$. This procedure assigns complex values to the variable y, which cannot be interpreted in terms of in phase and out of phase components. It also raises the question which of the two detours in the complex plane gives the correct result.

The integral (12.3) gives the jump of $\mathbf{u}$ across the critical layer, and it follows that below the critical layer a new component $\mathbf{u}_3$ must be added to the conventional solution that is constant and equal to

$$\mathbf{u}_3 = \pi \frac{U''(h)}{\alpha U'(h)} \mathbf{v}(h). \tag{12.4}$$

By the continuity equation we find that the "inviscid" solution must be supplemented by a contribution from the critical layer which takes the value

$$\mathbf{v}_3 = -i\pi \frac{U''(h)}{U'(h)} \mathbf{v}(h)(y - h), \qquad y < h. \tag{12.5}$$

The constant of integration has been set so that $\mathbf{v}_3$ vanishes at the critical layer. Above the critical layer $\mathbf{v}_3$ is zero.

By a proper choice of c_r we can guarantee that the real term corresponding to the "inviscid" solution vanishes at the wall. This is tantamount to the first step of the simplified analysis for a neutrally stable oscillation. However, it does not dispose of the term $\mathbf{v}_3$ and, with the best choice of c_r, we still have

$$\mathbf{v}_3(0) = i\alpha\pi \frac{hU''(h)}{U'(h)} \mathbf{v}(h). \tag{12.6}$$

The value of $\mathbf{v}(h)$ is also proportional to $\mathbf{v}(1)$, and we write

$$\mathbf{v}(h) = g\alpha h\, \mathbf{v}(1). \tag{12.7}$$

It can easily be seen by analogy with the "inviscid" solution that g is equal to the ratio $-\mathbf{u}(0)/\mathbf{u}(1)$ and that its value is near unity.

Thus, as we refer everything to $\mathbf{v}(1)$, we find that a small positive value of c_i does not change the solution inside the layer. To the contrary, the presence of c_i removes the difficulty at the critical layer.

The viscous solution can be treated as in Section 9; the result is a relation between $\mathbf{u}_2(0)$ and $\mathbf{p}(0)$. This relation is expressed by the function $\mathscr{F}(\eta)$ of (9.16). We can consider $\mathscr{F}$ as an analytic function of c and use analytic continuation if c becomes complex. We can also consider $\mathscr{F}$ as a constant but with a discontinuous function U and examine the effect of c_i. Because η increases with a real c, it follows that the curve giving $\mathscr{F}$ with a constant positive c_i will be nested inside the curve shown in Fig. 12.1.

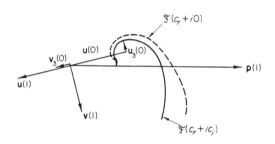

Fig. 12.1. Phase diagram for an unstable boundary layer oscillation. Compare with Fig. 11.1.

Thus every point located below the $\mathscr{F}$ curve and above the real axis will correspond to some unstable oscillation with particular values of α, ν, c_r, and c_i. The viscous solution also creates a quantity $\mathbf{v}_2(0)$ equal and opposite to $\mathbf{v}_3(0)$.

The phase diagram of Fig. 12.1 indicates the phases of the various components related to that of the pressure. The dotted line indicates $\mathscr{F}$ when c is purely real. Note that as a result of the rotations of the inviscid quantity $\mathbf{v}(1)$ and of the function $\mathscr{F}$ the contribution $\mathbf{u}_3$ must decrease. This means that η decreases. Because we still expect to find η in the range of 2 to ∞, this means that the unstable solutions will have diminishing phase velocities but rapidly increasing Reynolds numbers by (9.11) or (9.15). Exact results based on numerical calculations are given later.

It is also interesting to note that $\mathbf{v}_3(0)$ is almost out of phase with $\mathbf{p}(0)$, so that the oscillating layer actually exports energy to the viscous layer next to the wall, where dissipation takes place. This seems a waste of resources but, as we shall see in the study of flexible walls, the boundary layer oscillations always present this property.

(b) Unsteady Critical Layer, with Viscosity

If c_i is negative, the relations obtained for free-stream inviscid oscillations are still valid. The sign of $\overline{pv}$, now negative, indicates that energy is fed back to the mean flow by a process similar to a reversible thermodynamical transformation.

At the critical layer we may be tempted to use the inviscid approximation again. This is a fatal mistake. It changes the sign of the result of the integration (12.3), thus reversing that of $\mathbf{u}_3(0)$. This places the end of the vector $\mathbf{u}_1 + \mathbf{u}_3$ below the real axis in Fig. 12.1. Let us now consider the effect of a negative c_i on $\mathscr{F}$. By analytic continuation we find that $\mathscr{F}$ will now move away from the neutral curve. Thus it becomes impossible to find damped solutions in the usual way.

The mistake lies in the assumption that the inviscid solution, when taken through the critical layer, is a valid solution. In the limit of zero viscosity certain special effects occur which we now investigate. We start from the viscous vorticity equation and retain the term in c_i. The equation is simplified so that it reads

$$i\alpha\, U'(h)\left(y - h - i\,\frac{c_i}{U'(h)}\right)\mathbf{r} + U''(h)\,\mathbf{v}(h) = \nu\mathbf{r}''. \qquad (12.8)$$

We use the nondimensional variable $\eta = (y - h)/l$, with $l = (\nu/\alpha U'(h))^{1/3}$, and introduce the convenient parameter $L = c_i/lU'(h)$. The vorticity equation (12.8) can now be reduced to the form

$$(\eta - iL)\mathbf{r} - 1 = -i\mathbf{r}'', \qquad (12.9)$$

where the constant $iU''(h)\mathbf{v}(h)/\alpha lU'(h)$ has been replaced by unity for simplicity.

The corresponding inviscid solution is

$$\mathbf{r} = \frac{1}{\eta - iL}. \qquad (12.10)$$

We must now find the solution to (12.9), which reduces to (12.10) when η becomes large, on either side of $\eta = 0$.

The homogeneous form of (12.9), that is, the equation obtained by dropping the term -1, can be given in terms of Hankel functions. This solution leads to that of the complete equation by further integrations. We prefer to rely on the computer, using a method that may not be the most efficient but has the merit of being easily understood. An examination of (12.9) shows that if the real part of $\mathbf{r}$ is asymmetric the imaginary part must be symmetric. We assume that the solution has this symmetry and note that it agrees with the inviscid solution.

Let us begin with $L = 0$. We start the computations at $\eta = 6$ and proceed toward $\eta = 0$ in equal steps, $\Delta\eta = -0.01$.

The initial value of $\mathbf{r}$ is computed from the inviscid solution (12.10), and for the first try we also compute $\mathbf{r}'(6)$ from the same equation. The symmetry of the solution indicates that the real part of $\mathbf{r}$, as well as the imaginary part of $\mathbf{r}'$, should vanish at $\eta = 0$. The first attempt fails to fulfil these requirements, so we modify the complex value of $\mathbf{r}'(6)$ slightly and carefully observe the effect at $\eta = 0$. After a few attempts we can come close to a satisfactory solution. Hence we have an almost perfect match at $\eta = 6$ and note with satisfaction that our assumption on symmetry of the solution was justified. The symmetry guarantees that it will match the inviscid solution at $\eta = -6$. The real and imaginary parts of $\mathbf{r}$ are shown in Fig. 12.2 with a dotted curve that indicates the purely real inviscid solution. This solution is identical to the results given in Fig. 10.1 for a neutral wave except for a different choice for the value of $iU''(h)\mathbf{v}(h)$. Let us now reconsider the growing wave and treat $L = 1$ as an application. The procedure is the same, and the results are shown in Fig. 12.2. The real and imaginary parts of the inviscid solutions are now indicated by dotted lines.

The case $L = 10$ is another example of a growing oscillation in which the viscosity is so small that the nondimensional parameter L is large. As shown in Fig. 12.2, the solution varies smoothly and we note that it remains almost identical to the inviscid solution.

This sequence of results reveals that as the parameter L grows the viscous solution tends toward the inviscid solution.

Let us now examine the decaying waves, beginning with $L = -1$. The results are given in Fig. 12.3. Note that the viscous solution is not qualitatively different from the viscous neutral solution found with $L = 0$. The imaginary part of the inviscid solution, however, has a different behavior.

With $L = -3$, shown in Fig. 12.3, it was necessary to start the computations at $\eta = 10$. Between $\eta = 6$ and $\eta = 10$, the results are magnified 10 times to show clearly that they merge with the inviscid solution. Around the origin the solution oscillates vigorously, and it is evident that the two most important terms of (12.9) are $L\mathbf{r} = \mathbf{r}''$. Because L is negative, an oscillation is clearly possible.

The case $L = -5$ further illustrates the effect. As seen in Fig. 12.3 the amplitude of $\mathbf{r}$ becomes thousands of times larger than the inviscid solution. In the limit of $v \to 0$ with $c_i < 0$ we suspect that the solution cannot be represented by a power series near the origin. The solution may not even be analytic.

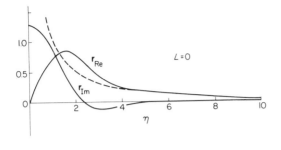

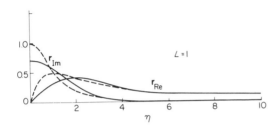

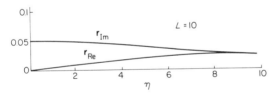

FIG. 12.2. Solutions of the vorticity equation through the critical layer for neutral and for growing waves. The solutions approach the inviscid solutions (dotted curves) as $L = c_i/lU'(h)$ becomes large.

The physical interpretation of this effect can be examined. We have limited ourselves to solutions that decay exponentially in time with the implicit assumption that they have so behaved for as long as we can remember. We are also examining flows with small viscosity, so that the vorticity fluctuations of long ago cannot easily diffuse. Away from the critical layer the transport process is fast enough to bring the fluid particles back to positions in which their net vorticity coincides with the mean vorticity. Thus the fluctuations disappear.

Near the critical layer, however, the transport term is slow, and diffusion of vorticity takes place. This causes some exchanges of vorticity which are similar to irreversible thermodynamic processes. The flow

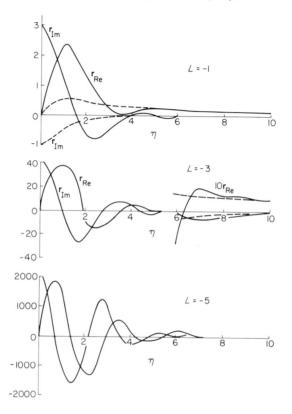

FIG. 12.3. Solutions of the vorticity equation through the critical layer for decaying waves. The solutions do not tend to the inviscid solution as $L = c_i/lU'(h)$ becomes large.

becomes spotted with vorticity fluctuations that persist for relatively long periods. The high values of $\mathbf{r}$ in Fig. 12.3 must therefore be regarded as vestiges of the past, slowly diffusing away.

The absolute magnitude of $\mathbf{r}$ in itself is meaningless because we have chosen $iU''(h)\mathbf{v}(h) = 1$. The results of Fig. 12.3 indicate that if the decay is allowed to proceed long enough the vorticity fluctuations inside of the critical layer will be thousands of times larger than those found outside of the critical layer.

It is also appropriate to remark that the effective thickness of the critical layer is approximately proportional to l when $|L| < 1$ and to $|L|$ itself when $|L| > 1$.

This terminates the discussion of the vorticity equation through the critical layer, and we are now ready to discuss the change in $\mathbf{u}$ across this layer. It should be clear that the inviscid solution is grossly inadequate if L is negative. For the viscous solutions, the computer can evaluate $\int_{-\infty}^{+\infty} \mathbf{r} \, d\eta$, and the results suggest that the value is independent of L and equal to $i\pi$. Of course, the numerical results become rapidly unreliable when L falls below -3, for then the solution oscillates vigorously.

The Cauchy theorem can be applied to the inviscid equation. It becomes necessary to make a detour about the origin in the complex plane of the variable η. The choice of two detours, giving the results $i\pi$ or $-i\pi$, leads to a delicate discussion (see Lin, 1955) which concludes that the limit of zero viscosity corresponds to one particular detour and that the other is meaningless. The integral of the solution of (12.8) becomes $i\pi$ for all values of L. In viscoelastic fluids v becomes a complex number and no difficulties are encountered as long as v is not purely imaginary. (See Section 60.) Although the use of a detour in the η-plane is perfectly rigorous and can even be performed with analog and digital computers, it leaves many students convinced that the results are not natural, for η is basically a distance from the wall and an essentially real quantity!

(c) THE DECAYING SOLUTION

Let us now return to the general problem of the decaying wave with the knowledge that $\mathbf{u}_3(0)$ is independent of both the sign and magnitude of L. The boundary condition on $\mathbf{u}$ leads to a diagram similar to Fig. 12.1, except that $\mathscr{F}(c)$ is on the convex side of the neutral curve, or $\mathscr{F}(c) = \mathscr{F}(c_r + i0)$. We can now find the value of c_i, which corresponds to a pair of values of α and v located outside of the neutral curve.

A final remark can be made. Let us consider a particular flow and a particular solution. The value of $U''(h)$ affects $\mathbf{u}_3$ and has a direct effect on c_i because it decides how far the vector sum $\mathbf{u}_1 + \mathbf{u}_3$ will reach above or below the neutral curve $\mathscr{F}(c_r)$. Let us now modify the program for our computations and arbitrarily reduce the value of U'' at the critical layer, leaving other values of U'' intact. This will reduce $\mathbf{u}_3$ and promote instability. Thus, although the term $\mathbf{u}_3$ is essential to oscillations from the point of view of energy production [Reynolds stresses, see (2.14)], it has the opposite effect as far as boundary conditions are concerned. This shows that such statements as "the critical layer is destabilizing" have a limited meaning and can lead to misunderstandings.

Chapter III

Machine Analysis of Viscous Flows

13. The Viscous Shear Layer

(a) GENERAL METHOD

The stability of the shear layer was examined in Section 5 without consideration of viscous effects. We now reconsider this problem and retain the viscous terms. Because the flow is unstable at large Reynolds numbers, we can expect that friction will be of minor importance except when the Reynolds number is low. Therefore we use a method suitable in the presence of large viscous terms. In particular, we follow a technique first devised by Brown (1961a). With the units of length and time such that $U_0 = 1$ and $L = 1$, we take the profile $U = \tanh y$ between $-3 < y < +3$. Outside this range U is constant and $U'' = 0$. The basic equation governing $\mathbf{v}$ can be obtained from (1.17), (1.18), and (1.19) by eliminating $\mathbf{u}$ and $\mathbf{p}$. It is known as the Orr-Sommerfeld equation and reads as follows:

$$(U - c)(\mathbf{v}'' - \alpha^2\mathbf{v}) - U''\mathbf{v} = -\frac{i\nu}{\alpha}(\mathbf{v}'''' - 2\alpha^2\mathbf{v}'' + \alpha^4\mathbf{v}). \qquad (13.1)$$

Note that if the viscous terms are dropped it reduces to the Rayleigh equation (5.7). Above the shear layer, when y is larger than 3, the coefficients of this equation are zero or constant and the solution has the form

$$\mathbf{v} = \sum_{n=1}^{4} A_n e^{p_n y}, \qquad (13.2)$$

with the exponents

$$p_1 = \alpha, \qquad\qquad p_2 = -\alpha,$$

$$p_3 = \alpha\left(1 + i\,\frac{1-c}{v\alpha}\right)^{1/2}, \qquad p_4 = -\alpha\left(1 + i\,\frac{1-c}{v\alpha}\right)^{1/2}. \qquad (13.3)$$

It is our convention that the square root will always be taken with a positive real part. The boundary conditions at infinity simply require that A_1 and A_3 be zero. In the region below the shear layer the same argument applies, and we write

$$\mathbf{v} = \sum_{n=1}^{4} B_n e^{q_n y}, \qquad (13.4)$$

with

$$q_1 = \alpha, \qquad\qquad q_2 = -\alpha,$$

$$q_3 = \alpha\left(1 + i\,\frac{1+c}{v\alpha}\right)^{1/2}, \qquad q_4 = -\alpha\left(1 + i\,\frac{1+c}{v\alpha}\right)^{1/2}. \qquad (13.5)$$

The conditions at minus infinity demand that B_2 and B_4 vanish.

We now use the computer to integrate this equation. Essentially, we must extend the upper solution through the shear layer and examine the conditions that permit a perfect match with the lower solutions. Because the upper solution contains two arbitrary constants, this requires two computing passes.

Some values of α and v, as well as some tentative values for c_r and c_i, must be chosen. We start at $y = +3$ and proceed in single precision with constant step size. We start the first pass with the choice $A_2 = 1$, $A_4 = 0$. In the free stream this corresponds to a simple pressure wave from below. With this choice of constants, the initial conditions are

$$\mathbf{v}(3) = e^{-3\alpha}, \qquad \mathbf{v}''(3) = p_2^2 e^{-3\alpha},$$
$$\mathbf{v}'(3) = p_2 e^{-3\alpha}, \qquad \mathbf{v}'''(3) = p_2^3 e^{-3\alpha}. \qquad (13.6)$$

As the computer proceeds toward $y = -3$ the solutions develop a characteristic oscillation roughly similar to $e^{p_3 y}$. We have seen that if U is constant the pressure fluctuations and the vorticity fluctuations will be completely uncoupled. In the shear layer the term $U''\mathbf{v}$ produces interaction between these processes. A pressure gradient generates a velocity fluctuation that disrupts the mean vorticity distribution. Furthermore, we have no assurance yet that some giant generators of vorticity fluctuations are not operating somewhere below the layer to produce a large term in B_4. The choice

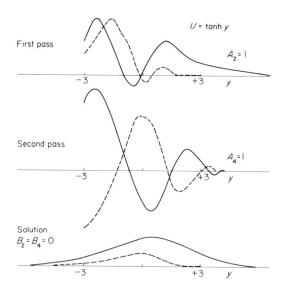

FIG. 13.1. Integration through a viscous shear layer, from $+3$ to -3 in two passes. Solid line for real part, broken line for imaginary part.

$A_4 = 0$ singles out a special solution such that no vorticity fluctuation reaches the upper free stream. It gives no guarantee as the computer proceeds downward that it will not encounter vorticity fluctuations of increasing magnitude, as shown in Fig. 13.1. This point is important. Among all the possible solutions that oscillate vigorously below the layer, we must find one that is well behaved as $y \to -\infty$.

At $y = -3$ we emerge with some values of $\mathbf{v}$ and its derivatives which we shall denote by $(d^n v / dy^n)_{\mathrm{I}}$; $n = 0, 1, 2, 3$. The Roman number denotes the first pass results. If we allowed the computer to proceed into the lower free stream, it would simply grind out some solutions of the form given by (13.4) with four particular constants which we denote as $(B_j)_{\mathrm{I}}$. Let us determine these values.

In general, we can obtain four relations between the four $d^n v / dy^n$ and the four quantities B_j. After inversion we obtain the following general relations:

$$B_1 = \frac{-e^{-\alpha y}}{2\alpha(q_3^2 - \alpha^2)} \left(\mathbf{v}''' + \alpha \mathbf{v}'' - q_3^2 \mathbf{v}' - \alpha q_3^2 \mathbf{v} \right), \qquad (13.7)$$

$$B_2 = \frac{e^{\alpha y}}{2\alpha(q_3^2 - \alpha^2)} \left(\mathbf{v}''' - \alpha \mathbf{v}'' - q_3^2 \mathbf{v}' - \alpha q_3^2 \mathbf{v} \right), \qquad (13.8)$$

$$B_3 = \frac{e^{-q_3 y}}{2q_3(q_3^2 - \alpha^2)} \, (\mathbf{v}''' + q_3 \mathbf{v}'' - \alpha^2 \mathbf{v}' - \alpha^2 q_3 \mathbf{v}), \qquad (13.9)$$

$$B_4 = \frac{-e^{q_3 y}}{2q_3(q_3^2 - \alpha^2)} \, (\mathbf{v}''' - q_3 \mathbf{v}'' - \alpha^2 \mathbf{v}' + \alpha^2 q_3 \mathbf{v}). \qquad (13.10)$$

The constants $(B_j)_{\mathrm{I}}$ are obtained by substituting the terms $d^n\mathbf{v}/dy^n$ in (13.7) to (13.10) at $y = -3$. These constants represent the amplitude and phase of the four basic oscillations whose combined existence in the lower free stream results in a single unit pressure wave of the upper free stream. The computer is instructed to determine and store the $(B_j)_{\mathrm{I}}$.

A second pass begins with $A_2 = 0$ and $A_4 = 1$ and corresponds to a unit vorticity wave diffusing from the shear layer upward. The initial conditions are similar to (13.6) with p_4 in place of p_2. During the integration the solution displays the same growing oscillations. At $y = -3$ we emerge with four quantities $(d^n\mathbf{v}/dy^n)_{\mathrm{II}}$. Just as in the first pass the computer determines the coefficients $(B_j)_{\mathrm{II}}$. The typical aspect of the second pass is shown in Fig. 13.1. Because the Orr-Sommerfeld equation is linear, a variety of solutions can be obtained by superimposing the results of the two preceding integrations. With two arbitrary values of A_2 and A_4, the oscillations of the lower free stream are given by the four following constants

$$B_j = A_2(B_j)_{\mathrm{I}} + A_4(B_j)_{\mathrm{II}}. \qquad (13.11)$$

We can choose $A_2 = 1$ without loss of generality and use the conditions $B_2 = 0$ to determine the desirable value of A_4. This choice of A_4 excludes all pressure waves that reach the shear layer from some source located far below. If we use these values of A_2 and A_4, the constant B_4 becomes

$$B_4 = (B_4)_{\mathrm{I}} - \frac{(B_2)_{\mathrm{I}}}{(B_2)_{\mathrm{II}}} \, (B_4)_{\mathrm{II}}. \qquad (13.12)$$

It indicates the amount of vorticity that must reach the oscillating layer from far below in order to produce an oscillation that presents the chosen values of α, ν, and c. The interesting oscillations of the flows are those for which B_4 vanishes. Therefore we must change the parameters α, ν, and c until the desired result is obtained.

It is convenient to keep α and ν constant and to vary c. Indeed, the complex number B_4 is an analytic function of the complex number c and the Cauchy relations are satisfied. This means that the change of B_4 produced by a small change of c_r is sufficient to determine by a simple multiplication the change of B_4 produced by any small complex change of

c. Thus, if we trace the curves of constant c_r and constant c_i in the complex plane of B_4, they will be orthogonal. The curves at constant α and constant c_r in the complex B_4 plane will not necessarily be orthogonal, and the search for an eigenvalue will be more time consuming. This search in c is described in the literature under the title of analytic continuation.

When an eigenvalue has been found, the corresponding solution can be computed by starting a third pass with the suitable value of A_4. The results then appear as shown in Fig. 13.1.

(b) AUTOMATED SEARCH FOR EIGENVALUES

The search for a zero for B_4 can be done by the machine proper. For this purpose, we describe a simple form of linear extrapolation. The computer starts with specified values of c and δc and computes $B_4(c)$ and $B_4(c + \delta c)$. Any other small change of c, such as Δc, would lead to

$$B_4(c + \Delta c) = B_4(c) + \frac{B_4(c + \delta c) - B_4(c)}{\delta c} \Delta c. \qquad (13.13)$$

Because we would like to have $B_4(c + \Delta c)$ as close to zero as possible, we choose

$$\Delta c = -\lambda \left(\frac{B_4(c + \delta c)}{B_4(c)} - 1 \right) \delta c, \qquad (13.14)$$

where the parameter λ is taken as unity if B_4 varies slowly with c. If the search meets with difficulties, it is advisable to reduce λ to 0.8 or 0.5 so that the computer will take small steps toward the goal.

The number c is replaced by $c + \Delta c$ and δc by $\mu \Delta c$. The computer is now ready to make another attempt. The role of the factor μ is to reduce the size of the exploratory steps as the target is approached. In practice we often use $\mu = \frac{1}{4}$. Thus the same subroutine is used for each subsequent attempt. After each attempt the computer verifies that Δc is less than some specified number, say, 10^{-3}. If this occurs, the computer prints SEARCH SUCCESSFUL and stops.

If this result is not obtained after N attempts, the computer prints SEARCH UNSUCCESSFUL. We have often used $N = 8$ and have found that if it did not converge in this time it would probably never do so.

(c) RESULTS OF NUMERICAL ANALYSIS

The shear layer was successfully treated by Betchov and Szewczyk (1963) who used the method described in the preceding sections. The

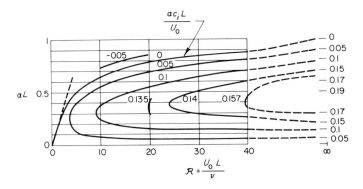

FIG. 13.2. Oscillations of a shear layer. $U = U_0 \tanh y/L$.

results are given in Fig. 13.2 for values of Reynolds numbers between 1 and 40. The rates of growth of the inviscid layer (infinite Reynolds number) are shown in Fig. 5.4. In these computations we note that if $\mathscr{R} > 40$ the quantities $(B_j)_I$ and $(B_j)_{II}$ become very large and the quantity B_4 of (13.12) is then a small difference between two large numbers. The truncation errors produce important fluctuations of B_4 and the computer wanders aimlessly in search of an eigenvalue.

The total growth of the numerical fluctuations varies with α, v, and the width of the shear layer. By taking advantage of the symmetry of the solution, however, we could gain a factor of two. The conditions $\mathbf{v}' = \mathbf{v}''' = 0$ should be applied at $y = 0$.

The case of a laminar boundary layer would not be essentially different from that of a half-shear layer, except that the boundary conditions should be $\mathbf{v} = \mathbf{u} = 0$ at $y = 0$. From the continuity equation we see that this amounts to $\mathbf{v} = \mathbf{v}' = 0$. Because we know from Section 2 that a boundary layer is certainly stable for $\mathscr{R} < 500$, we can anticipate serious difficulties. Indeed, as the computer proceeds from the edge of the layer toward the wall, in any of the necessary passes, the solution oscillates with an exponentially growing amplitude. This is caused by vorticity fluctuations rather than pressure fluctuations. The two passes therefore describe the same physical phenomenon, with different phases and magnitudes. The Jacobian of the solutions becomes relatively small, and, to comply with the boundary conditions, we must rely on a small difference between two large quantities. As shown by Brown (1959), the computations in single precision can be extended up to approximately $\mathscr{R} = 1000$. For a shear layer the growth factor is roughly $e^{\sqrt{0.7/2\alpha v}}$. For a boundary layer the same factor is smaller because the edge of the layer is located at $y = 1$ instead of $y = 6$. Thus v can be 50 times smaller in a boundary layer than in the shear layer before the errors become overwhelming. This corresponds to a boundary layer with a Reynolds number of 2000. A rough estimate of the growth factor given above for a shear layer leads to e^{27} or 10^{12}. In practice this factor is near 10^6, which indicates that the variation of the coefficients with y introduces an empirical factor of $\frac{1}{2}$ in the exponent. The accumulation of truncation errors contaminates the last two digits and reduces the accuracy to one part in 10^6. On the other hand, in double precision the computer would use 16 places. It would be necessary to decrease the step size to control the integration errors so that only the last four places would be lost to truncation errors. Thus a growth factor of 10^{12} is acceptable. This is twice the previous exponent or, for a laminar boundary layer, a maximum Reynolds number of 8000. In Section 14 we will see that special methods can be utilized to avoid this difficulty.

(d) EFFECT OF VISCOSITY ON DAMPED SOLUTIONS

An examination of Fig. 13.2 shows that as the Reynolds number increases the oscillations of a shear layer gradually merge with the results of the inviscid theory, as shown on the right side. However, this is true only for the unstable waves: it does not apply above the neutral line. Indeed, as shown in Fig. 13.2, the viscous analysis predicts negative values of c_i as indicated for $10 < \mathcal{R} < 20$. The inviscid equation has no solution for $\alpha > 1$.

It must be noted that the Rayleigh equation (8.4) is derived from a set of equations which is invariant to a reversal of the time variable and of all velocities. As a result, if it has a solution proportional to $e^{0.1t}$, it must have another solution proportional to $e^{-0.1t}$ and vice versa. Thus, for $\alpha = 0.5$, the solution corresponding to $c_i = 0.37$, as indicated in Fig. 5.4, is accompanied by another solution with $c_i = -0.37$. By the same argument any solution with a negative c_i, say for $\alpha = 2$, would lead to an unstable solution with $-c_i$. Furthermore, a numerical search at $\alpha > 1$ failed to show any solutions of the type $c = \pm a$, where a is real and positive. There are simply no continuous solutions of the inviscid problem for $\alpha > 1$. The two sets of inviscid eigenvalues are shown in Fig. 13.3 and labeled $v = 0$.

In the same figure we indicate numerical results for $\mathcal{R} = 20$ and $\mathcal{R} = 100$. They clearly suggest that the viscous theory merges with the inviscid theory for α less than unity. However, for $\alpha > 1$ something remarkable occurs.

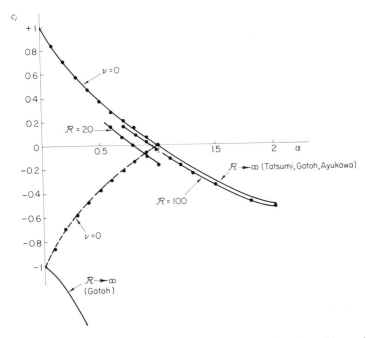

FIG. 13.3. Eigenvalues of the shear layer $U = \tanh y$ at various Reynolds numbers, in the limit $\nu \to 0$ and according to an inviscid theory. In all cases $c_r = 0$. The dots show numerical results.

The limit $\mathscr{R} \to \infty$ has been predicted theoretically by Tatsumi *et al.* (1964), with further refinements, such as the numerical results of Gotoh (1965). They used a delicate expansion in powers of c_i and found a special set of eigenvalues, starting from $c_i = 0$ at $\alpha = 1$ and approaching the asymptote $c_i = -0.376\alpha$ as $\alpha \to \infty$. As $\mathscr{R} \to \infty$, the eigenfunction becomes discontinuous at some point. Thus, if $\alpha = 2$ and $\mathscr{R} = 100$, a numerical study has shown that **v** is proportional to $e^{-\alpha y}$ for $y > 0$ and $Ae^{\alpha y}$ for $y < 0$. The constant A appears to be complex. In the vicinity of $y = 0$, **v** has a smooth behavior with a very large second derivative. As the viscosity vanishes, the solution approaches two different solutions of the inviscid equation in two adjacent portions of the flow. These functions are connected by a region in which the viscous terms retain an essential character. As the viscosity is decreased, the width of this region diminishes but it does not vanish. Through this intermediate region **v** is continuous, whereas **u**, proportional to **v′**, becomes discontinuous.

Clearly, the inviscid theory cannot predict the occurrence and locations of such viscous regions.

According to Gotoh (1965), another set of eigenvalues starts from $c_i = -1$ at $\alpha = 0$ and tends toward $c_i = -2.22\alpha$ as $\alpha \to \infty$. These results are indicated in Fig. 13.3.

It would be interesting to solve the time-dependent equation of motion with viscous terms, starting from the initial conditions which correspond to a negative eigenvalue of the inviscid equation.

(e) Effect of a Wall

In this section we examine the effect of a rigid wall placed on one side of the shear layer, say at the edge of the lower free stream. We use the profile $U = \tanh y$ between $-3 \leq y \leq 3$. This means that we impose the boundary condition $\mathbf{v} = 0$ at $y = -3$ instead of $y = -\infty$. The inviscid results appear in Fig. 5.6 and the viscous cases can be obtained without difficulty as long as the Reynolds number is not too large. Indeed, we must combine the results of the two passes of integration in such a way that $\mathbf{v}(-3)$ is zero and search for values of c to reach a combined solution in which $\mathbf{v}'(-3)$ is also zero. A fluid endowed with viscosity is prevented from moving along a rigid wall, and $\mathbf{u}$ must also vanish. In turn, this means $\mathbf{v}' = 0$ by application of the equation of continuity.

Some numerical results which show the neutral line are displayed in Fig. 13.4. The dotted lines correspond to the preceding results for a free

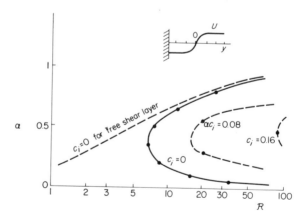

Fig. 13.4. Oscillation of a shear layer near a wall.

shear layer (see Fig. 13.2). We note that the wall has a more pronounced effect on long wavelengths than on short. This penalty on the oscillations at low α was already noticeable in the inviscid analysis (see Fig. 5.6). With the added effects of viscosity the values of c_i drop below the inviscid values. Note that if the Reynolds number is less than 6.7 all wavenumbers are damped. This is an example of a "critical Reynolds number." It marks the beginning of instability for some range of the wavenumber.

A few values of αc_i are indicated within the region of instability. For $\alpha > 1$ there are no inviscid solutions. At large Reynolds numbers, such as 100, we found the same type of strongly damped viscous solutions as shown in Fig. 13.3.

14. The Boundary Layer

(a) INTRODUCTION

The oscillations of the boundary layer occur at such high Reynolds numbers that the growth of the functions encountered during integration creates major difficulties. Roughly speaking, certain functions grow as fast as $e^{(\alpha \mathcal{R})^{1/2}}$ through the layer and this factor can be as large as 10^{40}. In single precision (eight digits), and with a variety of errors, it becomes difficult to find eigenvalues if the growth exceeds 10^6. On the other hand, in double precision and complex arithmetic, the machine time becomes excessive and the amplification cannot exceed 10^{12}, approximatively speaking.

The range of integration can be divided in two halves (cf. Nachtsheim, 1964). The solutions start at the outer edge and are matched with solutions which start at the wall—all in single precision. Again this seems limited to the level of 10^{12}. Various other methods have been used that require either the inversion of a large matrix (Thomas, 1953; Kurtz and Crandall, 1962; Gallagher and Mercer, 1962, 1964), or a combination of integration and power series (Powers, Heiche, and Shen, 1963). These methods become unreliable or expensive if the amplification exceeds 10^8.

Major progress in this matter was achieved by Kaplan (1964) who divided the layer into many small segments and purified the solution of the dangerous component at the end of each segment. In single precision, this method works with functions that grow by a factor 10^{40}. The method of Kaplan is akin to a general procedure of orthogonolization which is briefly outlined by Bellman and Kalaba (1965). This procedure has also been used by Wazzan, Okamura, and Smith (1966).

Their results are in good agreement with those of Kaplan, and amplifications are extended to the order of 10^{80}.

In this section, we present a modified form of Kaplan's scheme with
application to the Blasius boundary layer. Of all the methods, it seems to
be the easiest to understand, to program, and to use. The discussion of the
more sophisticated procedure of orthogonolization is consigned to Appendix III; for more complicated problems, this procedure may become
necessary.

(b) Local Modes

As a preliminary step to the understanding of this process, we define
four local modes $A_j(y)$. Let us consider a thin slice of the boundary layer
between $y - \Delta y$ and $y + \Delta y$. Within this region the coefficients $U(y)$ and
$U''(y)$ of the Orr-Sommerfeld equation can be regarded as constants
and therefore every solution can be expanded in the form

$$\mathbf{v}(y) = \sum_{j=1}^{4} A_j e^{p_j y}. \tag{14.1}$$

The exponential factors are roots of the equation

$$(U - c)(p^2 - \alpha^2) - U'' = -i\frac{v}{\alpha}(p^2 - \alpha^2)^2. \tag{14.2}$$

We have for large Reynolds numbers the approximate values

$$p_1 = \left(\alpha^2 + \frac{U''}{U - c}\right)^{1/2}, \qquad p_2 = -\left(\alpha^2 + \frac{U''}{U - c}\right)^{1/2}, \tag{14.3}$$

$$p_3 = \left(i\alpha\frac{U - c}{v}\right)^{1/2}, \qquad p_4 = -\left(i\alpha\frac{U - c}{v}\right)^{1/2}. \tag{14.4}$$

The factors A_j are related to the quantities $\mathbf{v}$, $\mathbf{v}'$, $\mathbf{v}''$, and $\mathbf{v}'''$ by four relations
similar to (13.7) to (13.10). Physically speaking, A_1 and A_2 represent the
amplitude and phase of two inviscid processes which are no longer ir-
rotational pressure waves. Because the term in U'' often exceeds the term in
α^2 of (14.3), A_1 and A_2 correspond to inviscid rotational oscillations with
p_1 and p_2 almost imaginary.

The terms A_3 and A_4 correspond to the familiar viscous vorticity
waves from above and below.

If we compare the set of four A_j of the slice centered on y with the set of
the adjacent slice centered on $y - 2\Delta y$, we will notice a difference due to

change in p_i and in $\mathbf{v}$, $\mathbf{v}'$, $\mathbf{v}''$, $\mathbf{v}'''$. Thus, as the coefficients U, U'' change from one slice to the next, the A_j vary. They represent the local amplitudes of four elementary processes which appear with different intensity and phase throughout the layer. Because we are integrating from $y = 1$ to $y = 0$, we are especially interested in the quantity A_4. The quantities $\mathbf{v}'''$ and $\mathbf{v}''''$ are stored in the computer, and therefore we can obtain a very good approximation to A_4. Neglecting the terms in A_1 and A_2, we can combine the two relations giving $\mathbf{v}'''$ and $\mathbf{v}''''$ to yield

$$A_4 = \frac{\mathbf{v}'''' - p_3\mathbf{v}'''}{2(p_3)^3} e^{p_3 y}. \tag{14.5}$$

(c) THE FIRST INTEGRATION PASS

We now obtain a basic solution, S_0, throughout the entire boundary layer which should contain as much as possible of the dangerous mode A_4. This solution will then be retained for use in future corrections of other solutions.

After choosing the step size Δy of the Runge-Kutta numerical integration procedure, we pick a number of stations, located at $y_1, y_2, \ldots, y_w$, as shown in Fig. 14.1. For example, we may take $\Delta y = 0.005$ and 10 stations 20 steps apart.

The solution S_0 corresponds to a pure vorticity wave in the free stream. To determine the initial conditions we must choose $A_1 = A_2 = A_3 = 0$ and $A_4 = 1$ at $y = 1$ and $\mathbf{v}(1)$ to $\mathbf{v}'''(1)$ in a manner analogous to (13.6). The computer proceeds to the wall in one uninterrupted pass. At each station, however, the values of $\mathbf{v}$, $\mathbf{v}'$, $\mathbf{v}'''$ and $\mathbf{v}''''$ are carefully stored in the memory. This function $\mathbf{v}$ will be noted as $(\mathbf{v})_{S_0}$.

The solution obtained in this manner may grow to very large amplitudes, as illustrated in Fig. 14.2. Sometimes, in fact, the value of $\mathbf{v}$ becomes larger than 10^{40} and the computer is unable to proceed. This difficulty can be remedied by changing the initial A_4 to a value, say, of $A_4 = 10^{-20}$.

(d) THE SECOND INTEGRATION PASS

We must now obtain another solution which is independent of the solution S_0 computed in the first pass. We start with a particular solution S_1, but, in order to control the outbreak of vorticity fluctuations, we shall not carry this new solution beyond the first station y_1. At y_1 we shift our interest to another solution S_2, which, at station y_1, is relatively free of

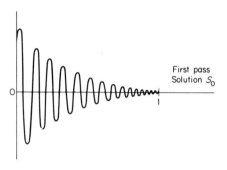

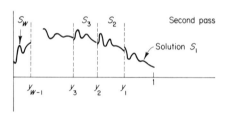

Fig. 14.1. Numerical integration by the method of Kaplan (from $y = 1$ down to $y = 0$, in two passes).

the unwanted mode. The procedure is repeated at each station, as shown in Fig. 14.1.

The computer starts at $y = 1$ with $A_1 = 1$ and $A_1 = A_3 = A_4 = 0$ and the associated initial conditions. It begins the integration and reaches y_1 with certain values of $\mathbf{v}$ and its derivatives. Later this function is noted as $(\mathbf{v})_{S_1}$. From y_1 to y_2 the computer determines another solution S_2 whose initial conditions at y_1 must now be found. We define S_2 as the sum of S_1 and some suitable fraction of S_0. Symbolically, this means

$$S_2 = S_1 + a_1 S_0. \tag{14.6}$$

Thus the new function $\mathbf{v}$ is related to the two others by

$$(\mathbf{v})_{S_2} = (\mathbf{v})_{S_1} + a_1(\mathbf{v})_{S_0} \tag{14.7}$$

Similar relations exist from the derivatives. The coefficient a_1 is determined

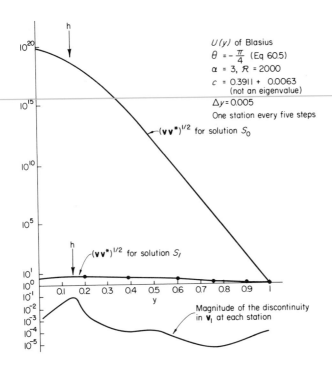

FIG. 14.2. An example illustrating the growth of solution S_0, the steadiness of solution S_1, and the corrections of **v**.

by the condition that the mode A_4 of solution S_2 must vanish at station y_1. This means that

$$a_1 = -\frac{(A_4)_{S_1}}{(A_4)_{S_0}}, \qquad (14.8)$$

where $(A_4)_{S_1}$ is obtained by applying (14.5) with the terms associated with solution S_1. Likewise, $(A_4)_{S_0}$ is generated from S_0.

The initial values for solution S_2 are now obtained by application of (14.7) at $y = y_1$, along with three similar equations for the derivatives. This connection between S_2 and S_1 is therefore uniquely established.

The computer now follows S_2 up to station y_2, where a similar procedure is pursued to eliminate the fourth mode. This procedure is repeated at each station to establish a link between the incoming and outgoing solutions. Hence the computer deals successively with several solutions by following every one between two adjacent stations only.

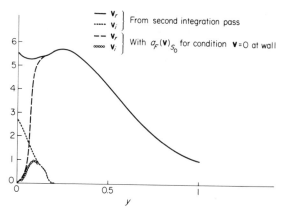

FIG. 14.3. Results of the second integration pass and the application of one wall condition. This case gives $v'(0) = -0.5 - i\,2.2$.

As shown in Fig. 14.2, the magnitude of $(\mathbf{v})_{S_i}$ remains of the order of unity, with small changes at the stations. The last solution reaches the wall and is formally related to S_1 and S_0 by

$$S_W = S_1 + (a_1 + a_2 + \cdots + a_{W-1})S_0; \qquad (14.9)$$

but note that some a_j may be so small and some components of S_0 so large that the computer cannot process any numerical relation accurately.

An example of the behavior of successive solutions is given in Fig. 14.3. It can be seen that the discontinuities of $\mathbf{v}$ are insignificant, whereas the discontinuities of $\mathbf{v}''$ are of some importance (see Fig. 14.4).

The result of the second pass of integration is therefore a solution corresponding to a unit pressure wave in the free stream combined with vorticity fluctuations vanishing at $y = \infty$, as shown by Eq. 14.9. The values of $\mathbf{v}$ and its derivative at the wall are known.

(e) BOUNDARY CONDITIONS

We shall now search for a final solution S_F to satisfy the wall boundary conditions $\mathbf{v} = \mathbf{v}' = 0$. We define a combination of the two passes such that

$$v(0) = (v(0))_{S_W} + a_F(v(0))_{S_0} = 0. \qquad (14.10)$$

The corresponding value of $\mathbf{v}'$ is given by the expression

$$v'(0) = (v'(0))_{S_W} - \frac{(v(0))_{S_W}}{(v(0))_{S_0}}(v'(0))_{S_0}. \qquad (14.11)$$

The effect of a_F is given in Fig. 14.3. This quantity is a function of α, v, and c. A search is now prepared to find the correct combination of parameters that will render $\mathbf{v}'(0)$ sufficiently small.

The eigenvalues of the Blasius boundary layer which were obtained by Kaplan are shown in Fig. 14.5. Later results based on the improved method of the orthogonalization procedure (see Appendix III) have been computed by Wazzan, Okamura, and Smith (1966) and are shown in Fig. 14.6. The rates of growth given in Fig. 14.6 are more accurate than those of Fig. 14.5 because of smaller tolerances in the search for eigenvalues. Wazzan *et al.* also treated certain velocity profiles of the Falkner-Skan variety.

(f) MISCELLANEOUS REMARKS

Some additional remarks pertinent to the aspects of this numerical method are worth noting. The eigenfunctions, for example, cannot be

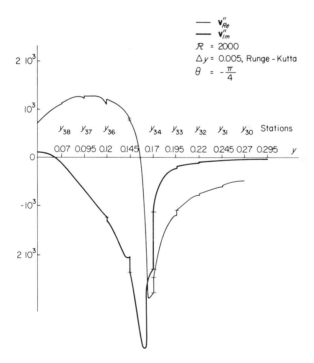

FIG. 14.4. Details of $\mathbf{v}''$ through critical layer, in the case of Fig. 14.2. Note the discontinuities of $\mathbf{v}''$.

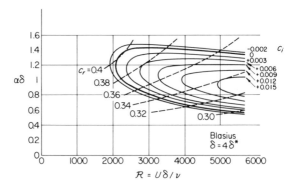

FIG. 14.5. Eigenvalues for the Blasius boundary layer (after Kaplan, 1964).

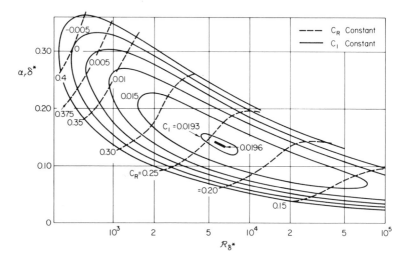

FIG. 14.6. Eigenvalues for the Blasius boundary layer (after Wazzan *et al.*, 1966).

obtained exactly. This is especially true of the higher derivatives of the eigenfunctions. However, if enough stations are used, the discontinuities in v and v' are negligible.

Because one step of integration following the procedure of Runge-Kutta approximates the solution by a fourth-order polynomial, it seems reasonable to space the stations several steps apart. Little machine time is added, however, if the stations are only one step apart. Indeed, the

Runge-Kutta procedure is more time-consuming than the transition from one solution to the next.

The last remark deals with the suppression of the dangerous mode at each station. Instead of reducing A_4 to zero as precisely as possible by (14.5), we can use such approximations as

$$A_4 = \mathbf{v}'''' - p_3\mathbf{v}''' \tag{14.12}$$

or even

$$A_4 = \mathbf{v}''''. \tag{14.13}$$

In his original work Kaplan stamped out the viscous terms completely. In effect, he wrote at each station

$$A_4 = (U - c)(\mathbf{v}'' - \alpha^2\mathbf{v}) - U''\mathbf{v} = 0. \tag{14.14}$$

More comments are given in Appendix III.

15. Channel Flows

Historically speaking the study of channel flows has been an inspiration and a challenge to two generations of applied mathematicians. By definition, a channel flow is one confined by two walls, and therefore the boundary conditions are applied at two finite values of y. As a result, the problem is well posed and the mathematical analysis is considerably simplified.

It should be realized that the channel flows we are discussing are those of the fully developed variety in which U is a function of y only. In practice, however, there is a long entrance region beginning with a flow of constant velocity between two thin boundary layers along each wall. The stability of the entrance flow has been studied by Tatsumi (1952).

Eventually the two boundary layers merge and the mean flow becomes truly independent of the variable x (except for the pressure, of course). There are two well-known examples of this system: (a) Poiseuille flow, in which both walls are at rest and the flow is driven by a constant pressure gradient, and (b) Couette flow, in which there is no pressure gradient and the motion of the walls is a parallel flow of one wall with respect to the other.

(a) POISEUILLE FLOW

The undisturbed Poiseuille flow is defined by

$$U(y) = U_0\left(1 - \left(\frac{y}{L}\right)^2\right). \tag{15.1}$$

We shall choose our units in such a way that U_0 and L are unity.

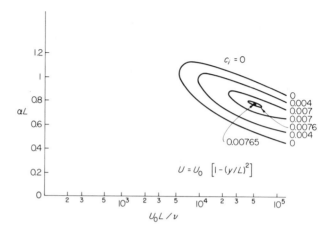

FIG. 15.1. Stability of plane Poiseuille flow (after Shen, 1954).

Answers to the problem were found by Shen (1954) who used an asymptotic power series. The results of this work are given in Fig. 15.1. He found that the minimum Reynolds number amounts to 5360 at $\alpha = 1.05$. Earlier Thomas (1953), in one of the first applications of the computer in the field of hydrodynamic stability, had found a minimum Reynolds number of 5780 at $\alpha = 1.02$. More recently, Nachtsheim (1964), using a modern computer and the method described at the beginning of Section 13, found a minimum Reynolds number of 5767 at $\alpha = 1.02$.

The method described in Section 14, due to Kaplan, can be used to study the stability of any channel flow, except that one should start with a wall instead of at the edge of a free stream. At the wall both $\mathbf{v}$ and $\mathbf{v}'$ will vanish, so that the first integration pass could start with $\mathbf{v}'' = 0$, and $\mathbf{v}''' = 1$. The second integration pass would then start with $\mathbf{v} = \mathbf{v}' = 0$, $\mathbf{v}'' = 1$ and $\mathbf{v}''' = 0$. The results of the two integration passes are combined at the other wall exactly as in Section 14.

The existence of higher modes for the eigenvalues has been disclosed by Grohne (1954) and, except for the lowest one corresponding to the work of Shen for example, they are strongly damped. These additional solutions are essentially similar to those of the plane Couette flow and are discussed in paragraph (c).

(b) SIMPLIFIED ANALYSIS OF PLANE POISEUILLE FLOW

We now discuss briefly the neutrally stable oscillations at a large Reynolds number. First, we examine the Rayleigh equation and note that it is possible to have both a symmetric and an antisymmetric solution. We suspect that the symmetric solution is less stable on the grounds that an antisymmetric solution is more exposed to viscous dissipation. Recognizing that we need a symmetric function in y with zero boundary conditions at either end and by taking advantage of the precedent of the boundary layer, we are led to the following conjecture:

$$v = a(1 - y^2). \tag{15.2}$$

From the continuity equation we can obtain

$$u = \frac{iv'}{\alpha} = -\frac{2iay}{\alpha}. \tag{15.3}$$

Because our description of v is more refined than that of u, we determine the pressure by integration of the v equation. Thus we find

$$p(-1) = -i\alpha a\left(\frac{8 - 10c}{15}\right), \tag{15.4}$$

which is the pressure at one wall.

We now examine the effect of viscosity along a wall. Because we expect large Reynolds numbers, we know that the oscillating viscous region along one wall will not interfere with its counterpart along the other wall. Thus the method of Section 9 becomes applicable, with the difference that the pressure gradient used to determine the viscous vorticity gradient is based on the equation

$$i\alpha p(-1) = vu''(-1). \tag{15.5}$$

As far as viscous effects are concerned, we can use the analogy of the oscillating boundary layer. The effects of friction at the critical layer can be treated, for example, as in Section 10. We now have all of the elements necessary to satisfy all the boundary conditions, namely (a) the inviscid flow, (b) the flow induced by the vorticity gained at the walls, and (c) the flow modification introduced at the critical layer. The results of this simplified analysis are in reasonable agreement with those shown in Fig. 15.1.

(c) PLANE COUETTE FLOW

For the Couette flow the mean velocity varies linearly between the two plates. We take as unity half the distance between the plates and half the velocity difference. Because $U'' = 0$ everywhere, we immediately notice

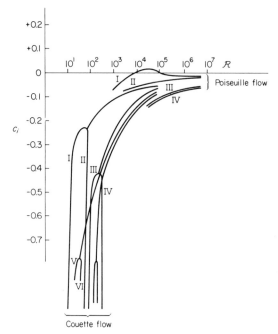

FIG. 15.2. The values of c as functions of $\mathscr{R}$ for the flows of Poiseuille ($\alpha = 0.87$) and Couette ($\alpha = 1$). Note the merging of modes at low Reynolds numbers and the unique positive portion on the Poiseuille curve for mode I.

that there is no production of vorticity fluctuations. In fact, the vorticity fluctuations are merely transported and diffused. Therefore we should not be surprised that all available data for this problem point toward complete stability of this flow to small disturbances. There is, however, little doubt that this flow will in reality become turbulent at large enough Reynolds numbers.

A major contribution was made by Grohne (1954) to the solution of this problem. He relied on the extensive tables of the Hankel function and found an infinity of modes which are all damped. These results are shown in Fig. 15.2. At low Reynolds numbers, typically $\mathscr{R} < 1000$, the perturbations do not propagate and vanish exponentially. At large Reynolds numbers these stationary modes coalesce and another type of solution appears. Essentially, we find one mode traveling with one plate and decaying at a slower rate and the other mode traveling with the other plate decaying with the same lower rate. At a Reynolds number of 10^4 the first 10 modes have phase velocities between 60 and 90% of the respective plate velocities; the values of c_i range between -0.05 and -0.02. In all, Grohne computed 12 modes for the Couette flow.

Later, Gallagher and Mercer (1962, 1964) reexamined the problem with the assistance of a computer. They made use of matrix methods and calculated the first few roots. In the worst case the matrix was of order 20. In the $\alpha\mathcal{R}$-plane the region in which c is complex begins to show a shape similar to that of the higher modes of a Poiseuille flow.

Finally, as $\mathcal{R} \to \infty$, the c_i of all modes becomes very small and tends asymptotically toward zero. It appears, moreover, that each mode retains its individuality.

16. Viscous Jets and Wakes

The instability of jets and wakes at high Reynolds numbers must agree with the findings of the inviscid studies. In the viscous shear layer we have found that the effects of viscosity are felt below a Reynolds number of approximately 50, and indeed there was no critical value of the Reynolds number found at all.

The case of the jet with $U = 1/\cosh^2 y$ has been singled out because it corresponds to a similarity solution of the viscous equations of the boundary layer type (cf. Schlichting, 1933c; Bickley, 1937; Tatsumi and Kakutani, 1958). Tatsumi and Kakutani investigated the stability of this jet in viscous flow. These authors used asymptotic power series and found that the viscosity has a stabilizing influence, beginning below a Reynolds number of 100. A critical Reynolds number was calculated to be 4.0 at $\alpha = 0.2$. More recently, Kaplan used his computational scheme, described in Section 14,

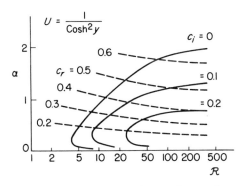

FIG. 16.1. Stability of a viscous symmetric jet (after Kaplan, 1964; Tatsumi and Kakutani, 1958).

to confirm these results. In addition, Kaplan computed the complete neutral line and mapped the region of instability; his work is shown in Fig. 16.1. Again, the machine computations could easily be applied to any other jet or wake and with a complex α rather than a complex ω.

When α becomes small, say $\alpha < 0.1$, both theoretical and numerical analyses encounter various obstacles because the ratio v/α becomes large. These difficulties should not be taken too seriously, for $\alpha = 0.1$ corresponds to a wavelength of $\lambda = 60$, and therefore the fact that the basic profile spreads gradually with x renders the Orr-Sommerfeld equation invalid for such long wavelengths.

Note added in proof: A recent reference which is devoted specifically to machine analysis of the Orr-Sommerfeld equation should be noted. This work is due to Landahl (1966) and allows for great diversity in its use. For example, the boundary layer over flexible (see Section 57) as well as solid boundaries is contained among the list of possibilities. Moreover, the program is so fixed that it permits combinations of the standard varieties by insertion or removal of the appropriate element.

Part Two

The General Problem

Chapter IV

General Analysis of Oscillating Flows

17. Three-Dimensional Governing Equations

After examining in detail the stability of two-dimensional flows in Part One we now consider the general case of a three-dimensional flow and extend the preceding results.

The general equations governing an incompressible three-dimensional flow are

$$\frac{\partial u}{\partial t} + u\frac{\partial u}{\partial x} + v\frac{\partial u}{\partial y} + w\frac{\partial u}{\partial z} + \frac{\partial p}{\partial x} = \varphi_x + \frac{\partial \tau_{xx}}{\partial x} + \frac{\partial \tau_{xy}}{\partial y} + \frac{\partial \tau_{xz}}{\partial z}, \quad (17.1)$$

$$\frac{\partial v}{\partial t} + u\frac{\partial v}{\partial x} + v\frac{\partial v}{\partial y} + w\frac{\partial v}{\partial z} + \frac{\partial p}{\partial y} = \varphi_y + \frac{\partial \tau_{yx}}{\partial x} + \frac{\partial \tau_{yy}}{\partial y} + \frac{\partial \tau_{yz}}{\partial z}, \quad (17.2)$$

$$\frac{\partial w}{\partial t} + u\frac{\partial w}{\partial x} + v\frac{\partial w}{\partial y} + w\frac{\partial w}{\partial z} + \frac{\partial p}{\partial z} = \varphi_z + \frac{\partial \tau_{zx}}{\partial x} + \frac{\partial \tau_{zy}}{\partial y} + \frac{\partial \tau_{zz}}{\partial z}, \quad (17.3)$$

$$\frac{\partial u}{\partial x} + \frac{\partial v}{\partial y} + \frac{\partial w}{\partial z} = 0, \quad (17.4)$$

where τ_{ij} are the components of a stress tensor produced by friction forces (divided by the density) and φ_i is the component of a body force and again p is the kinematic pressure.

All investigations of flow stability, which have used this set of equations as a starting point, have, by necessity, been limited to problems that can be reduced to a system of ordinary differential equations. Almost the only method for dealing with the kind of partial differential equations that describe general three-dimensional flows requires the massive use of a computer. Some two-dimensional results in this respect have been obtained by Fromm and Harlow (1963). They have the additional advantage, though, that they are not limited to linearized perturbations but rather give exact solutions to the two-dimensional problem.

We now introduce the restrictions that lead to the desired level of difficulty: a system of ordinary differential equations. This system stipulates that all coefficients of the fluctuation equations must be functions of one independent variable only, and we choose the variable y for this purpose. Thus the mean components U, V, and W, which will be coupled to the perturbations by nonlinear terms, must be functions of y alone.

Because the mean flow must satisfy the equation of continuity, V can at most be a constant. Because P is not coupled to fluctuations, it could vary linearly with x or z, as, for example, in viscous channel flows.

We have introduced the friction stress tensor because we wish to include the possibility of a variable viscosity. The presence of a body force is necessary to deal with centrifugal and Coriolis forces; it may be used further for studies in magnetohydrodynamics.

In accordance with our policy of aiming at a system of ordinary differential equations, we assume that the perturbations are small, making it possible to linearize the equations. The variation with x, z, and t is expressed by a factor $e^{i(\alpha x + \beta z)}e^{-i\omega t}$. It must be understood that α, β, and ω could be complex, in which case the assumption of small perturbation breaks down on some boundary in space or time. Most experimental studies (Klebanoff *et al.*, 1962) have been performed with a real frequency ω, a complex α, and a real β. In this respect, a point of rigor must be raised. By application of Laplace transforms, a broad class of linear perturbations can be expressed by a sum of complex exponentials in x and z. However, unstable perturbations with no Laplace transforms may exist. In practice none has been encountered, but it must be recognized that such instabilities would slip completely through the net of our analysis. As an illustration, perhaps we can imagine a perturbation growing as e^{t^2}.

An interesting but little noted fact about the z-dependence can now be made. The basic equations of motion are invariant to changes of sign of both the variable z and the velocity component w. Thus, if a particular oscillation proportional to $e^{i\beta z}$ is a solution to the problem, another oscillation exists in which β is replaced by $-\beta$. Assuming that β is real, we can combine the oscillation in $e^{i\beta z}$ and $e^{-i\beta z}$, so that the sum vanishes at certain places. A case in point would be boundary conditions on w at two walls perpendicular to the z-axis, just as in standing waves. If β is complex, two solutions will vanish at $+\infty$ and $-\infty$. However, it is impossible to form a linear combination of the complex function w which has more than one zero. The existence of a solution $e^{i\beta z}$ does not necessarily imply the existence of a solution in $e^{-i\beta^* z}$ without a change in α or ω.

With $\mathbf{u}$, $\mathbf{v}$, $\mathbf{w}$, $\mathbf{p}$, τ, φ to indicate the complex amplitude functions (all functions of y) in the velocity, pressure, stress, and body force, the linearized set of equations is

$$\left(U - \frac{\omega}{\alpha}\right) i\alpha\mathbf{u} + V_0\mathbf{u}' + i\beta W\mathbf{u} + U'\mathbf{v} = -i\alpha\,\mathbf{p} + \varphi_x$$

$$+ i\alpha\tau_{xx} + \tau'_{xy} + i\beta\tau_{xz}, \quad (17.5)$$

$$\left(U - \frac{\omega}{\alpha}\right) i\alpha\mathbf{v} + V_0\mathbf{v}' + i\beta W\mathbf{v} = -\mathbf{p}' + \varphi_y$$

$$+ i\alpha\tau_{yx} + \tau'_{yy} + i\beta\tau_{yz}, \quad (17.6)$$

$$\left(U - \frac{\omega}{\alpha}\right) i\alpha\mathbf{w} + V_0\mathbf{w}' + i\beta W\mathbf{w} + W'\mathbf{v} = -i\beta\mathbf{p} + \varphi_z$$

$$+ i\alpha\tau_{zx} + \tau'_{zy} + i\beta\tau_{zz} \quad (17.7)$$

$$i\alpha\mathbf{u} + \mathbf{v}' + i\beta\mathbf{w} = 0. \quad (17.8)$$

The primes denote derivatives with respect to y.

18. Equivalent Transformations

We now introduce a set of transformations, first used by Squire (1933), that reduces a three-dimensional problem to an equivalent two-dimensional system. This technique is of general value in a large variety of stability problems, including flows with body forces, variable viscosity, compressibility, and even magnetohydrodynamic forces. Moreover, the method permits the use of the two-dimensional computer program to obtain the eigenvalues of the full three-dimensional problem. In the limited case of incompressible flows it leads to the famous Squire theorem.

Let us consider the standard incompressible three-dimensional problem, with constant viscosity and no body forces. We then introduce the quantities

$$\bar{\alpha}^2 = \alpha^2 + \beta^2, \quad (18.1)$$

$$\bar{\alpha}\,\bar{\mathbf{u}} = \alpha\mathbf{u} + \beta\mathbf{w}. \quad (18.2)$$

Now, by adding the product of Eq. (17.5) by α to the product of (17.7) by β we have in this new notation

$$i\alpha\left(U + \frac{\beta}{\alpha}W - c\right)\bar{\alpha}\,\bar{\mathbf{u}} + V_0\bar{\alpha}\,\bar{\mathbf{u}}' + \alpha\left(U' + \frac{\beta}{\alpha}W'\right)\mathbf{v} + i\bar{\alpha}^2\mathbf{p} = \nu\bar{\alpha}(\bar{\mathbf{u}}'' - \bar{\alpha}^2\bar{\mathbf{u}}),$$

$$(18.3)$$

in which we substituted directly for the viscous terms and used the fact that $\omega = \alpha c$. Note that this equation corresponds to the fluctuation of the component parallel to the wavenumber vector.

Going further, we define the additional relations

$$\bar{\mathbf{v}} = \mathbf{v}, \tag{18.4}$$

$$\frac{\bar{\mathbf{p}}}{\bar{\alpha}} = \frac{\mathbf{p}}{\alpha}, \tag{18.5}$$

$$\frac{\bar{V}_0}{\bar{\alpha}} = \frac{V_0}{\alpha} \tag{18.6}$$

$$\bar{U} = U + \frac{\beta}{\alpha} W, \tag{18.7}$$

$$\bar{c} = c. \tag{18.8}$$

Then, after substituting (18.4) to (18.8) into (18.3) and (17.6) we find the corresponding two equations in terms of the new variables:

$$i\bar{\alpha}(\bar{U} - c)\bar{\mathbf{u}} + \bar{U}'\bar{\mathbf{v}} + i\bar{\alpha}\,\bar{\mathbf{p}} = \bar{V}_0\bar{\mathbf{u}}' + v\,\frac{\bar{\alpha}}{\alpha}\,(\bar{\mathbf{u}}'' - \bar{\alpha}^2\bar{\mathbf{u}}), \tag{18.9}$$

$$i\bar{\alpha}(\bar{U} - \bar{c})\bar{\mathbf{v}} + \bar{\mathbf{p}}' = \bar{V}_0\bar{\mathbf{v}} + v\,\frac{\bar{\alpha}}{\alpha}\,(\bar{\mathbf{v}}'' - \bar{\alpha}^2\bar{\mathbf{v}}). \tag{18.10}$$

Close inspection of (18.9) and (18.10) reveals that they are analogous to the equations governing a two-dimensional oscillation except for the presence of the factor that multiplies the viscous terms. This leads us to the conclusion that the stability of a three-dimensional wave is governed by the same equation as that of a two-dimensional wave but having a viscosity given by

$$\bar{v} = v\,\frac{\bar{\alpha}}{\alpha}, \tag{18.11}$$

and we emphasize that α, $\bar{\alpha}$ are taken as real only.

Thus the eigenvalues of a three-dimensional flow, in the presence of W and V components, can be obtained by using the same method of integration as that used for a two-dimensional oscillation.

The determination of the eigenfunctions, however, requires some additional labor. Indeed, if $\bar{\mathbf{u}}$ is known, the calculation of either $\mathbf{u}$ or $\mathbf{w}$ requires some additional integration. It is necessary to integrate (17.5) or (17.7), where $\mathbf{v}$ and $\mathbf{p}$ are already known from the integrations leading to the determination of the eigenvalues. Once **w** is known, **u** is immediately made available by (18.2). The root of this difficulty lies in the presence of the term $U'v$ in the equation for **u**.

In the new notation the three-dimensional continuity equation reads

$$i\bar{\alpha}\bar{u} + \bar{v}' = 0. \tag{18.12}$$

By elimination of $\bar{p}$ between (18.9) and (18.10), the equation in $\bar{v}$ can be formed. We find

$$(\bar{U} - \bar{c})(\bar{v}'' - \bar{\alpha}^2\bar{v}) - \bar{U}''\bar{v} = i\frac{V_0}{\bar{\alpha}}(\bar{v}''' - \bar{\alpha}^2\bar{v}') - \frac{i\bar{v}}{\bar{\alpha}}(\bar{v}'''' - 2\bar{\alpha}^2\bar{v}'' + \bar{\alpha}^4\bar{v}).$$

$$\tag{18.13}$$

The term involving V_0 is seldom retained because the majority of mean flows are confined by rigid walls which would not allow any mean flow in the y-direction. It corresponds to the transport of vorticity fluctuations by the mean flow along the y-axis.

If the body forces indicated in the basic equations (17.5) to (17.8) are of the conservative variety, it should be immediately obvious that they will not contribute to (18.13). Indeed (18.13) is essentially a vorticity equation, and potential forces have no effect.

Further applications of this transformation are discussed in the appropriate specialized sections.

19. The Squire Theorem

The Squire theorem states that when $V_0 = W = 0$ the lowest value of the critical Reynolds number occurs when $\beta = 0$. Indeed, the equation governing a three-dimensional oscillation is the same as that of a two-dimensional oscillation, except that the viscosity must be replaced by $\bar{v} = v(\bar{\alpha}/\alpha)$. If α and β are real, the presence of β, in effect, raises the viscosity. Thus the minimum Reynolds number for instability is higher for an oblique wave than for a pure two-dimensional wave.

Because β has the same effect as an increase in viscosity, it can have a destabilizing effect in the vicinity of the upper branch of the neutral curve. This matter is explained in detail in Section 22.

The net result can be summarized as follows: if α and β are real, the obliquity of the wave promotes stability.

An important comment is now in order. In many experimental situations the growth of the oscillations occurs along the x-direction and not in time. Thus α and eventually β are complex, whereas $\omega = \alpha c$ is real. The transformation is still useful, although it may introduce complex viscosity and complex coefficients. The Squire theorem still holds, for it is concerned strictly with the neutral case in which α, c, and ω are purely real.

The rate of growth, however, is affected by the complex character of $\alpha/\bar{\alpha}$. (See viscoelastic effects, Section 60.)

Finally we note that the presence of W modifies the basic flow, and an inflection of the combined profile $\bar{U}$ could occur where none occurs in the U profile. Indeed, the linear combination of U and W expressed by (18.7) could look very different from the ordinary profile given by U.

Hence the inviscid instability might not occur except for waves of sufficiently large obliquity. This remark is further pertinent to the work of Gregory *et al.* (1955) on the stability of three-dimensional boundary layers.

20. Some Properties of Inviscid Oscillations

(a) SIGNIFICANCE OF AN INFLECTION POINT

In the preceding sections we learned the importance of the term $U''\mathbf{v}$ in the analysis of flow oscillations. The purpose of this section is to explore the point more thoroughly. The discussion herein is concerned with two-dimensional oscillations, but the Squire theorem (Section 18) can always be invoked to extend our conclusions to three-dimensional oscillations.

We recall that for a neutral oscillation both α and c are purely real. Moreover, instability occurs if the imaginary part of c is positive and vice versa. We now neglect the viscous stresses, and the Orr-Sommerfeld equation reduces to the Rayleigh equation:

$$(U - c)(\mathbf{v}'' - \alpha^2\mathbf{v}) - U''\mathbf{v} = 0. \tag{20.1}$$

Let us multiply this equation by $\mathbf{v}^*$, the complex conjugate of $\mathbf{v}$, divide by the factor $(U - c)$, and integrate the result between some limits y_1 and y_2. With some regrouping of the first term we arrive at

$$\int_{y_1}^{y_2} [(\mathbf{v}'\mathbf{v}^*)' - \mathbf{v}'\mathbf{v}^{*\prime} - \alpha^2\mathbf{v}\mathbf{v}^*]\, dy - \int_{y_1}^{y_2} \frac{U''\mathbf{v}\mathbf{v}^*}{(U - c)}\, dy = 0. \tag{20.2}$$

The first term can be integrated exactly:

$$\int_{y_1}^{y_2} (\mathbf{v}'\mathbf{v}^*)'\, dy = \left(\mathbf{v}'\mathbf{v}^*\right)_{y_1}^{y_2}. \tag{20.3}$$

We must now decide whether the limits y_1 and y_2 are located at walls or at $\pm \infty$, so that either **v** or **v′** or both will vanish at the limits. From the positive definite nature of the two following terms it follows that

$$\int_{y_1}^{y_2} \frac{U''\mathbf{v}\mathbf{v}^*}{(U - c)} \, dy = -k^2, \qquad (20.4)$$

where k is a real number and α is assumed real.

In this equality all terms are real except $c = c_r + ic_i$. Let us consider only the imaginary part of (20.4), which reads

$$c_i \int_{y_1}^{y_2} \frac{U''\mathbf{v}\mathbf{v}^*}{(U - c_r)^2 + c_i^{\,2}} \, dy = 0. \qquad (20.5)$$

From this expression we can infer that either c_i is zero or the integral must vanish. If c_i is zero, we might expect some exceptional neutral solution, but the very existence of the integral often becomes questionable. When c_i is not zero, (20.5) cannot be satisfied unless U'' has at least one zero. This result was first obtained by Rayleigh (1880).

A stronger version of the Rayleigh theorem on the consequence of the inflection point in the mean velocity profile was noted by Høiland (1953). Let us accept $c_i \neq 0$ and define the value of the mean velocity at the point of inflection as U_i. Then, if we multiply (20.5) by the factor $(c_r - U_i)/c_i$ and add the result to the real part of (20.4) we will find that the integral

$$\int_{y_1}^{y_2} \frac{U''(U - U_i)\mathbf{v}\mathbf{v}^*}{(U - c_r)^2 + c_i^{\,2}} \, dy$$

must be negative. This can only be true if $U''(U - U_i)$ is negative in the interval of integration. It becomes apparent that only certain shapes of profiles are permissible. But even more important, an unstable oscillation can exist only if the mean vorticity has an absolute maximum within the flow.

Let us now return to (20.2) and consider the case of c purely real and α complex. The imaginary part of the equation reduces to

$$2\alpha_r\alpha_i \int_{y_1}^{y_2} \mathbf{v}\mathbf{v}^* \, dy = 0. \qquad (20.6)$$

From these results we can conclude that the problem of an inviscid flow has no solutions unless its velocity profile has at least one inflection point. This refers to oscillations with a real α. Oscillations with a purely real c and a complex α are excluded.

As pointed out before, many experimental conditions occur with a real frequency $\omega = \alpha c$ and a complex α. Thus we find that both α and c are complex. With the proviso that ω is purely real, which implies that the oscillation is periodic in time, the imaginary part of (20.2) leads to the relation

$$\int_{y_1}^{y_2} \frac{U'' \mathbf{v} \mathbf{v}^*}{(U - c_r)^2 + c_i^2} \, dy = 2 \frac{\alpha_r^2}{c_r} \int_{y_1}^{y_2} \mathbf{v} \mathbf{v}^* \, dy. \qquad (20.7)$$

We can now understand a fundamental difference between flows having an inflection of the velocity profile, such as shear layers, jets, and wakes, and flows without inflection, such as ordinary boundary layers and channel flows. Those with inflections can be unstable at sufficiently high Reynolds numbers, and, in general, the effect of the viscosity appears to be the addition of damping, as shown by Figs. 13.2 and 16.1. Those without inflections are either stable at all Reynolds numbers (Couette flow) or their instability occurs only when the viscosity can have the remarkable destabilizing influence as shown by the results in Section 7.

Hence, for real α, an inflection point is necessary for instability of an inviscid flow and the collection of available facts even suggests that an inflection is sufficient. This last point, however, has not been rigorously proved.

(b) LOCATION OF THE CRITICAL LAYER

We now demonstrate that if α is real and c is complex the phase velocity must lie somewhere between the extreme values of the mean velocity. For this purpose we define the auxiliary function

$$\mathbf{G} = \frac{\mathbf{v}}{(U - c)}. \qquad (20.8)$$

By replacing $\mathbf{v}$ with $\mathbf{G}$, and after some rearrangement, Eq. (20.1) takes the following form:

$$[\mathbf{G}'(U - c)^2]' - \alpha^2 (U - c)^2 \mathbf{G} = 0. \qquad (20.9)$$

Let us multiply (20.9) by $\mathbf{G}^*$ and integrate between the limits y_1 and y_2 to find

$$\left[\mathbf{G}^* \mathbf{G}(U - c)^2 \right]_{y_1}^{y_2} - \int_{y_1}^{y_2} (U - c)^2 (\mathbf{G}' \mathbf{G}^{*\prime} + \alpha^2 \mathbf{G} \mathbf{G}^*) \, dy = 0. \qquad (20.10)$$

The limits are now placed at walls or at infinities and only the integral remains. Separation into real and imaginary parts gives

$$\int_{y_1}^{y_2} [(U - c_r)^2 - c_i^2](\mathbf{G}'\mathbf{G}^{*\prime} + \alpha^2 \mathbf{G}\mathbf{G}^*) \, dy = 0, \tag{20.11}$$

$$c_i \int_{y_1}^{y_2} (U - c_r)(\mathbf{G}'\mathbf{G}^{*\prime} + \alpha^2 \mathbf{G}\mathbf{G}^*) \, dy = 0. \tag{20.12}$$

Although (20.11) places a constraint on c_i, (20.12) points out that $(U - c_r)$ must change sign somewhere between the limits of integration. This demonstrates the point. The height at which $(U - c_r)$ vanishes, $y = y_c$, defines the critical layer. This result was first demonstrated by Rayleigh (1913).

c) REYNOLDS STRESSES

In the general discussion of the energy of the oscillation leading to (2.12) we met a term representing a direct exchange of energy between the mean flow and the oscillation. This exchange term is defined by (2.14), and by analogy with (1.29) we can express it directly in terms of complex amplitudes:

$$\mu_1 = -U'\overline{\tilde{u}\tilde{v}} = -\tfrac{1}{4}U'(\mathbf{u}^*\mathbf{v} + \mathbf{u}\mathbf{v}^*). \tag{20.13}$$

The Reynolds stresses divided by the density are defined as

$$\sigma(y) = -\overline{\tilde{u}\tilde{v}} = -\tfrac{1}{4}(\mathbf{u}^*\mathbf{v} + \mathbf{u}\mathbf{v}^*), \tag{20.14}$$

and we now explore their behavior.

Using the continuity equation (1.17), we obtain

$$\sigma = \frac{i}{4\alpha}(\mathbf{v}'^*\mathbf{v} - \mathbf{v}'\mathbf{v}^*). \tag{20.15}$$

Let us differentiate (20.15) and eliminate $\mathbf{v}''$ with the help of (20.1) or its conjugate. This leads to

$$\sigma' = \frac{c_i}{2\alpha} \frac{U''}{(U - c_r)^2 + c_i^2} \mathbf{v}\mathbf{v}^*. \tag{20.16}$$

If we now let $c_i \to 0$, σ' vanishes except near the critical layer, where it diverges.

With a small c_i we can consider that $\mathbf{v}$, U'', and U' are nearly constant through the critical layer, and the integration of (20.16) gives

$$\left(\sigma\right)_{h-\Delta y}^{h+\Delta y} = \frac{\pi}{2\alpha} \frac{U''(h)}{U'(h)} (\mathbf{vv}^*)_{y=h}. \tag{20.17}$$

In Section 10 we saw that for a neutrally stable oscillating boundary layer $\mathbf{u}$ and $\mathbf{v}$ are out of phase above the critical layer and therefore $\sigma = 0$. Below the critical layer a new component $\mathbf{u}_3$ has appeared [see Eq. (10.17)] which is in phase with $-\mathbf{v}$. Thus, when $c_i = 0$, the viscous terms that we are neglecting in this section guarantee the continuity of the solution. However, a rapid change takes place at the critical layer and σ undergoes the jump indicated by (20.17). This is illustrated in Fig. 21.1. At the wall σ vanishes because of the viscous effects. The characteristic thickness of such regions where σ vanishes is the length $l = (v/\alpha U'(h))^{1/3}$.

It is now appropriate to remark that the function $e(y, t)$ of (2.12) is the local energy density and therefore $U'\sigma$ indicates the local supply of energy. Thus in a neutrally stable boundary layer this supply of energy occurs only between the wall and the critical layer.

It can be shown with the use of the continuity equation that the integral of the term μ_2 of (2.15) over a box containing one wavelength of the boundary layer is identically zero. Thus, although the term in μ_2 may redistribute the energy, it does not produce or dissipate any. The energy produced by the Reynolds stresses is therefore transformed into heat by the viscous term μ_3.

Between the critical layer and the wall the inviscid equation (20.16) indicates that σ remains constant. Because the boundary conditions require both $\mathbf{u}$ and $\mathbf{v}$ to vanish, it is quite clear that the viscous terms must intervene and bring σ down to zero. One can show that σ builds up in every viscous region along a wall. Thus a critical layer must exist so that σ can return to zero. This is another reason why a boundary layer, without inflection point, cannot have a neutral oscillation at infinite Reynolds number.

21. Some Properties of Viscous Oscillations

(a) The Critical Reynolds Number

As shown in the particular case of a Blasius boundary layer (see Fig. 8.1) and of a few other cases (see Fig. 11.3), certain flows are stable for all real values of α as long as the Reynolds number remains less than some imit $\mathcal{R} = \mathcal{R}_c$, which is often called the critical Reynolds number. If $\mathcal{R} > \mathcal{R}_c$, c_i becomes positive in a certain range of values of α between the two branches of the neutral curve.

Before World War II it was generally felt that turbulence occurs as soon as the laminar flow becomes unstable. By turbulence we mean a flow that is so complicated, so rich in fine details, that all hopes of finding analytic or numerical solutions become futile. Thus the prediction of $\mathcal{R}_c$ was the main goal of many investigations. Today, it is recognized that $\mathcal{R}_c$ marks only the beginning of a regime in which certain disturbances are amplified while being washed downstream. These disturbances are amenable either to mathematical analysis or to numerical computations. When they become sufficiently large certain nonlinear effects suddenly open the door to random fluctuations and the flow becomes turbulent. The Reynolds number characteristic for the location of this turbulent transition is known as the transitional Reynolds number. We now demonstrate certain inequalities contributed by Synge (1938) concerning $\mathcal{R}_c$.

Let us consider a two-dimensional flow, with a real α and a complex c. The boundary conditions $\mathbf{v} = \mathbf{v}' = 0$ are applied at y_1 and y_2 so that our results are applicable to boundary layers, channel flows, shear flows, jets, and wakes.

The Orr-Sommerfeld equation (13.1) is multiplied by $\mathbf{v}^*$ and integrated as before. The various terms are integrated by parts and regrouped. We use the following condensed notations:

$$I_0{}^2 = \int_{y_1}^{y_2} \mathbf{v}\mathbf{v}^* \, dy, \tag{21.1}$$

$$I_1{}^2 = \int_{y_1}^{y_2} \mathbf{v}'\mathbf{v}'^* \, dy, \tag{21.2}$$

$$I_2{}^2 = \int_{y_1}^{y_2} \mathbf{v}''\mathbf{v}''^* \, dy, \tag{21.3}$$

$$Q = \int_{y_1}^{y_2} [(U\alpha^2 + U'')\mathbf{v}\mathbf{v}^* + U\mathbf{v}'\mathbf{v}'^*] \, dy + \int_{y_1}^{y_2} U'\mathbf{v}^*\mathbf{v}' \, dy. \tag{21.4}$$

This leads to the following integral of the Orr-Sommerfeld equation:

$$I_2{}^2 + 2\alpha^2 I_1{}^2 + \alpha^4 I_0{}^2 = \frac{i\alpha}{\nu}[-Q + c(I_1{}^2 + \alpha^2 I_0{}^2)]. \tag{21.5}$$

The real part of this relation is

$$I_2{}^2 + 2\alpha^2 I_1{}^2 + \alpha^4 I_0{}^2 = -\frac{\alpha}{\nu}\left[\frac{Q^* - Q}{2i} + c_i(I_1{}^2 + \alpha^2 I_0{}^2)\right]. \tag{21.6}$$

It follows from (21.4) that

$$Q - Q^* = \int_{y_1}^{y_2} U'(v^*v' - vv'^*) \, dy. \tag{21.7}$$

Taking the absolute value, we have the following inequality:

$$|Q - Q^*| \leq 2 \int_{y_1}^{y_2} |U'| \, |v| \, |v'| \, dy. \tag{21.8}$$

With U'_{max} for the maximum value of $|U'|$ and an application of the Schwarz inequality, (21.8) becomes

$$|Q - Q^*| \leq 2U'_{\text{max}} I_0 I_1. \tag{21.9}$$

It now follows from (21.6) that

$$\frac{\alpha c_i}{v}(I_1{}^2 + \alpha^2 I_0{}^2) \leq \frac{\alpha}{v} U'_{\text{max}} I_0 I_1 - (I_2{}^2 + 2\alpha^2 I_1{}^2 + \alpha^4 I_0{}^2). \tag{21.10}$$

At first glance this inequality suggests that as v becomes very large c_i is forced to stay negative. This conclusion is apparently correct for boundary layers and channel flows, which indeed have critical Reynolds numbers. In the free shear layer the asymptotic calculations of Esch (1957) by means of power series and the numerical calculations of Betchov and Szewczyk (1963) have shown that the neutral line reaches the origin with $\alpha v = $ constant. (See Fig. 13.2.) With $U = \tanh y$ the calculations gave $\alpha v = 0.15$. Using the example of $\mathcal{R} = 1$, it appears that if $\alpha > 0.15$ the negative terms of the right-hand side of (21.10) will exceed the term in U'_{max}. However, if $\alpha < 0.15$, the solution becomes so smooth that the term $I_2{}^2$, originating from second derivatives, no longer controls the right-hand side of (21.10). This can be examined in detail by using a parabola to approximate v inside the shear layer and the exponentials $e^{-\alpha y}$ and $e^{\alpha y}$ for the free-stream solutions. Then (21.10), in $\mathcal{R} = 1$, guarantees that c_i will be negative for $\alpha > 1.5$. A comparison with Fig. 13.2 shows the weakness of the inequality.

A general method of evaluating the upper limit of $I_2{}^2$ has been given by Synge (1938) for boundary layers and channel flows and by Lessen (1952) for flows away from walls. This subject has also been reviewed by Shen (1964). It is believed that inequalities of this type are generally weak.

A rough estimate for the critical Reynolds number of a boundary layer has been made by Lin (1945), and we now present a nearly equivalent formula based on our simplified theory. In Section 11 we found that the highest value of α, allowing for a neutral solution, corresponds to the first intercept between a curve at constant α in Fig. 11.2 and the function $1 + \mathscr{F}(\eta)$.

The highest point of the neutral curve corresponds roughly to $\eta = 4$. An examination of the results in Fig. 11.2 suggests that the critical Reynolds number corresponds to $\eta = 3.5$ or to the largest phase angle of the function $1 + \mathscr{F}(\eta)$. Thus the critical Reynolds number corresponds to a point such that the phase of the left-hand side of Eq. (11.1) is maximum. We then obtain

$$-\pi h \frac{U''(h)}{U'(h)} = 0.42. \tag{21.11}$$

For a given profile this relation permits us to obtain h. From the relation $\eta = h/l = 3.5$ we can obtain l. Because $l = (v/\alpha U'(h))^{1/3}$, the corresponding value of v can be determined. Assuming that $\alpha = 1$, we find that the minimum Reynolds number is approximately

$$\frac{U_0 L}{v} = (3.5)^3 \frac{U_0}{h U'(h)} \left(\frac{L}{h}\right)^2. \tag{21.12}$$

If we use the real part of (11.1) to provide a relation between α and c, we obtain

$$\frac{U_0 L}{v} = \frac{(3.5)^3 (0.28) U_0^2 L^2}{h^4 U'(h)}. \tag{21.13}$$

For example, the parabolic profile gives $h = 0.118$ and $\mathscr{R}_c \approx 14$. The profile $U = 1 - e^{-3y}$ gives $h = 0.042$ and $\mathscr{R}_c = 200$. The Blasius profile gives $h = 0.2$ and $\mathscr{R}_c = 2000$.

These results are acceptable and they show that the above relations are valid over a wide range of situations.

(b) REYNOLDS STRESSES AT A WALL

At this point we consider the behavior of the viscous stresses near the wall for a neutrally stable oscillation. In this vicinity the analysis corresponds closely to that already discussed in Section 9, in which we described the viscous correction. The equation governing the vorticity in this region was given in (9.1). A suitable solution for this equation was found to be

$$\mathbf{r} = m \exp\left[-(1 - i)\left(\frac{\alpha c}{2v}\right)^{1/2} y\right]. \tag{21.14}$$

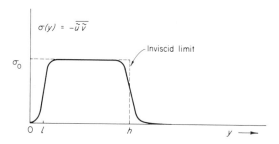

FIG. 21.1. The Reynolds stresses between the wall and the critical layer in an oscillating boundary layer. Idealized example, with $h \gg l$.

Using the continuity equation, the definition of the vorticity, and integrating twice, we find

$$\mathbf{u} = -(1 + i)m\left(\frac{v}{2\alpha c}\right)^{1/2}\left\{1 - \exp\left[-(1 - i)\left(\frac{\alpha c}{2v}\right)^{1/2}y\right]\right\}, \qquad (21.15)$$

$$\mathbf{v} = (1 + i)m\left(\frac{v}{2\alpha c}\right)^{1/2}\left\{y + (1 + i)\left(\frac{v}{2\alpha c}\right)^{1/2}\left(\exp\left[-(1 - i)\left(\frac{\alpha c}{2v}\right)^{1/2}y\right] - 1\right)\right\}, \qquad (21.16)$$

to be the amplitude components of the velocity in this same region near the wall. Likewise, we find that the Reynolds stress corresponding to this solution is

$$\sigma = -\overline{\tilde{u}\tilde{v}} = \tfrac{1}{4}(\mathbf{u}^*\mathbf{v} + \mathbf{u}\mathbf{v}^*). \qquad (21.17)$$

Note that $\mathbf{u}^*\mathbf{v}$ and $\mathbf{u}\mathbf{v}^*$ vary exponentially with the decay length $\delta = (v/2\alpha c)^{1/2}$. We saw in Section 4 that for an inviscid neutral oscillation σ is zero above the critical layer and jumps to a positive constant value σ_0 below it. When the effect of the vorticity diffusing from the wall, as well as the boundary conditions, are taken into account, the total stress behaves qualitatively as shown in Fig. 21.1. In the event that c_i is negative, similar calculations with the same boundary conditions will result in a sign reversal in σ; this is also true if the sign of α is changed.

22. Stability Surfaces

(a) NEUTRAL SURFACES

In Section 18 we examined the stability of three-dimensional oscillations and found that under very general conditions the calculations of the eigenvalues and eigenfunctions were identical with those of two-dimensional waves. In this section we use this relationship and examine the stability of oscillations of the form $e^{i(\alpha x + \beta z)}e^{-i\alpha ct}$, where α and β are real wavenumbers. As before, the complex amplitudes are functions of y.

We assume that W is zero for simplicity and reconsider this matter at the end of the section.

In place of a neutral curve in the $\alpha\mathscr{R}$-space, we must now think of a neutral surface in the $\alpha\beta\mathscr{R}$-space, defined by the set of values of α and β which correspond to $c_i = 0$ for a given Reynolds number $\mathscr{R} = U_0 L/v$.

It follows from the Squire theorem (Section 19) that such eigenvalues correspond to an eigenvalue

$$\bar{\alpha} = (\alpha^2 + \beta^2)^{1/2} \tag{22.1}$$

and to a Reynolds number

$$\bar{\mathscr{R}} = \frac{\overline{U}L}{v} = \frac{\alpha}{\bar{\alpha}}\mathscr{R}. \tag{22.2}$$

Thus, once the neutral curve has been plotted for $\beta = 0$, a simple rescaling of both coordinates will give the neutral curve for some constant value of $\bar{\alpha}$. To simplify this task it is convenient to plot $\bar{\alpha}$ as a function of $\bar{\alpha}\bar{\mathscr{R}} = \alpha\mathscr{R}$, for this quantity no longer varies with β. A typical curve is shown in Fig. 22.1. Because $\bar{\alpha}$ plays the role of a radius in the $\alpha\beta$-space, it follows that the neutral surface can be generated by rotating the neutral curve about the $\alpha\mathscr{R}$-axis, as shown in Fig. 22.2. Thus a cut at constant $\alpha\mathscr{R}$ will show two circles generated by the upper and lower branches.

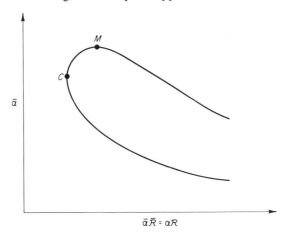

FIG. 22.1. The neutral curve, shown as a function of $\alpha\mathscr{R}$.

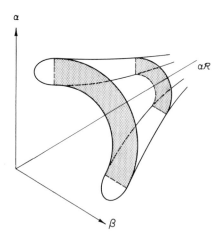

FIG. 22.2. The neutral surface, generated by a rotation about the $\alpha\mathscr{R}$ axis.

Because $\mathscr{R}$ is characteristic of a particular flow, and not $\alpha\mathscr{R}$, we must now obtain the curves at constant $\mathscr{R}$ in the $\alpha\beta$-plane. This requires some knowledge of the relation between $\bar{\alpha}$ and $\alpha\mathscr{R}$, and we define the slope of the neutral curve on a particular branch as

$$\gamma = \frac{\partial\alpha}{\partial(\alpha\mathscr{R})} \qquad (22.3)$$

where the derivative is taken at $c_i = 0$ and for $\beta = 0$. Anywhere in the $\alpha\beta$-plane we have the relation

$$\Delta\bar{\alpha} = \gamma\,\Delta(\alpha\mathscr{R}). \qquad (22.4)$$

Then, along a curve taken at constant $\mathscr{R}$, we have

$$\alpha\,\Delta\alpha + \beta\,\Delta\beta = \mathscr{R}\gamma\,\Delta\alpha. \qquad (22.5)$$

Between the points C and M of Fig. 22.1 the slope γ is positive; elsewhere it is always negative. This means that, according to (22.5), the curve at constant $\mathscr{R}$ will move away from the origin as α decreases, except for the points related to the C-M interval.

We can rewrite (22.5) in the form

$$\beta\,\Delta\beta = -(\alpha - \mathscr{R}\gamma\bar{\alpha})\,\Delta\alpha, \qquad (22.6)$$

and, if γ is negative, it means that α decreases when $|\beta|$ increases.

In Fig. 22.3 we show some results obtained by Criminale (1960) and Criminale and Kovásznay (1962) for a Blasius boundary layer.

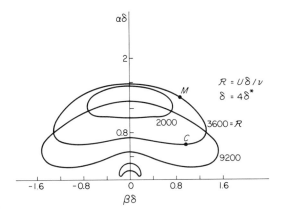

FIG. 22.3. Neutral curves in $\alpha\beta$ plane, sometimes called *kidney curves* (Blasius boundary layer).

As described in the example of $\mathcal{R} = 3600$, point M occurs at maximum distance from the origin, and at point C the curve is parallel to the β-axis.

As $\mathcal{R}$ becomes very large, the kidney curve shrinks, but the maximum value of the ratio β/α increases continuously.

In fact, if we define $\tan\theta = \beta/\alpha$, we find approximately that

$$\tan\theta_{max} \to \mathcal{R}/\mathcal{R}_c.$$

Similar results have been obtained by Watson (1960) for plane Poiseuille flow.

A remarkable feature of these curves is that at sufficiently large Reynolds numbers a wave of given α and zero β may be stable, whereas the wave with the same α and some finite β may be unstable.

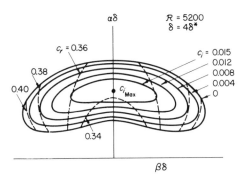

FIG. 22.4. Amplification of oblique waves in a Blasius boundary layer.

(b) AMPLIFIED CASES

Inside each curve of Fig. 22.3 we can plot the curves at constant c_r or at constant c_i. Some typical results are given in Fig. 22.4 for a Blasius layer. Any point of the $\alpha\beta$-plane thus corresponds to an oblique wave propagating with velocity c_r in the direction θ, where $\theta = \tan^{-1}(\beta/\alpha)$.

Chapter V

The Initial Value Problem

23. Formulation of the Problem

The oscillations of a laminar flow that occur in reality can be regarded as a sum of Fourier components. As long as the amplitudes are small enough, these components do not affect each other and proceed independently, according to the linearized equations. So far, we have considered one Fourier component at a time; that is, our solutions have always been related to the real part of functions proportional to $e^{i\alpha(x-ct)}e^{i\beta z}$. This means that a spectral analysis of the motion would show only one line corresponding to the wave vector with components α and β. Practically speaking, spectral analysis shows broader spectra. Because the Fourier components do not interact, each part of the spectrum evolves independently in the course of time. This means that the spectrum at any time is the result of the initial conditions at some origin in time and of the effect of the factor $e^{-i\alpha ct}$.

In this chapter we illustrate the role of the initial conditions and point out certain selective properties of laminar flows. In particular, we consider a completely stationary Blasius boundary layer at time $t = 0$ and apply a well-defined localized perturbation. As time increases, we follow the evolution of the developing flow by taking into account the propagation and growth or decay of every Fourier component. For large values of time we examine the physical nature of the perturbation. The method used for this problem, however, should be recognized as general and applicable to any other system.

Before treating the example in specific detail, we need to recall the general linearized equations for a three-dimensional perturbation of a laminar flow with only one mean velocity component, $U(y)$, in the x-direction:

$$\tilde{u}_t + U\tilde{u}_x + U'\tilde{v} + \tilde{p}_x = v\,\nabla^2\tilde{u}, \tag{23.1}$$

$$\tilde{v}_t + U\tilde{v}_x + \tilde{p}_y = v\,\nabla^2\tilde{v}, \tag{23.2}$$

$$\tilde{w}_t + U\tilde{w}_x + \tilde{p}_z = v\,\nabla^2\tilde{w}, \tag{23.3}$$

$$\tilde{u}_x + \tilde{v}_y + \tilde{w}_z = 0, \tag{23.4}$$

where ∇^2 indicates the three-dimensional Laplace operator, which, in the above equations, is

$$\nabla^2 = \frac{\partial^2}{\partial x^2} + \frac{\partial^2}{\partial y^2} + \frac{\partial^2}{\partial z^2}.$$

For $t < 0$ we use the trivial solution in which $\tilde{u} = \tilde{v} = \tilde{w} = \tilde{p} = 0$. In other words, the flow is completely steady before the onset of the disturbance. A very short time before $t = 0$ we imagine that unspecified external forces are applied with the result that at $t = 0$ the four functions $\tilde{u}(x, y, z, 0)$, $\tilde{v}(x, y, z, 0)$, $\tilde{w}(x, y, z, 0)$, and $\tilde{p}(x, y, z, 0)$ have known expressions. In principle, these functions can be quite arbitrarily specified as long as it is ensured that two equivalent constraints are met. These initial expressions must satisfy the continuity equation (23.4), as well as the pressure equation [see (2.1)], which, in three dimensions, takes the following form

$$\nabla^2\tilde{p} = -2U'\tilde{v}_x. \tag{23.5}$$

We now determine the fluctuations for positive times. Such a determination requires more information than has been needed up to now, and it is not an easy task, practically speaking. In order that it may be clearly understood, we have attempted to cover all the essential points in the following order:

1. Rewrite the equations in a convenient form for solving.
2. Transform $\tilde{v}$ to facilitate the problem.
3. Use a set of eigenfunctions for the initial values.
4. Solve successively for $\tilde{v}$, $\tilde{p}$, $\tilde{u}$, and $\tilde{w}$.

(a) A CONVENIENT FORM

Equation (23.1) is brought down directly and written as

$$\tilde{u}_t + U\tilde{u}_x - v\,\nabla^2\tilde{u} = -\tilde{p}_x - U'\tilde{v}. \tag{23.6}$$

Next we take the derivative of (23.1) with respect to x, the derivative of (23.2) with respect to y, and the derivative of (23.3) with respect to z. Addition of the three resulting equations and use of (23.4) yield the result

$$\nabla^2 \tilde{v}_t + U \nabla^2 \tilde{v}_x - U'' \tilde{v}_x - \nu \nabla^2 \nabla^2 \tilde{v} = 0. \tag{23.7}$$

Then (23.3) is copied exactly and written in an analogous fashion to (23.6),

$$\tilde{w}_t + U\tilde{w}_x - \nu \nabla^2 \tilde{w} = -\tilde{p}_z. \tag{23.8}$$

Finally, (23.5) is used as the last relation. This equation is, in essence, a replacement for continuity (23.4). Thus we have

$$\nabla^2 \tilde{p} = -2U'\tilde{v}_x. \tag{23.9}$$

We can immediately see that (23.7) of this new set of equations is independent of the others and homogeneous in the dependent variable $\tilde{v}$. This is an important fact that is employed later to obtain the final solutions.

(b) Transformation of $\tilde{v}$

Because (23.7) is linear and has coefficients that are functions of y only, it is appropriate to use a Fourier transformation in the variables x and z. Thus we define in the usual way the relation

$$\hat{v}(\alpha, \beta; y, t) = \frac{1}{(2\pi)^2} \int_{-\infty}^{+\infty} \int_{-\infty}^{+\infty} \tilde{v}(x, y, z, t) e^{-i(\alpha x + \beta z)} \, dx \, dz. \tag{23.10}$$

By performing the operation indicated by (23.10) on the equation that governs $\tilde{v}$ we have only a minor simplification, for we are still left with a partial differential equation. Thus we must now concentrate our attention on the time dependence in order to bring about a familiar form that will render itself solvable.

In the preceding sections we saw that elementary solutions can be found for various wavenumbers, provided that the time dependence is in the form of $e^{-i\alpha ct}$, where c takes on only the special eigenvalues. As examples we recall the inviscid jet in which we found that for a certain range of the wavenumber (specifically $\alpha < 1$) there are two unstable solutions corresponding to symmetric and axisymmetric oscillations and that each solution has its own complex eigenvalue. On the other hand, in viscous flows such as Couette flow and Poiseuille flow we reported the existence of a large number of modes which are all damped except for the first one in the

Poiseuille flow system. Fortunately, well-known techniques enable us to make a connection between these elementary solutions and the problem that we are now pursuing.

(c) The Set of Eigenfunctions

In general, a differential equation, such as that of Orr and Sommerfeld (13.1), which is constrained by sufficient boundary conditions, admits an infinite but discrete set of solutions. Each solution is called a mode and has a particular corresponding eigenvalue. We use the subscript n to label the various modes v_n and the notation c_n for the eigenvalues.

Now let us examine the transformed equation for $\tilde{v}$, which, as we have already mentioned, can be obtained by the operation indicated in (23.10). The dependent variable of this equation becomes $\hat{v}$. We note that, except for the time dependence, it is almost the same as the Orr-Sommerfeld equation. Indeed, this fact is the basis for solving this intermediate equation.

A solution is obtained if we decompose $\hat{v}$ into a sum over the many different modes. To do so we write

$$\hat{v}(\alpha, \beta; y, t) = \sum_{n=1}^{\infty} A_n \, v_n(\alpha, \beta; y)e^{-i\alpha c_n t}, \qquad (23.11)$$

where the coefficients A_n represent the amplitudes of the various modes. These coefficients can be determined by using the known initial conditions at $t = 0$. At once we see that this step forms the link with the elementary system, for the modes (or eigenfunctions) of the expansion (23.11) must satisfy the Orr-Sommerfeld equation. To check this conclusion we merely substitute (23.11) into the intermediate equation. It is recognized that this general procedure is not unique to fluid mechanics. In fact, it is quite common in many other fields.

Let us progress further. We have seen that each mode has an arbitrary amplitude and phase. By choosing the various eigenfunctions and judiciously selecting their amplitude and phase they can be processed into a set of orthogonal and normalized functions. For example, for the purpose we might impose the equations

$$\frac{1}{4} \int_{y_1}^{y_2} (v_n v_m{}^* + v_n{}^* v_m) \, dy = \delta_{mn}, \qquad (23.12)$$

$$\frac{1}{4} \int_{y_1}^{y_2} (v_n v_m{}^* - v_n{}^* v_m) \, dy = 0. \qquad (23.13)$$

The integration limits correspond to the boundaries of the particular flow under consideration; δ_{mn} is the Kronecker delta and the asterisk denotes the complex conjugate. Acceptance of this form allows us to evaluate the A_n's in the expansion for $\hat{v}$. First, we multiply (23.11) by v_n^* and integrate between the limits at time $t = 0$. By virtue of (23.12) we obtain

$$\int_{y_1}^{y_2} \hat{v}(\alpha, \beta; y, 0) \, v_n^*(\alpha, \beta; y) \, dy = 2A_n; \qquad (23.14)$$

and, just as for other functions and transforms, if the initial perturbation $\hat{v}(\alpha, \beta; y, 0)$ is symmetric in y, the A_n corresponding to asymmetric eigenfunctions will be zero. As a result these modes will not appear in the final solution. This outline covers the essentials well enough, but nothing has been said about the validity of the expansion or the availability of the important set of eigenfunctions.

Actually, relatively little attention has been given to the initial-value analysis in fluid mechanics. As a result the requirements that we seek for this problem are available in part only. In spite of the fact that the methods of arriving at the results are well known, the computations have not been readily obtainable. Even the simple orthogonality and normalization relations given as illustrations in (23.12) and (23.13) have been oversimplified. Specifically, an expansion theorem which ensures the validity of (23.11) has been established in a fairly recent work by Schensted (1960). (A partial review of this same subject has also been provided by Reid, 1965.) On the other hand, the set of eigenfunctions of the expansion in (23.11) has never been computed for any particular flow. Even more, as we have already mentioned, the status of the set of eigenvalues for any flow has been only moderately examined.

Because it is pertinent to the discussion at hand, we find it appropriate to mention other salient points of the Schensted work. Essentially, it is an application of classical means of applied mathematics to the initial value problems of stability theory; the reader is referred to the original work for the complete details.

First, it is pointed out that the Orr-Sommerfeld equation which governs the eigenfunctions is not a self-adjoint differential equation. In keeping with our notation the equation would be written as

$$(U - c_n)(D^2 - \bar{\alpha}^2)v_n - U''v_n = \frac{-i\nu}{\alpha}(D^2 - \bar{\alpha}^2)^2 v_n, \qquad (23.15)$$

where $\bar{\alpha} = (\alpha^2 + \beta^2)^{1/2}$ and the operator $D = d/dy$.

Schensted demonstrates the details of the expansion process for arbitrary initial conditions by use of the adjoint equation to (23.15) and its solutions. The adjoint equation is easily computed and, if we denote g_n as the dependent variable for this equation, we will have

$$(D^2 - \bar{\alpha}^2)[(U - c_n)g_n] - U''g_n = \frac{-i\nu}{\alpha}(D^2 - \bar{\alpha}^2)^2 g_n. \qquad (23.16)$$

By using (23.15) and (23.16) Schensted then finds the orthogonality relationship between the v_n and g_n. This is

$$\int_{y_1}^{y_2} v_n(D^2 - \bar{\alpha}^2)g_m \, dy = \int_{y_1}^{y_2} g_m(D^2 - \bar{\alpha}^2)v_n \, dy = 0. \qquad (23.17)$$

if $n \neq m$. When $n = m$ we have the accompanying normalization requirement which is given by

$$\int_{y_1}^{y_2} v_n(D^2 - \bar{\alpha}^2) g_n \, dy = 1. \qquad (23.18)$$

The limits of integration again correspond to the flow boundaries. Finally, with these equations as bases, the validity to the expansion of (23.11) was established, but now the A_n are correctly evaluated by the integral

$$A_n = \int_{y_1}^{y_2} \hat{v}(\alpha, \beta; y, 0)(D^2 - \bar{\alpha}^2)g_n(\alpha, \beta; y) \, dy. \qquad (23.19)$$

Let us now add a few specialized remarks. Some authors have discussed the possibility that the set of eigenvalues might be continuous and not discrete. This would mean that the right-hand side of (23.11) should be an integral of the ordinary Laplace transform variety in place of the indicated sum.

A continuous set must be excluded, as pointed out by Lin (1961). We can arrive at the same conclusion by observing that when we use the computer to find eigenvalues we must perform a search in the complex plane of the function B_4 in Section 13. In other words, most values of c do not satisfy the boundary conditions; the complex function B_4 may have many zeros, but they are isolated from one another.

There is, however, one example of a continuous set of eigenvalues. In the example of Couette flow we can see in Fig. 15.2 that if the Reynolds number tends toward infinity the various c_i tend toward zero. Thus in the limit of zero viscosity we find a "continuous" set of zeros.

The system we have not previously mentioned is that of the laminar boundary layer. All the evidence at hand indicates that the first mode is unstable and that all the others are damped, but none of the higher modes has been computed. Thus, when t is large enough, the first term of the sum indicated in (23.11) is the only important one. Therefore it is not necessary to know the initial conditions in every detail, for only A_1 is important. This point is equally pertinent to plane Couette and Poiseuille flows for which several of the higher modes have been computed (cf. Grohne, 1954).

(d) SUCCESSIVE SOLUTIONS

From the initial values we can determine the A_n necessary for the long-time behavior of the solution and obtain $\hat{v}(\alpha, \beta; y, t)$ from (23.11). By using the inverse transform of (23.10) we can find $\tilde{v}$ from $\hat{v}$ or

$$\tilde{v}(x, y, z, t) = \int_{-\infty}^{+\infty} \int_{-\infty}^{+\infty} \hat{v}(\alpha, \beta; y, t)e^{i(\alpha x + \beta z)} \, d\alpha \, d\beta. \tag{23.20}$$

We can now use (23.9) to determine $\tilde{p}$; the integration will introduce an arbitrary function which is a solution of the homogeneous equation. Because it is well known that such solutions of the Laplace equation must have singularities or sources within the range of integration, such arbitrary solutions can be eliminated by specifying the boundary conditions or by referring back to the initial conditions at $t = 0$.

When both $\tilde{v}$ and $\tilde{p}$ are known, $\tilde{u}$ or $\tilde{w}$ can be determined by using (23.6) and (23.8). The homogeneous solutions again appear as "integration constants" and can be eliminated by referring to conditions at $t = 0$. Thus a full solution is obtained.

Because the pressure equation (23.5) is a consequence of the continuity equation (23.4), our full solution will automatically comply with this requirement.

More details on this topic are given by Criminale (1960).

24. An Example of the Initial Value Problem

In a Blasius boundary layer the effect of a localized initial perturbation has been examined in some detail by Criminale and Kovásznay (1962). Other examples of initial value problems can be found in the literature: Benjamin (1962), or Lamb's (1932) review of the illuminating problem of a fishing line in a moving stream which was treated by Rayleigh (1911). In view of its importance for the problem of transition from oscillating laminar flow to turbulence, we now review the boundary layer case.

At the outset we must specify $\tilde{v}$ at $t = 0$. In order to have a localized perturbation, we introduce a small length σ and we assume

$$\tilde{v}(x, y, z, 0) = f(y) \exp\left[-\frac{(x^2 + z^2)}{2\sigma^2} \right], \qquad (24.1)$$

where $f(y)$ is unspecified. A Gaussian function was chosen merely to facilitate the mathematics and actually, in the limit, contains the ideal localized disturbance or Dirac delta function. By introducing this expression into (23.10) we find that the Fourier transform of (24.1) at time $t = 0$ is

$$\hat{v}(\alpha, \beta; y, 0) = \frac{\sigma^2}{8\pi} f(y) \exp\left[-\frac{\sigma^2}{2}(\alpha^2 + \beta^2) \right]. \qquad (24.2)$$

This expression should now be used to determine the various A_n.

In a Blasius layer we know only of one unstable mode; and, because we are concerned primarily with the ultimate fate of a disturbance, we assume that it is the only one and discontinue the use of the subscript n. Such a drastic reduction of the infinite set of modes to one brings about only a limited simplification, however, for the details of the single eigenfunction (as a function of the wavenumber and the variable y) is not well known. This drawback led Criminale and Kovásznay to introduce additional approximations.

In Section 8 we saw that $\mathbf{v}(y)$ can be approximated by a parabola between the wall and the edge of the layer and that it drops exponentially beyond. Using this form for $\mathbf{v}$, we show in a calculation of the integral of $\mathbf{v}^2$ that the result varies very slowly with $\bar{\alpha}$ (or α) in the range $0.3 < \bar{\alpha} < 2$, which is wider than the range in which the eigenvalue has amplification. Thus the normalization constant is roughly independent of $\bar{\alpha}$. With this fact in mind, a casual inspection will show that if f is constant, or examined at a particular value of y only (this will mean that the initial disturbance was injected at this y-location), the integral leading to the single A is almost independent of the value of $\bar{\alpha}$. For the sake of simplicity we choose this particular shape with $f = 1$, which leads to the approximate result,

$$A = \frac{\sigma^2}{16\pi} e^{-\sigma^2 \bar{\alpha}^2/2}. \qquad (24.3)$$

It now follows from (23.11) that

$$\hat{v} = Ae^{-i\alpha ct}, \qquad (24.4)$$

but we cannot proceed beyond this point without referring to the specific relation between c, α, and β.

For piecewise calculations relying on the use of a computer, the exact shape of the amplification surfaces (Figs. 22.3 and 22.4) could be retained. Following Criminale and Kovásznay, we approximate the kidney curves by a set of ellipses and write

$$c_i = c_{i(\max)} - b_1(\alpha_{\max} - \alpha)^2 - b_2\beta^2. \tag{24.5}$$

For the phase velocity we use the following parabolic approximation:

$$c_r = c_{r(\max)} - a_1(\alpha_{\max} - \alpha) + a_2\beta^2. \tag{24.6}$$

In these expressions the subscript max refers to the point at the center of the kidney curve at which c_i is maximum.

Thus $\hat{v}$ is expressed in terms of exponentials, including such factors as $e^{i\alpha^3}$. This means that we must again use machine computations to obtain $\tilde{v}$, with (23.20) in any exact manner. Because $\mathbf{v}(y)$ has a known behavior, we explore only the variations with x and z and examine some aspects of $\tilde{v}(x, 1, z, t)$ or at any other y location.

Criminale and Kovásznay inverted the integral given in (23.20) for the case $\tilde{v}(x, 1, z, t)$ using numerical methods and the functions (24.5) and (24.6) for c. Once this was done, the data could be presented for different values of time after the initial instant ($t = 0$) of introduction of the disturbance as seen when looking down on the flat plate. (In other words, a

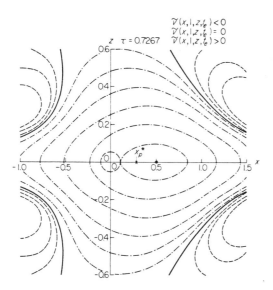

FIG. 24.1. Development of Gaussian pulse disturbance originally located at origin (shown as dashed circle for $t = 0$) at an early time ($t = t_e$) after introduction. View is vertically down on flat plate and nondimensional time, $\tau = \alpha_{\max} U_0 t$.

sequence of movie clips with the camera pointing vertically downward.) The result is a display of constant contours of $\tilde{v}$ as a function of x and z. One view corresponding to a relatively early time after the initial disturbance is shown in Fig. 24.1. For reference, the initial disturbance contour appears as a circle about the origin. It can be seen that the first effect is to spread the peak of the locally concentrated energy over a broader area and, in effect, decrease its amplitude. A similar view for much later time can be seen in Fig. 24.2. We note that the disturbance has progressed to a developed wave packet in the sense that there are now negative $\tilde{v}$ contours which cross the x axis. For the earlier times, the "wrinkling" was off this axis completely. Again, the initial circle contour would be centered about the origin of introduction if drawn on this scale.

A search for the location of the maximum value of $\tilde{v}$ from the contours as a function of time was made by these authors and found to always occur

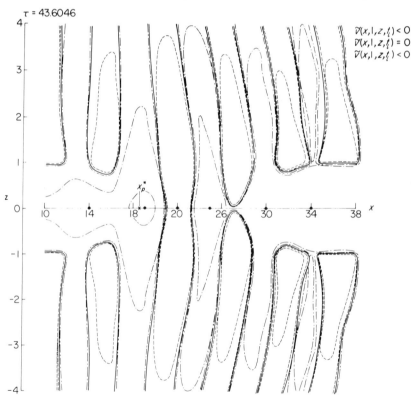

FIG. 24.2. Later development of pulse $(t = t_c)$ looking vertically downward on flat plate. Original circular pulse is now far upstream by comparison; $\tau = \alpha_{max} U_0 t$.

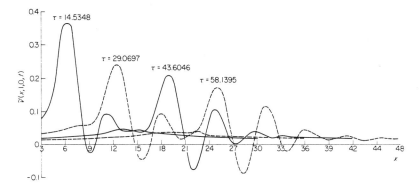

FIG. 24.3. Maximum amplitude of pulse as determined from cross plots of vertical views. Location is always for $z = 0$; $\tau = \alpha_{max} U_0 t$.

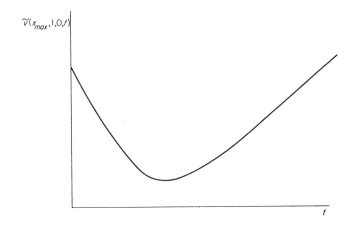

FIG. 24.4. Maximum of pulse as a function of time.

on the x axis. Thus, a cross sectional cut along the x axis will indicate major features of the developed $\tilde{v}$ as a function of x for a fixed time and $z = 0$. Figure 24.3 illustrates this point very well.

Close inspection of the three sets of curves reveals that it is only locally that contour lines can be considered aligned perpendicular to the direction of the mean flow; generally speaking there is a noticeable curvature which corresponds to the three-dimensionality present. Criminale and Kovásznay were able to compute that it is in the limit of $t \to \infty$ that a true two-dimensional system appears and, in this case, the surviving wavenumber —the one of maximum amplification—becomes that of Tollmien and Schlichting. This conclusion has been summarised in the plot of Fig. 24.4

where the maximum value of $\tilde{v}$ is depicted as a function of time. The asymptotic form of the wave packet tends to be an ellipse with the major axis along the x axis. Thus, the boundary layer is indeed a strong filter.

Finally, Criminale and Kovásznay showed that the maximum of $\tilde{v}$ is essentially the same as if a particle had been released at the disturbance location and subsequently traveled downstream on the most amplified wave. The points marked by x_p^* on Figs. 24.1 and 24.2 denote such a point at that instant of time. A comparison of these values with those of the group revealed that the group leads this single wave and gives the impression of waves entering from the front and leaving from rear of the wave packet.

Chapter VI

Experiments

25. Experiments Confirming Theory

Since the late 1920's, a continuous program of experimental studies of fluid oscillations and turbulence has been pursued at the National Bureau of Standards in Washington, D.C. The first efforts resulted in reliable hot-wire anemometers (e.g., Dryden, 1948) and various probings of turbulence. Soon after the end of World War II the work was directed more specifically toward the oscillations of laminar boundary layers. In 1943 Schubauer and Skramstad obtained the first laboratory confirmation of the calculations of Tollmien and Schlichting. The agreement between theory and experiment was sufficiently close to convince physicists and mathematicians of the soundness of their approach. In the subsequent years the investigations became deeper and resulted in the important findings reported by Schubauer and Klebanoff (1955, 1956), Klebanoff and Tidstrom (1959), and Klebanoff et al. (1962).

In this section we describe the experimental setup used by these workers and summarize the results, insofar as the fluctuations are small enough to agree with a linear theory. This naturally leads to a review of nonlinear processes. It goes without saying that the lead in progress toward an understanding of this difficult subject has been equally shared by both theoreticians and experimentalists.

The general layout of the facilities at the Bureau of Standards is sketched in Fig. 25.1. A rigid flat plate with a sharp edge is placed in a wind tunnel. The walls of the tunnel are far enough apart to be of insignificant influence. They are slightly divergent so that the pressure along the useful side of the plate is constant. The incoming stream of air has a low turbulence level and a constant velocity. Along the useful side of the plate a Blasius boundary layer develops to a thickness, δ, that is proportional to

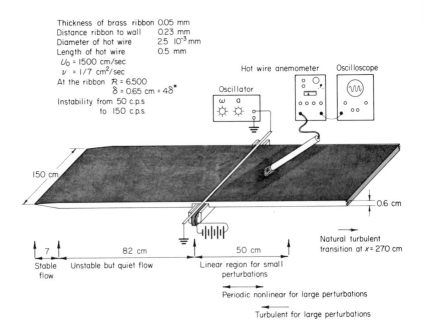

FIG. 25.1. Experimental installation to study boundary layer oscillations.

$\sqrt{x}$. We define δ for the Blasius layer as four times the boundary layer displacement thickness δ^*. At $y = \delta$, this choice gives $U/U_0 = 0.9999$ and $U''(\delta)/U''_{\max} = 0.0061$. Thus the value of U'' is negligibly small at the edge of the layer. (With $\delta/\delta^* = 2.88$, we have $U/U_0 = 0.99$ and $U''(\delta)/U''_{\max} = 0.246$, which is not at all negligible!) Normally, the layer begins to oscillate without any external prompting, and it becomes turbulent at some distance x_T from the leading edge. According to the theory, it should become unstable to small perturbations at some distance x_C corresponding to $\mathscr{R}_C$.

To verify the theory, an artificial perturbation can be introduced and, if it stimulates an unstable wave, the artificial oscillation will outgrow the natural waves and be easily observed. The artificial perturbation is produced by a thin metallic ribbon stretched close to the wall and traversed by an oscillating electric current. On the opposite side of the wall an electromagnet produces a constant magnetic field. In this way the ribbon oscillates to and fro from the wall. The amplitude and frequency of its motion can be regulated.

The u-component of the air stream can be measured by a hot-wire anemometer parallel to the edge of the wall. Both mean velocity and fluctuations can be observed. The other velocity components can be sensed by a combination of oblique hot wires (Kovásznay *et al.*, 1962). The frequency ω is dictated by the oscillator. By comparing the phase of the signal with that of the oscillator and moving the wire parallel to the x-axis, at a constant value of the ratio y/δ (thus slightly changing y), the wavelength can be measured to correspond with α_r. The value of α_i is deduced from measurements of the amplitude of the signal.

At any position x and z the amplitude and phase of every component of the velocity can be examined. The oscillator provides the reference for phase measurements.

It is a genuine experience to sit by the wind tunnel and slowly change the frequency of the vibrating ribbon, while at the same time watching the signal from a probe located in the boundary layer. The Reynolds number is constant, and if the frequency is lower than some value, say, 50 cps, we will be below the neutral curve and the oscilloscope will show only insignificant natural perturbations. As the frequency exceeds 50 cps, a sinusoidal signal appears, which reaches a maximum amplitude at a value, perhaps, of 100 cps. For higher frequencies it declines and vanishes at 150 cps, which means that the upper branch of the hairpin curve has been crossed.

The experimental measurements of neutral oscillations are indicated in Fig. 25.2 and compared with a curve obtained by digital computations. This curve was first computed by Brown (1959) and later independently confirmed by Kurtz and Crandall (1962), Kaplan (1964), and Betchov (1965). The neutral curves, calculated by using the Tietjens functions and power series (e.g., Tollmien, 1947), generally fall to the left of the experimental points and show considerable scatter.

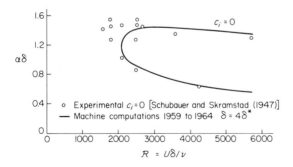

FIG. 25.2. Experimental and theoretical results, for neutrally stable oscillations of the Blasius layer.

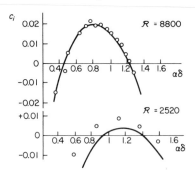

FIG. 25.3. Experimental and calculated values of c_i. (After Schubauer and Skramstad, 1947, and Kaplan, 1964.)

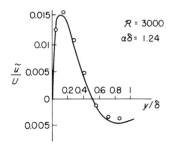

FIG. 25.4. Experimental and calculated values of u. (After Schubauer and Skramstad, 1947, and Kaplan, 1964.)

In Fig. 25.3 we compare the measured rate of growth c_i with digital computations.

The measurements of eigenfunctions are also in good agreement with calculated functions (see Fig. 25.4). The oscillations produced by the ribbon always displayed some variations with z, which were eventually traced back to imperfections in the incoming wind-tunnel flow. In particular, the antiturbulence screens introduce weak but large-scale disturbances. These three-dimensional variations can influence the position of turbulent transition.

To clarify this question Klebanoff *et al.* (1962) minimized pre-existing three-dimensional disturbances by using screens of lower solidity. They purposely introduced a controlled three-dimensional perturbation by placing pieces of thin tape at regular intervals on the wall immediately under the vibrating ribbon. This assumes control of the phase of the periodic variations in z.

In principle, the presence of the tape could alter the magnitude of the initial perturbation; it could also create a local modification of the mean flow and cause substantial changes in the rate of amplification. A local inflection, for example, could be very effective.

It appears that the main effect of the pieces of tape is to make the Reynolds number a function of z, with a weak maximum downstream of each piece of tape. These mean flow modifications are too small to be measured, even immediately behind the ribbon. However, they have been inferred from observation of the amplitude of the velocity fluctuation, examined as a function of z, at some constant x. If the frequency is high, corresponding to an eigenvalue near the upper branch of the neutral curve, the amplitude of the signal varies in the z direction with a minimum downstream of the pieces of tape. If the frequency is low, thus near the lower branch, the signal has a maximum amplitude downstream of the piece of tape. This is exactly what we would observe if $\mathscr{R}$ varied with z, for at a constant and low frequency the factor αc_i increases with $\mathscr{R}$; conversely it decreases at high frequencies. By changing the frequency we can place the regions of maximum activity either downstream of the pieces of tape or downstream of the intermediate gaps. The linear theory is still valid, as proved by the fact that the amplitude grows exponentially with x. The oscillation can be described by an expression such as

$$v = [\tilde{v}(x, y) + \hat{v}(x, y) \cos \beta z] \cos \omega t. \qquad (25.1)$$

26. Experiments Related to Subsequent Development of Instability

If the hot wire is placed sufficiently far from the ribbon or if the amplitude of the ribbon is large enough, the measurements show evidence of nonlinear processes. The velocity fluctuations are no longer sinusoidal in time, the rate of growth with x becomes linear rather than exponential, and the mean profile shows sone systematic alterations. Of these three effects only the last has become the object of a successful theory.

As we have already stated, an increase in the frequency of the ribbon changes the amplitude in the wake of the pieces of tape from a maximum to a minimum. If the amplitude of the fluctuations is large enough, either because the hot wire is sufficiently far from the ribbon or because a large current passes through it, we can observe the effect of the frequency on the mean flow disturbances. Klebanoff *et al.* found that the transversal modifications of the mean flow were influenced by the frequency, just as in the case of small fluctuations. This is a certain sign that the mean-flow disturbances are created by the time-dependent fluctuations.

The mean profile is depleted where the oscillation is most vigorous. Furthermore, we observe a component W of the mean flow which varies with y but does not extend into the free-stream region.

The alteration of the mean flow corresponds to longitudinal vortices whose strength grows with x. A theory that analyzes such possibilities is discussed in Section 27.

It is in these regions in which the mean flow has become warped and in which the oscillation is a sum of two main contributions and of higher harmonics that a radically new phenomenon appears. More precisely, these new events occur in the region in which the stationary longitudinal vortices carry the fluid away from the wall and at such times when $\tilde{u}$ is a minimum. Then the hot-wire anemometer shows a sharp spike in the u-component as a function of the time. Further downstream these spikes appear in groups of two, three, or more, and soon the entire boundary layer has become turbulent (Klebanoff *et al.*, 1962; Kovásznay *et al.*, 1962).

If the ribbon is not vibrated, the same general process is also observed, but it is now initiated by unspecified perturbations. As x increases, we observe successively the region of weak two-dimensional oscillations, the region of mixed two- and three-dimensional waves, the appearance of longitudinal vortices, and transition to turbulent flow. However, the whole structure is subject to unpredictable changes, probably related to the initial perturbations. Transition to turbulence becomes intermittent in space and time. When the ribbon is energized, the entire process becomes more regular and stationary. The spikes characteristic of turbulence transition were first observed in this ribbon-governed condition at the appropriate values of z. It was then actually possible to detect the presence of these spikes in previous recordings of hot-wire signals, taken without vibration of the ribbon. It helps to know what you are looking for.

A theory that deals with the concentration of vorticity in a three-dimensional flow and offers an explanation for the formation of the first spike has been proposed by Stuart (1965). Subsequent rolling up of the region of concentrated vorticity would account for multiple spikes.

A different process of transition was observed by Tani and Komoda (1962) in a flow containing a substantial component of vorticity perpendicular to the wall.

Note added in proof: An independent and often overlooked confirmation of the Tollmien-Schlichting theory has also been made in the laboratory by Liepmann (1943). Moreover, Liepmann already detected and measured the existence of a w-component of velocity.

Chapter VII

Nonlinear Effects

27. Theory of Longitudinal Vortices

In the experimental studies made by Klebanoff *et al.* (1962) on the oscillations of the boundary layer it was noticed that the mean flow was affected by the presence of oscillations. As already mentioned, the main effect is the appearance of vorticity oriented in the x-direction and periodically varying with z. The magnitude of the stationary vortices grows with x, and they are confined entirely within the boundary layer. Thus the mean flow acquires components $V(x, y, z)$ and $W(x, y, z)$, as illustrated in Fig. 27.1. These observations prompted Benney and Lin (1960) to construct an appropriate theory which we now summarize. For additional details the reader is referred to the papers of Benney (1961, 1964).

To begin, we consider a growth in time rather than a growth along the direction of the x-axis. This provides mathematical simplicity and corresponds roughly to the point of view of an observer moving with the phase velocity c_r. In turn, this means that all factors will either be independent of x or periodic in x. The steps leading to the development are the following:

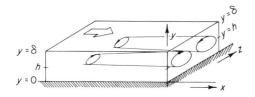

Fig. 27.1. Stationary longitudinal vortices occurring in an oscillating boundary layer and produced by nonlinear effects. (Measurements of Klebanoff.)

135

1. Establish a general vorticity equation.
2. Postulate a single combination of waves.
3. Evaluate the production of longitudinal stationary vorticity.
4. Treat the case of the shear flow.
5. Treat the case of the boundary layer.

(a) GENERAL VORTICITY EQUATION

Let us use the Cartesian tensor notation u_i, $i = 1, 2, 3$, to indicate the Eulerian velocity components u, v, and w. The Navier-Stokes equations read as follows:

$$\frac{\partial u_i}{\partial t} + u_k \frac{\partial u_i}{\partial x_k} + \frac{\partial p}{\partial x_i} = v \frac{\partial^2 u_i}{\partial x_j \, \partial x_j} \tag{27.1}$$

$$\frac{\partial u_i}{\partial x_i} = 0, \tag{27.2}$$

where summation over repeated indices is implicit. The components of the vorticity vector are defined by equations such as

$$r_1 = \frac{\partial u_2}{\partial x_3} - \frac{\partial u_3}{\partial x_2}. \tag{27.3}$$

The three vorticity equations can now be obtained with the help of the continuity equation. In fact, they can be put in the form of

$$\frac{\partial r_i}{\partial t} = \frac{\partial u_i}{\partial x_p} r_p - u_p \frac{\partial r_i}{\partial x_p} + v \frac{\partial^2 r_i}{\partial x_j \, \partial x_j}. \tag{27.4}$$

(b) WAVE COMBINATION POSTULATE

We now postulate that the oscillating layer can be described by the following expressions:

$$u_i = U_i(y, z, t) + \tilde{u}_i(x, y, t) + \hat{u}_i(x, y, z, t). \tag{27.5}$$

The term $U_i(y, z, t)$ represents the mean flow in a boundary layer or in a free shear layer. It is considered to be a slowly varying function of time. Its dependence on y is typical of a boundary layer or shear layer flow and the z-dependence enters only in the form of longitudinal vortices, as shown in Fig. 27.1.

The term $\tilde{u}_i$ represents an ordinary two-dimensional oscillation periodic in x with wavenumber α and periodic in t with frequency αc_r.

For simplicity we assume that $c_i = 0$, although this limitation is unnecessary. The last term, $\hat{u}_i$, represents a three-dimensional oscillation with the same wavenumber in α as $\tilde{u}_i$ and the same frequency. This wave is also assumed to be neutrally stable. The assumption that both oscillations have the same α and the same frequency $\omega = \alpha c_r$ can perhaps be strengthened by remarking that in nonlinear mechanics it is well known that coupled oscillators often do "lock in." In this respect the coupling through variation of the mean flow may be important.

We now pass to a frame of reference that moves with the velocity c_r so that $\tilde{u}$ and $\hat{u}$ no longer oscillate in time.

Recognizing that we intend to handle nonlinear terms, we find it advisable to avoid the use of complex functions. We have seen that a periodic boundary layer oscillation can be described by the following expressions:

$$\tilde{u}_1 = -\frac{\mu}{\alpha}[F'(y) \sin \alpha(x - ct) + K'(y) \cos \alpha(x - ct)]. \tag{27.6}$$

$$\tilde{u}_2 = \mu[F(y) \cos \alpha(x - ct) + K(y) \sin \alpha(x - ct)], \tag{27.7}$$

$$\tilde{u}_3 = 0. \tag{27.8}$$

The functions F and K are the real and imaginary parts of the eigenfunction, $\mathbf{v}$, and the parameter μ is used for scaling purposes. The origin of the x-axis can be selected so that $K(y)$ will vanish above the critical layer in the region where the Reynolds stress is zero.

Analogously, the terms in $\hat{u}_i$ can be written as

$$\hat{u}_i = A_i(x, y, t) \cos \beta z + B_i(x, y, t) \sin \beta z. \tag{27.9}$$

The experimental observations suggest that only one set of the modes appears, and this is perhaps caused by lateral boundary conditions or initial values. Suitable expressions for each of the components therefore are

$$\hat{u}_1 = \lambda[g(y) \sin (\alpha(x - ct) + \varphi) + l(y) \cos (\alpha(x - ct) + \varphi)] \cos \beta z, \tag{27.10}$$

$$\hat{u}_2 = \lambda[f(y) \cos (\alpha(x - ct) + \varphi) + k(y) \sin (\alpha(x - ct) + \varphi)] \cos \beta z, \tag{27.11}$$

$$\hat{u}_3 = \lambda[h(y) \cos (\alpha(x - ct) + \varphi) + g(y) \sin (\alpha(x - ct) + \varphi)] \sin \beta z. \tag{27.12}$$

The parameter λ is again an adjustable scale and φ is the phase difference between the $\hat{u}_i$ and $\tilde{u}_i$ systems.

Above the critical layer l, k, and g vanish. Furthermore, in the free stream we have $\beta g = \alpha h$.

Using these relations for the velocity components, we can evaluate the vorticity fluctuations. We will have

$$\tilde{r}_1 = \tilde{r}_2 = 0, \tag{27.13}$$

$$\tilde{r}_3 = -\frac{\mu}{\alpha}\left[(F'' - \alpha^2 F)\sin \alpha(x - ct) + (K'' - \alpha^2 K)\cos \alpha(x - ct)\right], \tag{27.14}$$

$$\hat{r}_1 = -\lambda \sin \beta z[(h' + \beta f)\cos(\alpha(x - ct) + \varphi)$$
$$+ (g' + \beta k)\sin(\alpha(x - ct) + \varphi)], \tag{27.15}$$

$$\hat{r}_2 = \lambda \sin \beta z[(\beta l + \alpha g)\cos(\alpha(x - ct) + \varphi)$$
$$+ (\beta g - \alpha h)\sin(\alpha - ct) + \varphi)], \tag{27.16}$$

$$\hat{r}_3 = \lambda \cos \beta z[(l' - \alpha k)\cos(\alpha(x - ct) + \varphi)$$
$$+ (g' + \alpha f)\sin(\alpha(x - ct) + \varphi)]. \tag{27.17}$$

(c) PRODUCTION OF LONGITUDINAL VORTICITY

We now have the information necessary to evaluate the production of longitudinal vorticity in the oscillating layer. From (27.4), and neglecting the viscous terms, we have

$$\frac{\partial r_i}{\partial t} = \frac{\partial R_i}{\partial t} = \frac{\partial u_i}{\partial x_p}r_p - u_p\frac{\partial r_i}{\partial x_p} \tag{27.18}$$

where R_i is the vorticity of the mean flow. Substitution of (27.5) into this vorticity equation would generate a large number of terms which fall in six distinct groups. The first group would contain products of the type $(\partial u_1/\partial x_3)R_3$ and would represent the interaction of the mean flow with itself. As long as the longitudinal vortices are small, these terms can be neglected. We are presently interested in the infancy of these vortices and not in their fully developed status. The terms of the second group are all proportional to μ. As long as R_1 and R_2 are negligible, they cancel each other, for we have assumed that these waves are neutrally stable.

The terms of the third group are proportional to λ and vanish in the limit in which $R_2 = R_3 = 0$, for the $\hat{u}_i$ are precisely the solutions of linearized equations that specify that the vorticity is independent of t. The next group is proportional to μ^2 and vanishes altogether because the terms are strictly two-dimensional.

The fifth group is proportional to λ^2. This results in terms that are either independent of z or vary like $\cos 2\beta z$. The effect of these terms has been studied by Benney. For the purpose of this discussion they are not relevant.

Finally, the sixth group collects terms proportional to $\lambda\mu$. Each varies with wavenumber β and contributes to the production of longitudinal vorticity. Hence, recalling that $\tilde{r}_1 = \tilde{r}_2 = 0$ and omitting all groups but the last, we have for the equation governing the longitudinal vorticity the relation

$$\frac{\partial R_1}{\partial t} = \frac{\partial \tilde{u}_1}{\partial x_1}\hat{r}_1 - \tilde{u}_1\frac{\partial \hat{r}_1}{\partial x_1} + \frac{\partial \tilde{u}_1}{\partial x_3}\tilde{r}_3 + \frac{\partial \tilde{u}_1}{\partial x_2}\hat{r}_2 - \tilde{u}_2\frac{\partial \hat{r}_1}{\partial x_2}. \quad (27.19)$$

Let us now examine this equation above the critical layer where four functions specify the oscillations. Indeed, this will leave us with only

$$\frac{\partial R_1}{\partial t} = \lambda\mu\{[F'(h' + \beta f) + F(h'' + \beta f')]\cos(\alpha(x - ct) + \varphi)\cos x(x - ct)$$

$$+ [F''h + F'(h' + \beta f) - \alpha\beta gF]\sin(\alpha(x - ct) + \varphi)\sin\alpha(x - ct)\}\sin\beta z. \quad (27.20)$$

Taking the average of (27.20) with respect to x and t, we are left with

$$\frac{\partial R_1}{\partial t} = \lambda\mu[Fh'' + F''h + 2F'(h' + \beta f) + F\beta(f' - \alpha g)]\frac{\cos\varphi}{2}\sin\beta z; \quad (27.21)$$

but it is possible to simplify this relation. Referring to the continuity equation, we see that for the $\hat{u}_i$ oscillation

$$\alpha g + f' + \beta h = 0. \quad (27.22)$$

Using this fact and $\beta g = \alpha h$, we see that (27.21) in the free stream becomes

$$\frac{\partial R_1}{\partial t} = 0. \quad (27.23)$$

Hence there is no production of longitudinal vorticity in the free stream.

(d) THE SHEAR LAYER

Let us take $U = \tanh y$ at a large Reynolds number and use the following expression to approximate the amplitude of the two-dimensional fluctuations:

$$F = \frac{1}{\cosh y}, \qquad c = 0. \quad (27.24)$$

This expression is exact if $\alpha = 1$. For the three-dimensional amplitudes w
start with the inviscid approximations:

$$f = \frac{1}{\cosh y}. \tag{27.25}$$

$$g = -\frac{\sinh y}{\cosh^2 y} + \frac{\beta^2}{\sinh y}, \tag{27.26}$$

$$h = -\frac{\beta}{\sinh y}, \tag{27.27}$$

which are exact if $\alpha^2 + \beta^2 = 1$ by the Squire theorem. We can easil
determine the pressure fluctuations

$$\tilde{p} = -\frac{\alpha\mu}{\cosh y} \sin \alpha x, \tag{27.28}$$

$$\hat{p} = -\frac{\alpha\lambda}{\cosh y} \sin \alpha x \cos \beta z. \tag{27.29}$$

Note that h, and therefore $\hat{w}$, become infinite at $y = 0$. A little reflectio
suggests that $\hat{w}$ should be the solution of the viscous equation

$$U \frac{\partial \hat{w}}{\partial x} + \frac{\partial \hat{p}}{\partial z} = \nu \frac{\partial^2 \hat{w}}{\partial y^2}, \tag{27.30}$$

where $\hat{p}$ can be obtained from f. We note the similarity of this equatio
with the vorticity equation encountered in the study of boundary laye
oscillations. Thus a viscous critical layer exists whose thickness is propor
tional to $(\nu/\alpha U'(0))^{1/3}$, and h is asymmetric, whereas g has a symmetric peak
centered on $y = 0$. The viscosity also removes the infinity from the functio
$g(y)$.

It is then shown by substitution into (27.21) that $\partial R_1/\partial t$ is asymmetri
with y. Thus the mean flow is modified by the system of alternate vortice:
sketched in Fig. 27.2. The phase angle which best promoted the growth o
these longitudinal vortices is $\varphi = 0$.

(e) THE BOUNDARY LAYER

The boundary-layer case was treated by Benney (1964) in the approximation that U' is constant from the wall to the edge of the free
stream. This leads to $F \approx \sinh y$ if $y < 1$ and $F \approx e^{-\alpha y}$ for $y > 1$. A similar
expression is given for f. In a better approximation we could use a parabolic

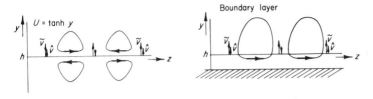

FIG. 27.2. Stationary longitudinal vortices in an oscillating shear layer and boundary layer. (Theory of Benney and Lin.)

expression for F and f. The resulting expressions for $\tilde{u}$ and $\hat{u}$ must be corrected for viscous effects at the critical layer and at the wall. Similar corrections are necessary for $\hat{w}$, which tends to diverge at the critical layer, just as in the preceding case. When all the results are assembled, we find essentially the flow sketched in Fig. 27.2. The production of mean vorticity has a negligible effect below the critical layer. The vorticity is similar to that of a shear layer in the region $c < U < 1$. In the free stream we find a larger potential fluctuation than in the free shear layer.

Consequently, we see that in free shear or boundary layers the nonlinear interaction between two elementary waves can produce stationary and streamwise vorticity. According to Klebanoff *et al.* (1962), these vortices set the stage for turbulent transition. It is surmised that similar vortices exist in fully developed flows (Townsend, 1956), but no theory has yet been proposed.

28. Theory of Nonlinear Critical Layer

The experiments have indicated that three-dimensional effects occur before the nonlinear oscillatory terms become significant. However, this should not discourage us from studying the effect of nonlinearity in a purely two-dimensional flow. Indeed, there is the danger that the three-dimensional effects may distract our attention and delay the full understanding of salient nonlinear processes. It is in this light that we want to review the work of Meksyn and Stuart (1951), who examined the scale of the nonlinear terms in an oscillating-plane Poiseuille flow.

When the amplitudes are such that the nonlinear terms become significant, the entire problem of stability must be reconsidered. The nonlinear terms contain partial derivatives such as $\tilde{v}(\partial \tilde{u}/\partial y)$, so that the failure of the linearized theory can occur if for some small amplitude some derivative becomes very large. For example, when the viscosity is allowed to vanish, while $\tilde{v}$ retains some prescribed small amplitude, we know that $\partial\tilde{u}/\partial$ becomes very large at the critical layer. Thus we can expect that the non-linear terms will assume control of the solution in place of the viscous terms. To our knowledge the details of this process at the critical layer have not been given.

Essentially Meksyn and Stuart recognized that the eigenvalues were very sensitive to the value of U'' at the critical layer and that any nonlinear effect on the basic profile would first manifest itself at the critical layer. We consider the case of a laminar boundary layer and discuss the non-linear effects on the curvature of the profile near the critical layer. For this purpose we take U as a slowly varying function of the time. The vorticity equation for the mean motion combining these points can be written as

$$U_{yt} + \overline{\tilde{u}\tilde{u}_{yx}} + \overline{\tilde{v}\tilde{u}_{yy}} = \nu U_{yy}. \tag{28.1}$$

It includes the transport of the vorticity fluctuation by the $\tilde{u}$ fluctuation and the production by $\tilde{v}$ in the presence of $\tilde{u}_{yy}$.

Among all the terms in (28.1) we concentrate our attention on $\overline{\tilde{v}\tilde{u}_{yy}}$ because we expect rapid variations near the critical layer; hence the highest derivatives in y will in fact be the dominant terms.

Differentiating (28.1) once more with respect to y and retaining only the terms with the highest derivatives, we are led to

$$\frac{d}{dt} U_{yy} = -\overline{\tilde{v}\tilde{u}}_{yyy} + \nu U_{yyy}. \tag{28.2}$$

The inviscid solution, which gives a gross description of the neutral oscil-lation of the layer, has the property that $\tilde{v}$ and $\tilde{u}$ are 90° out of phase. Thus it cannot contribute to (28.2). Near the wall the viscous effects create some components of $\tilde{u}$ in phase with $\tilde{v}$, which may contribute, but the largest term in $\tilde{u}$, which is in phase with $\tilde{v}$, comes from the critical layer.

In Section 10 we have seen that in a region in which $\tilde{v}$ is positive (ascending jet, Fig. 11.1) a negative component of $\tilde{u}$ appears below the critical layer. This term is induced by the vorticity stagnating in a zone of thickness proportional to $l = (\nu/\alpha U_y(y_c))^{1/3}$. The magnitude of the term is given by (10.18) and by the approximation relating $\tilde{u}$ to $\tilde{v}$ from the con-tinuity relation. Thus we have the approximation

$$(\tilde{u}_{yyy})_{y=h} = \left(\frac{\pi U_{yy}\tilde{v}}{\alpha U_y l^3}\right)_{y=h} = \frac{\pi (U_{yy}\tilde{v})_c}{\nu}. \tag{28.3}$$

At the critical layer we can use (28.3) and substitute into (28.2). Neglecting the term in U_{yyy}, we find

$$\frac{1}{U_{yy}} \frac{d}{dt} U_{yy} = -\frac{\pi \overline{\tilde{v}_c^2}}{v} \tag{28.4}$$

Under these circumstances it can be clearly seen that the oscillation tends to reduce the curvature at the critical layer. Referring to the vector diagram of Fig. 11.1, we see that this effect is destabilizing. By reducing U_c'' it becomes possible to have oscillations at higher α and consequently at lower Reynolds numbers, for η remains roughly constant.

According to (28.4), a fluctuation $\tilde{v} = 0.01 U_0$, at a Reynolds number of 3000 causes a major change of U_{yy} in one unit of time. In this time the free stream advances by one boundary-layer thickness, δ. Thus the entire structure of the classical theory begins to collapse, for the mean flow can no longer be treated as nearly constant in time or in the space variable x, depending on the point of view.

Meksyn and Stuart found a similar flattening of the profile in the plane Poiseuille flow and a reduction of the critical Reynolds number for amplitudes of the fluctuation up to 0.04. For higher amplitudes their calculations indicate an increase in the critical Reynolds number, but it is not certain that the assumptions used are still justified. In their approach U becomes a function of y and $\tilde{v}^2$. They find an actual inflection of U for the amplitude of 0.04 in the fluctuation.

The stability of the plane Couette flow for perturbations of finite amplitude has been examined recently by Kuwabara (1966) following the general procedure of Meksyn and Stuart together with the mathematical method of Galerkin. The results are shown in Fig. 28.1 where E is

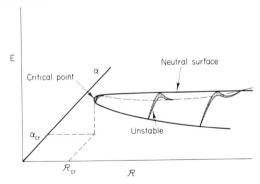

FIG. 28.1.   Nonlinear stability surface for plane Couette flow. E is the perturbation energy. (After Kuwabara, 1966.)

the energy of the perturbation. There is no instability if $\mathscr{R} > 45,000$. Above this critical value, and for a certain region of the $\alpha - \mathscr{R}$ plane, instability appears only in a rather narrow range of energy. The amplitude of the oscillation is sufficiently large to cause a large distortion of the mean velocity profile and produce a pronounced inflection in the center of the channel.

Another approach in this general direction considers a temporary inflection in the velocity profile of the boundary layer (Betchov, 1960a) Here, if the amplitude of the oscillation is large enough, the actual velocity denoted by $u = U + \tilde{u}$, will present an inflection at certain times and at certain locations. Thus a local shear layer is created which may have its own instability. Such inflections occur near the outer edge of the boundary layer and their existence is long enough to allow considerable amplification of a secondary disturbance. The phase velocity would be near $0.7U_0$ and the characteristic wavelength would perhaps be $\alpha\delta = 4$.

The critical layer can therefore become an inflection point and temporary inflections can become the critical layers of higher harmonics.

29. Analytic Studies of Nonlinear Processes

In this and the two succeeding sections we present various contributions to the understanding of oscillations when the nonlinear terms become of major importance. The studies have been performed either by pure analysis or by numerical computations. It should be understood that in each case the scope is so limited that it is not possible to give a general and comprehensive view of the situation.

This particular section is concerned with qualitative studies only. We discuss first the contributions related to boundary layer oscillations and then those dealing with flows away from walls.

(a) THE STUART METHOD

A method based on Fourier analysis and dealing with certain nonlinear effects has been developed by Stuart. For details the reader is referred to Stuart (1960, 1962). A two-dimensional incompressible flow is described by its stream function ψ. The vorticity has but the one component: $r = -\nabla^2\psi$ and the vorticity equation reads

$$\frac{\partial r}{\partial t} + \frac{\partial \psi}{\partial y}\frac{\partial r}{\partial x} - \frac{\partial \psi}{\partial x}\frac{\partial r}{\partial y} = \nu\,\nabla^2 r. \tag{29.1}$$

To study the stability, the stream function is expressed in a power series

$$\psi = \phi_0(y, t) + \phi_1(y, t)e^{i\alpha(x-ct)} + \phi_1^*(y, t)e^{-i\alpha(x-ct)}$$
$$+ \phi_2(y, t)e^{2i\alpha(x-ct)} + \phi_2^*(y, t)e^{-2i\alpha(x-ct)} + \cdots, \quad (29.2)$$

in which the presence of the complex conjugate guarantees that ψ will be real. Note that α and c are real and that the aperiodic variations in time are still incorporated in the various $\phi_i(y, t)$.

The various components are assumed to have the same phase velocity c to promote simplicity. We can always hope that the various components will be synchronized by the nonlinear processes, for it often occurs with coupled nonlinear oscillators. We define the operator for the Orr-Sommerfeld equation as

$$L(\alpha) = \left(U - c - \frac{i}{\alpha}\frac{\partial}{\partial t}\right)\left(\frac{\partial^2}{\partial y^2} - \alpha^2\right) - U'' + \frac{i\nu}{\alpha}\left(\frac{\partial^4}{\partial y^4} - 2\alpha^2\frac{\partial^2}{\partial y^2} + \alpha^4\right),$$
$$(29.3)$$

where α is a variable parameter. By substituting the series (29.2) into (29.1) and grouping the terms that do not vary with x or those with the wavenumbers α and 2α, we obtain the set of equations

$$\frac{\partial^2 U}{\partial t\,\partial y} + i\alpha\frac{\partial^2(\phi_1'\phi_1^* - \phi_1^{*'}\phi_1)}{\partial y^2} + 2i\alpha\frac{\partial^2(\phi_2'\phi_2^* - \phi_2^{*'}\phi_2)}{\partial y^2} + \cdots - \nu\frac{\partial^3 U}{\partial y^3} = 0,$$
$$(29.4)$$

$$L(\alpha)\phi_1 = \phi_2'(\phi_1'' - \alpha^2\phi_1)^* + 2\phi_2(\phi_1'' - \alpha^2\phi_1)^{*'}$$
$$- 2\phi_1^{*'}(\phi_2'' - 4\alpha^2\phi_2) - \phi_1^*(\phi_2'' - 4\alpha^2\phi_2)'$$
$$+ \cdots + O(\phi_2^*\phi_3), \quad (29.5)$$

$$L(2\alpha)\phi_2 = -\tfrac{1}{2}(\phi_1'\phi_1' - \phi_1\phi_1'')' + \cdots + O(\phi_1^*\phi_3). \quad (29.6)$$

These equations are completed by their complex conjugates. The term in ϕ_1 in (29.4) is responsible for the effect already examined at the critical layer. Note also that (29.4) can be integrated once immediately. These equations do not include an equation for $V(y, t)$, which contains terms of the order of ϕ_1^2, ϕ_2^2, and therefore tends to modify another important component of the mean flow. For small amplitudes we must find the solution of the linear problem, denoted as $\varphi_1(y)e^{\alpha c_i t}$. To find some way of separating the variables, Stuart writes

$$\phi_1(y, t) = A(t)\,\varphi_1(y) + B(t)\,\varphi_2(y), \quad (29.7)$$

$$\phi_2(y, t) = A^2(t)\,\varphi_3(y). \quad (29.8)$$

In order to write (29.5) in a form in which the variations with the time can be completely separated from the variations with y, Stuart takes $B = A^2 A^*$ and arrives at the relation

$$\frac{d}{dt}(AA^*) = 2\alpha c_i (AA^*) - 2\alpha K (AA^*)^2, \tag{29.9}$$

where K is a number that must be determined from the shape of the functions $\varphi_i(y)$. The integration of (29.9) is possible and gives

$$AA^* = \frac{c_i C e^{2\alpha c_i t}}{1 + K C e^{2\alpha c_i t}}, \tag{29.10}$$

where C is an arbitrary real constant. If $c_i > 0$, the small perturbation is already unstable, and, if $K > 0$, it grows exponentially until it approaches a ceiling located at c_i/K. The case $K < 0$ with $c_i < 0$ seems meaningless. Let us now examine the situation $c_i < 0$. For small amplitude we are outside the neutral curve and the oscillation will decay. For $t \to -\infty$ we have the amplitude $\sqrt{c_i/K}$, and K must therefore be negative. The differential equation (29.9) now shows that if $AA^* < c_i/K$ the first term on the right-hand side is leading and the amplitude decays. However, if $AA^* > c_i/K$, the second term dominates and, as expected, the amplitude goes to infinity. Therefore we have a critical threshold for the initial perturbation.

In view of the numerous assumptions this result can be interpreted only as evidence that there may be some metastable solutions. Such solutions are damped for small initial values, but they diverge if the initial amplitude is large enough.

Stuart's equations are related to the general approach of Krylov and Bogoliubov (1943) and other authors who concentrate their attention on a few key Fourier components while assuming that the others are somehow adjusting to some unimportant level. In the event that the system resonates at some unexpected frequency or wavenumber, serious errors can occur. Thus the method is limited to those situations in which we have a good preview of the solution. As examples of the pitfalls, we might think of the possibility of subharmonic or superharmonic resonance, as well as to situations in which the nth harmonic becomes unstable—if and when the fundamental reaches some critical amplitude (intermittency).

As pointed out by Stuart, the method may be of value in free shear layers, jets, and wakes. The relations between the functions φ_i, however, are so complicated that they have not been persued.

Other contributions in the same vein have been made by Raetz (1959) and Watson (1962).

(b) THE ECKHAUS METHOD

The behavior of the solutions of a large class of problems which includes those encountered in the theory of flow stability has been the object of a major contribution by Eckhaus (1962a, 1962b, 1963, 1965). His first paper considers a function of two variables alone, say $f(y, t)$, which obeys an equation that gives $\partial f/\partial t$ in terms of f, $\partial f/\partial y$, $\partial^2 f/\partial y^2$ and includes some nonlinear terms. Boundary conditions are given at two values of y. Initially, $f(y, 0)$ is given. Eckhaus considers the set of eigenfunctions $\varphi_n(y)$ of the linearized problem, corresponding to solutions of the type

$$f = \sum_{n=1}^{\infty} A_n \varphi_n e^{\omega_n t}. \tag{29.11}$$

The φ_n are a set of orthonormal eigenfunctions and the ω_n are the eigenfrequencies. The A_n are merely integration constants. In general, the nonlinear solution is taken by Eckhaus in the form of the series

$$f = \sum_{n=1}^{\infty} A_n(t) \varphi_n(y), \tag{29.12}$$

which yields a generalized form of (29.9) for each value of n:

$$\frac{dA_n}{dt} - \omega_n A_n = K_n(t). \tag{29.13}$$

In simple cases this leads to results similar to those of Stuart or Landau and Lifshitz (1953). In more sophisticated cases the growth of one mode may stabilize or destabilize another mode.

As an example and a test of the method, Eckhaus considers the model of turbulence proposed by Burgers (1948). This problem is remarkable by its similarity to the flow equations. The eigenfunctions φ_n are simply sine waves, and one by one they become unstable as a parameter $\mathscr{R}$, analogous to the Reynolds number, exceeds certain critical values such as π^2, $4\pi^2$, and $9\pi^2$.

The amplitudes of the first few Fourier components, according to Eckhaus, agree with those obtained by Burgers. It would be interesting and instructive to integrate this problem with a computer.

In a second paper Eckhaus extends his method to two-dimensional problems and considers solutions whose wavenumbers are close to the neutral curves. It appears that, by starting from general initial conditions, the oscillation does not become periodic in the x-direction unless the Reynolds number is sufficiently above the critical Reynolds number.

(c) THE FREE SHEAR LAYER

The shear layer, with $U = \tanh y$, has been examined by Michalke (1965), beginning with the linear solutions and continuing in the light of Stuart's method. The flow is considered inviscid.

According to the Helmholtz theorem, each fluid particle preserves its vorticity. Because the maximum vorticity of the unperturbed flow is $r_{max} = 1$, the vorticity in the oscillating flow should never exceed this value. If we now refer to the linearized inviscid theory, we find that the most unstable solutions occur near $\alpha = 0.5$. We also find that the vorticity fluctuation can be roughly approximated by $\tilde{r} = Ae^{-y^2}e^{\alpha c_i t} \cos \alpha x$, in which A is a small amplitude. Obviously the superposition of this fluctuation and that of the mean vorticity will in places exceed the prescribed maximum. When the nonlinear terms are considered, it is found that r never exceeds unity.

However, if $\alpha x = 2m\pi$ $(m = 0, 1, 2, \ldots)$ the region of high vorticity is broad, whereas it is narrow if $\alpha x = (2m + 1)\pi$.

The curves of constant vorticity are sketched in Fig. 29.1, in the xy-plane. The separate profile gives the vorticity distribution.

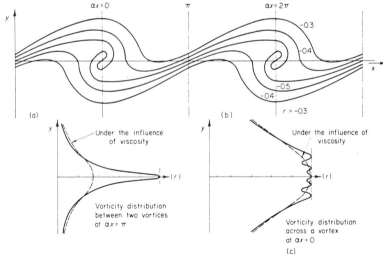

FIG. 29.1. Distribution of vorticity in the oscillating shear layer with $U = \tanh y$. In (a) the flow does not oscillate. With (b) we find oscillations at $\alpha x = \pi$. (From Michalke, 1965.)

The effects of viscosity are indicated by a dotted line; they show that diffusion cannot increase r above the maximum value, although it may increase the vorticity of some particles.

The results sketched in Fig. 29.1 have been obtained by Michalke (1965) by application of the Stuart method. The viscous terms were dropped, but the third harmonic was retained. A computer was used to integrate the equations giving the functions φ_i. The constant K in (29.9) is related to the first Landau constant and, according to Schade (1964a, 1964b), we have

$$K = -\frac{16}{3\alpha\pi}.$$
(29.14)

30. Numerical Experiments with Distributed Vortices

(a) INTRODUCTION

With the advent of the computer, the traditional division between theoretical and experimental physics is becoming somewhat outdated. It is now possible to experiment with theoretical fluids or plasmas which do not occur in nature but offer us instead an opportunity to sharpen our intelligence.

In the preceding chapters we have introduced the computer as a mathematical instrument ideally suited for integrating a system of differential equations. From a historical point of view, however, some of the earliest applications of computers occurred in the field of nuclear physics and were akin to stochastic processes (Monte Carlo method). In the next two sections we describe some studies in stability theory which fall into the category of numerical experiments.

Before discussing specific cases, we wish to make a few general remarks on what is implied by the difference between a "numerical integration" and a "numerical experiment." The classical image of the theoretician is that of a man using fewer than 100 symbols (some Greek, some Roman) and arriving at a few key equations. By using subscripts, indices, and implicit summations he may be able to handle a much greater number of terms. At times the theoretician can formulate general statements about the problem without reference to any specific solution: for example, the Squire theorem. This is a purely theoretical result. (To date machines have not discovered new theorems. Yet the possibility of creative mechanical mathematics should not be dismissed, as indicated by the work of Wang,

1960, for example.) In some rare cases the solution can be given in a closed form, but more often the solution is expressed in terms of specific expressions, such as Bessel functions or a suitable power series. Only then is it necessary to use a computer to determine these special functions. In general, the results of such efforts terminate in a few well-behaved curves whose general aspect can be stored in the human memory (e.g., stability curves). Thus, once it is agreed that the Orr-Sommerfeld equation provides an acceptable basis of operation, the plotting of rate of growth factors can be completely mechanized by a program of numerical integrations.

The established image of an experimentalist is that of a man using various measuring instruments and trying to obtain reproducible readings. In practice there is always a wide assortment of parasitic effects and disturbances which affect the results. The operator can change many parameters, and a dialogue takes place between him and the instruments. An experiment is successful if it leads to such reproducible results and to the conclusion that the parasitic effects are unimportant.

Nowadays there is a third kind of scientist who uses a computer to perform experiments. He starts from a set of equations having 10^N variables with N up to 4 under present limitations. On theoretical grounds he has already simplified these equations as far as possible and obtained some guarantee that the ordinary theoretical method could not handle the problem. Using the computer, he determines one or, more often, a large set of solutions. A variety of solutions can often be determined by making relatively insignificant changes in the initial conditions. Just as the experimentalist, he must examine the influence of various limiting factors, such as step size, boundary conditions, or round-off errors. The results of these numerical experiments are sometimes given in picture form, which in turn leads to some simple conclusion such as: there is a row of vortices. At other times the computer can average over a set of solutions or over an interval of time and print out the numerical value of some useful coefficient to indicate the heat transfer or the production of neutrons.

Numerical experiments can reveal the properties of flows which are strictly limited to two dimensions. This will lead to a better understanding of the role of the third dimension in real flows. Thus it will help to bridge the actual gap that separates theory from experiment. Moreover, a numerical experiment can be a testing ground for a new theory and a path to the discovery of new concepts. It can be just as fruitful or as wasteful as a purely theoretical effort or a complicated experimental program. Thus scientific effort must now be regarded as falling into three categories: theory, numerical experiment, and pure experiment. The boundaries between these regions are not sharp. We are aware of a situation in which an experimentalist used a computer to process information obtained by his probe, according to theoretical equations (linearization of hot-wire signals). There are two striking examples of this new approach pertinent to this area that warrant discussion.

(b) The Streaklines in an Oscillating Shear Flow

We have said very little about the actual path of fluid particles in an oscillating flow. In general, we are concerned with the velocity of whatever particles happen to be at some location at some time; this is the Eulerian point of view. The experimentalist can measure Eulerian velocities by using a hot-wire anemometer or other device, but he can also observe the location of some specially tagged particles at a given time. This is done by injecting smoke, ink, heat, ions, or bubbles into the moving fluid to provide information from a different point of view, usually named after Lagrange, although Euler apparently used both.

When the flow is stationary, the location of a set of tagged particles (often referred to as streaklines) coincides with the streamlines, and the interpretation of the measurements is not difficult.

In an oscillating flow such as the unstable wake behind an obstacle the streaklines often show marked irregularities. They do occasionally back up or form little spirals. This has often been interpreted as evidence of the presence of isolated vortices and even of turbulence. A remarkable clarification of this question has been given by Hama (1962) who examined the streaklines of a simple flow with the assistance of a computer. The mean velocity is given by

$$U = 1 + \tanh y, \tag{30.1}$$

so that the velocity is 1 at the point of inflection. For the fluctuation Hama chose the neutral inviscid case that corresponds to $\alpha = 1$ in Fig. 5.4:

$$\tilde{u} = \frac{a \sinh y}{\cosh^2 y} \sin (x - ct), \tag{30.2}$$

$$\tilde{v} = \frac{a}{\cosh y} \cos (x - ct), \tag{30.3}$$

where a is the arbitrary amplitude and $c = 1$.

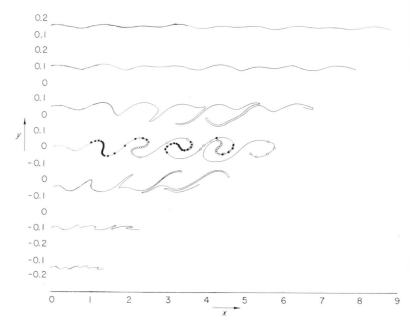

FIG. 30.1. Streaklines in an oscillating shear flow. (From Hama, 1962.)

The computer is programmed to follow the motion of a large number of particles, originally released from the same location at regularly spaced intervals of time. The locations of particles released at $x = 0$ and $y = 0$ after some time t are indicated in the middle of Fig. 30.1 by two kinds of dots. The streakline is the curve interpolating between the dots. The streaklines obtained with other altitudes of the injection point are also shown in this figure, without indications of the dots. The amplitude is $a = 0.04$.

At first sight we are tempted to say that the flow oscillates with an amplitude growing in the x-direction or with a complex α. We could also note the concentration of particles at the centers of the spirals and speak of discrete vortices. The truth is much simpler.

Let us first compare the streaklines at $y = 0.15$ and at $y = -0.15$. The difference in apparent wavelength is due to the different mean velocities: large at 0.15, small at -0.15. Thus the streaklines are not invariant to a Galilean translation, and we must proceed with caution. Let us now examine the results at $y = 0$. For an observer moving with the phase velocity $c = 1$ the Eulerian velocity is stationary in time; the streamlines appear as shown in Fig. 30.2. The closed streamlines are often referred to as " cats

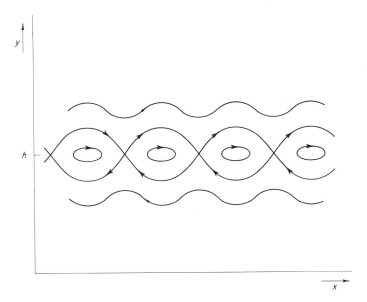

Fig. 30.2. Streamlines of an oscillating shear flow as seen by an observer moving with wave velocity.

eyes." The particle released at $x = 0$ and at $y = 0$ may find itself in the center of an eye, exposed to very little velocity. The next particle will be released somewhere to the left of the center and will start a small orbiting motion. Successive particles are launched on wider orbits until one particle, released exactly between the eyes, will drift far away. As time advances, these particles will perform a variable number of rotations, giving to the streakline the observed shape of a spiral. Thus the apparent amplification is simply an accumulation of various displacements. The particles released slightly off the critical layer will orbit inside the eye or drift away. This explains the long filaments formed by the streaklines at $y = 0.05$.

The computer can also indicate the path of a single particle, as a camera recording the flight of a shooting star. In the vicinity of the critical layer, and with the camera at rest, each pathline appears as a sine wave. However, the different orbits correspond to different periods so that the pathlines have different wavelengths! Thus the initial condition is all-important, and a set of particles which is released by a process whose phase is not related to that of the oscillation of the flow will soon show random properties.

All of these results are almost unchanged if the computer takes $\tilde{u} = 0$, but retaining $\tilde{v}$, as given by (30.3). This shows that the horizontal displacements of a particle are determined by its altitude $y(t)$ in the variable mean flow $U(y)$, rather than by the fluctuation $\tilde{u}$.

Hama's observations illustrate clearly the fact that a flow can have a simple Eulerian structure, yet can present intricate properties in Lagrange variables. Thus the experimentalist joins the theoretician in his preference for the Eulerian approach. It is also obvious that the information obtained by a technique of flow visualization must be interpreted with great caution. In practice, this information is often misleading and worthless.

(c) Nonlinear Numerical Solutions

As the various types of computers have become available, it has been possible to integrate nonlinear ordinary differential equations and nonlinear partial differential equations in two, and later in three, variables. One of the first applications to fluid dynamics with the variables x, y, and t was made by Evans and Harlow (1958); a compressible flow past a cylinder shows the bowed shock wave and a turbulent wake. A variety of programming codes has been developed and applied to numerous problems (see Alder and Fernbach, 1964). We confine our attention to a single example which illustrates the potential of modern computers, such as the IBM 7090. This work, done by Harlow and Fromm (1964), deals with viscous two-dimensional flow behind an obstacle. Additional details and refinements are discussed by Fromm (1964).

Essentially, the flow is described by the stream function $\psi(x, y, t)$ which is specified at about 1500 points and forms a rectangular lattice representing the flow from left to right, between two parallel walls. The vorticity equation can be written as

$$\frac{\partial r}{\partial t} + \frac{\partial r}{\partial x}\frac{\partial \psi}{\partial y} - \frac{\partial r}{\partial y}\frac{\partial \psi}{\partial x} = \nu\left(\frac{\partial^2 r}{\partial x^2} + \frac{\partial^2 r}{\partial y^2}\right). \tag{30.4}$$

The stream function can be determined from the vorticity by the relation

$$r = \frac{\partial^2 \psi}{\partial x^2} + \frac{\partial^2 \psi}{\partial y^2}. \tag{30.5}$$

The integration of (30.5) takes into account the boundary conditions by specifying that both components of the velocity must vanish at the walls and on the obstacle, which is represented by a rectangular block. The walls move with the mean velocity, which gives them as little effect as possible. On the left the flow should enter with constant horizontal velocity, but it is sometimes convenient to take periodic boundary conditions. Whatever appears at the downstream end is reintroduced upstream of the obstacle.

This is equivalent to studying the flow over a periodic array of obstacles, for the flow starts from a laminar condition and the ends are sufficiently far from the obstacle; this periodicity is unimportant as long as the flow does not last so long that the wake from one obstacle reaches the next.

The partial differential equations (30.4) and (30.5) are replaced by finite difference equations. At $t = 0$ the computer is loaded with some simple solutions such as $\psi = y$ and the modification of the vorticity is determined for a first advance in time. Of course, the initial increment of r occurs only at the obstacle. Knowing $r(x, y, t_1)$, we find that it is necessary to integrate (30.5) to determine the new stream function. This requires several iterations, and it is necessary to adopt a particular weaving pattern and to specify that the iteration cease when the modifications of ψ fall below a certain threshold. The result is $\psi(x, y, t_1)$, and the vorticity $r(x, y, t_2)$ can now be determined. With about 30 iterations and 1000 steps in time, Fromm used up to four hours of computer time, or more than 10^9 elementary operations.

Starting from a perfectly symmetric flow, it is found that the vorticity appears in two symmetric lines, trailing behind the obstacle. This wake, however, is unstable, and after a relatively long time the lines oscillate and the vorticity forms separate islands, as shown in Fig. 30.3. The initial perturbation responsible for this breakdown comes either from the weaving pattern or from the threshold of error. If the numerical experiment is repeated, all results are exactly reproducible, down to the last digit. This does not mean that they are free from error but only that chance plays no part in the numerical process.

To save time it is advisable to stimulate the oscillation by applying one single impulse to the flow, say just ahead of the obstacle. This starts the oscillation, and from then on the vorticity is shed alternately by each side of the obstacle.

The computer can feed a variety of information to a high-speed plotter so that the various aspects of the flow can be displayed. In Fig. 30.3(a) we have instantaneous streamlines for an observer fixed to the obstacle. The Reynolds number based on the width of the obstacle is 200. In Fig. 30.3(b) we have streamlines seen by the observer moving with the phase velocity.

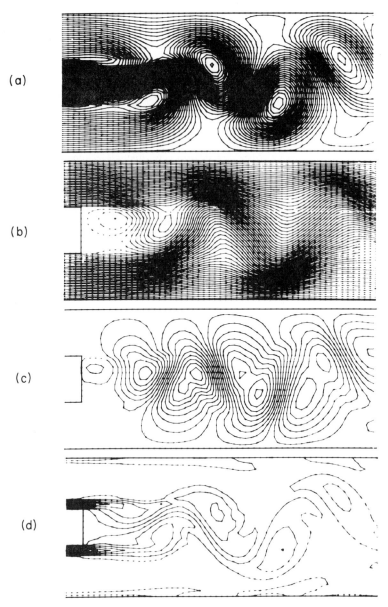

FIG. 30.3. Numerical solutions of a time-dependent viscous flow. (Results of Fromm and Harlow, 1963; Fromm, 1964.) The Reynolds number is 200. (a) Streamlines, with obstacle in motion. (b) Streamlines, with obstacle at rest. (c) Contours of constant v velocity component. (d) Vorticity contours.

The lines of constant v, displayed in Fig. 30.3(c), show a generally symmetric dependence on y. The lines of constant vorticity are shown in Fig. 30.3(d). Note the gradual formation of islands and the appearance of little boundary layers along the walls in response to the fluctuations of u.

Although these Eulerian pictures of v and r are relatively simple, they already show some evidence of random fluctuations. It is not known whether these irregularities are due to numerical errors or whether they are a true part of the solution. This problem is of some importance, for it amounts to determining whether the solution is sensitive to noise. In reality there are no physically perfect flows: the walls always have some roughness, the incoming flow is irregular, and the fluid itself is agitated by thermal motion. It may be that the arithmetic noise of the computer is even smaller than the generalized noise of the actual flow.

The computer can determine the drag on the obstacle or the coefficient of heat transfer by performing additional operations. Thus it furnishes the same kind of information as a wind tunnel, subject to special limitations but also less restricted by material properties.

Further work by Fromm (1964) explores the mean heat transfer and momentum transfer in such nonlinear oscillating flows.

Flows around rotating cylinders have been studied numerically by Thoman and Szewczyk (1966) who used cells of variable size and shape to fit the flow and the boundaries.

31. Numerical Experiments with Discrete Vortices

This section is concerned with the linear and nonlinear oscillations observed in two-dimensional flows, behind bluff bodies or sufficiently far downstream of thin obstacles. We begin with a review of the observed facts and continue with a study of the pertinent numerical experiments and analytic calculations. The main object is the well-known von Kármàn alley of vortices.

(a) THE OBSERVED FACTS

A most spectacular demonstration of the variety and beauty of hydrodynamics can be performed in a small smoke tunnel with a stroboscopic light. Sometimes the water-towing tank and camera are preferred. An ordinary cylinder is exposed broadside to the flow, and, when the Reynolds number falls in the range of 20 to 300, the flow becomes oscillatory; we also observe that the tagged particles form a variety of arabesques.

Essentially the boundary layers on either side become separated and form two trails of vorticity. These regions oscillate and start to roll up. At some distance the amplitude is so large that the two regions can no longer be distinguished. Instead, the flow forms isolated vortices, located on two staggered rows. Eventually the vorticity is diffused by a laminar process or it disintegrates in a turbulent wake.

It is true that any picture obtained with smoke or ink must be interpreted with caution, for a situation that appears to be intricate in Lagrangian variables may have a simple Eulerian description (see Section 30). One way to obtain Eulerian information is to search for maps of the pressure. In compressible flows, in the range of low Mach numbers, the correspondence between pressure and density is unique, and optical devices can be used to give pressure maps. Indeed, the gas density controls the index of refraction and techniques based on ray optics or actual interference can be used.

As an example, in Fig. 31.1 we show a Schlieren photograph of the wake behind a thin plate at Mach number 2.2. With a rigidly held plate, the wake is somewhat irregular, and in this particular experiment a small vibration has been imposed to the plate at the natural frequency of the wake to enhance the periodicity. The frequency is 34 kcps and the photographic exposure is 1 μsec. The dark areas indicate negative values of $\partial p/\partial z$ with positive value in the light areas. Thus, toward the right, the pressure is below average at the center of each of the isolated circular vortices. The wake begins with a spacially growing sinusoidal oscillation. It then forms spirals which become the centers of nearly circular disturbances on two staggered rows.

The existence of this alley of vortices was first reported by Bénard (1908). Its stability was established in 1912 by von Kármàn. Behind a bluff body the initial stages are telescoped, and the vortices are alternatively shed on either side without the preliminary linear oscillation. There is further evidence that at very large Reynolds numbers the turbulent wake of a cylinder behaves in a similar fashion. Apparently the eddy viscosity assumes the usual role of the molecular viscosity (Roshko, 1954).

The frequency of the von Kármàn alley of vortices has been measured by various means, in particular with hot-wire anemometers (Roshko, 1954; Kovásznay, 1949). With radius R and free-stream velocity U_0, the results generally indicate $f = \omega/2\pi = 0.1 U_0/R$. There is weak dependence on the Reynolds number. Instruments that sense the frequency in order to determine the mean velocity have been built.

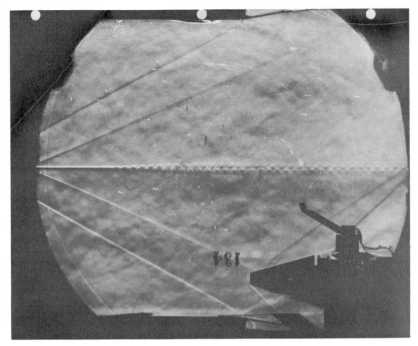

Fig. 31.1. Wake behind a thin plate at Mach 2.2. (Courtesy of Kendall, Jet Propulsion Laboratory.)

(b) The Single Layer

We now examine the stability of a single layer as a preliminary to the study of the von Kármàn formation. In Section 5 we saw that the simple shear layer is unstable to small perturbations. As long as the wavelength is large enough compared with the width of the shear layer, we can rely on Helmholtz's simpler model. It reduces to the instability of two adjacent parallel streams (see Fig. 4.4).

When the amplitude of the oscillation becomes comparable with the width or with the wavelength, the linearized theory collapses. Stuart's theory (see Section 29) and Michalke's contribution (Section 29) are applicable.

A different approach was proposed by Rosenhead (1931), who started from the assumption that the flow can be described by a single layer of isolated and concentrated vortices. A typical example is given in Fig. 31.2. Thus the vorticity vanishes except at certain points at which it become infinite. However, a large number of point vortices is found in each portion of the layer. When the point vortices are grouped in bunches or are well separated, we speak of a row of vortices. The flux of the vorticity

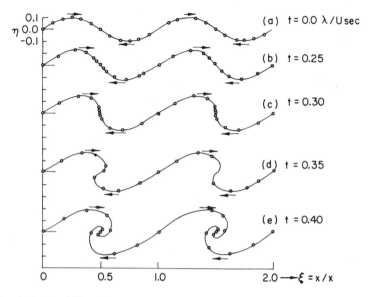

FIG. 31.2. Instability of a single layer of point vortices. (From Rosenhead, 1931.)

through the xy plane at each point vortex remains constant and identical with the circulation γ around the point. The initial distance between vortices is a, and therefore the mean flow induced far from the single row is $U_0 = \pm \gamma/a$. At point x, y, a vortex located at x_j, y_j will induce the velocity given by

$$u = \frac{\gamma}{2\pi} \frac{(y - y_j)}{r^2}, \tag{31.1}$$

$$v = \frac{-\gamma}{2\pi} \frac{(x - x_j)}{r^2}, \tag{31.2}$$

with

$$r^2 = (x - x_j)^2 + (y - y_j)^2.$$

In a two-dimensional flow we know that the vorticity accompanies the fluid particle. Thus the velocity of a point vortex is identical to that of the fluid.

The velocities induced by each vortex on any particular one must be added. Such calculations show that an initial disturbance in the location of the vortices leads to larger displacements. Typical results are shown in Fig. 31.2. Note that the vorticity tends to accumulate at the centers of the spirals at the expense of the ligaments. Thus the layer transforms into a row. When high-speed computers became available, the problem was re-examined by Birkhoff and Fisher (1959). They found that the rolling-up process can be quite irregular and questioned the notion of a smooth rolling-up process. Further work by Hama and Burke (1960) showed that the irregularities tend to appear when the steps of integration in time become too small. It is possible that when the computer uses too many steps the truncation errors act as a source of random perturbations. Moreover, the rolling-up flow could develop its own instabilities.

A good way to settle this question would be to test the reversibility of the computations by reversing the program and observing the unrolling. The arithmetic errors involve a loss of information equivalent to irreversibility. Thus a flow that does not reverse is a flow spoiled by numerical truncations or by defective methods of integration. In this respect we note that a routine like that of Runge-Kutta is not exactly reversible.

(c) The Single Row

In the last paragraph the distance between vortices was made small to simulate a continuous distribution of vorticity. The case of finite separation is interesting because it gives a preview of what takes place in the Kármàn alley.

Let us consider the vortices labeled $j = -2, -1, 0, +1, +2$ in Fig. 31.3. In the unperturbed state the velocities induced at the point $j = 0$ by vortices $j = 1$ and -1 are exactly equal and opposite.

Let us now pin down all vortices except the one labeled $j = 0$ which will now receive a small initial displacement in the upward direction. This displacement is illustrated by a dotted line in Fig. 31.3 and is drawn very much out of proportion. The velocities induced by the two nearest neigh-

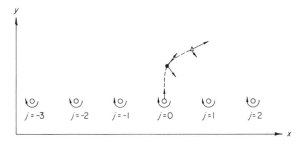

FIG. 31.3 A single row of vortices and the velocities induced on a displaced vortex by its nearest neighbors.

bors rotate and a net resultant to the right becomes apparent. Thus the vortex drifts to the right. Again the orientation of each induced velocity varies, but the dominant effect is simply the reduction in the distance from the vortex $j = +1$. As a result, the initial upward displacement is reinforced.

If the initial displacement is horizontal and to the right, a similar argument shows that the vortex tends to orbit around the point $j = 1$.

With $\tilde{x}$ and $\tilde{y}$ denoting changes in the position of the moving vortex, the analysis based on (31.1) and (31.2) leads to the system

$$\frac{d\tilde{x}}{dt} = A\tilde{y}, \tag{31.3}$$

$$\frac{d\tilde{y}}{dt} = A\tilde{x}, \tag{31.4}$$

where A is given by a sum of real terms, taking the pairs $j = \pm 1, \pm 2$, and so on. Obviously this problem has an unstable solution. Similar results are obtained when all vortices are allowed to move in accordance with fluid dynamics. This problem, discussed in Lamb (1932), reduces to the above system of linearized equations, except that A now depends on the wavenumber of the perturbation.

A similar instability occurs when the circulation of the vortices changes in sign with the parity of j. The odd vortices support one instability, whereas the even vortices sustain another.

(d) THE DOUBLE LAYER

Two parallel shear layers with a smooth distribution of the mean velocity, such as $U = 1/\cosh^2 y$, were examined in Section 6. We also saw that a continuous velocity distribution with discontinuous derivatives (polygonal contours) can be treated by algebraic methods (Section 6). The results indicate a range of unstable wavelengths with a broad maximum of the rate of growth. These results correspond to the phenomenon observed in the left side of Fig. 31.1, where the perturbations are of small magnitude. There is nothing to announce the formation of a double row of vortices. Thus we must use a nonlinear analysis to cover the transition from a parallel laminar flow to a double row of vortices. The stability of the double vortex row is examined later, and the analysis reverts to a linearized approximation, starting from a flow that is no longer parallel.

To study the nonlinear behavior of two adjacent layers of vorticity, Abernathy and Kronauer (1962) used a large computer to follow the

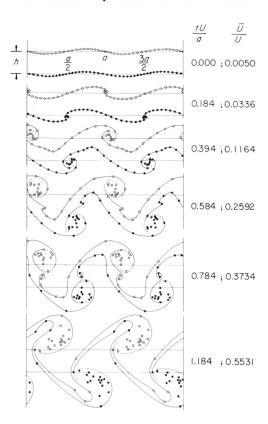

FIG. 31.4. Instability of a double layer of point vortices. (From Abernathy and Kronauer, 1962.)

motion of a number of point vortices. In principle, each induces on every other the velocities specified by (31.1) and (31.2), where x and y must be replaced by differences between the coordinates of the two interacting vortices. Typical results are shown in Fig. 31.4. The unperturbed flow has a velocity U above and below the double layer. Between the layers the mean velocity is taken as $-U$. This implies a relation between the strength of the individual vortex, the spacing between vortices, and U.

At $t = 0$ a particular disturbance, which can be symmetric or anti-symmetric and has an arbitrary wavelength is imposed. At first each layer behaves independently, but, as the vortices of one layer approach those of the opposite layer, a complicated interaction takes place. This can be seen in Fig. 31.4, beginning at $t = 0.584$.

The clockwise vortices are indicated by circles, the counterclockwise by dots. It can be seen that at times such that $t = 1.184$ some entrainment has taken place and both kinds of vortices are mixed. Thus the double layer evolves into a periodic array of clouds, which form a double alley of larger vortices. The net strength of a cloud is not proportional to the number of vortices, for they are not all of the same polarity. The entrainment simply weakens the large clouds.

It must also be noted that the final double row is wider than the original double layer. The presence of viscosity would smooth the vorticity distribution within each cloud so that the final stage would be similar to the structure observed on the right-hand side of Fig. 31.1.

Several other details have been studied by Abernathy and Kronauer, such as the mean velocity of the general pattern, the number of trapped vortices, and the role of the initial wavelength. Essentially, longer wavelengths produce the same general results, except for smaller clouds in the middle region with a somewhat greater degree of irregularity.

(e) THE DOUBLE ROW

The stability of double rows of vortices has been studied by von Kármàn and a detailed account is given by Lamb (1932).

Let an infinite number of equal vortices be placed on the line $y = 0$ at regular intervals $x = 0$, $x = \pm a$, $x = \pm 2a$, and so on. Each vortex is considered to have a circulation γ. The velocity field induced by these vortices is represented by an infinite series, each term according to (31.1) and (31.2). This series can be summed, and the result is

$$u = \frac{\gamma}{2a} \frac{\sinh(2\pi y/a)}{\cosh(2\pi y/a) - \cos(2\pi x/a)}, \qquad (31.5)$$

$$v = \frac{-\gamma}{2a} \frac{\sin(2\pi x/a)}{\cosh(2\pi y/a) - \cos(2\pi x/a)}. \qquad (31.6)$$

The streamlines are given by the relation $dx/u = dy/v$, which can easily be integrated; the result is

$$\cos\left(\frac{2\pi x}{a}\right) = \cosh\left(\frac{2\pi y}{a}\right) - K. \qquad (31.7)$$

The integration constant K is characteristic for each streamline. For $K = 0$ it marks the location of the vortices. For $-2 < K < 0$ the streamlines are closed. For $K < -2$ the streamlines are open and wavy. The borderline is given by $K = -2$, where saddlepoints occur halfway between the vortices. Along this borderline the maximum value of y occurs immediately above a vortex; that is, $\cosh(2\pi y_{max}/a) = 3$ or $y_{max}/a = 0.2805$. It has often been said that the closed streamline resembles the eye of a cat, and the ratio y_{max}/a is therefore characteristic of this particular structure.

Let us now place a second row of vortices, each with circulation $-\gamma$ located at $y = -ba$ and $x = c$, $x = c \pm a$, $x = c \pm 2a$, and so on.

For every value of b and c the velocity induced by the upper row on the lower row is equal to the velocity induced by the lower row on the upper row. Therefore the entire configuration moves with a uniform velocity in the xy-plane. This constant velocity is not parallel to the x-axis, except when $c = 0$ or $c = a/2$. These values define two special cases, denoted as the symmetric double row and the staggered double row.

In a system of coordinates moving with the system of vortices we observe a flow stationary in time but decidedly not parallel. Although it is necessary to use a nonlinear analysis to follow the transition from an unstable parallel flow to a nonparallel flow, it is nevertheless interesting to reconsider the question of stability to small perturbations.

The velocity induced by a vortex located at (x_i, y_i) on a vortex located at (x_j, y_j) is specified by (31.1) and (31.2). For small variations in the positions of both vortices the following expressions can be derived:

$$\Delta u_i = \frac{\gamma}{2\pi r^4} \{[(x_j - z_i)^2 - (y_j - y_i)^2](\Delta y_j - \Delta y_i)$$
$$- 2(x_j - x_i)(y_j - y_i)(\Delta x_j - \Delta x_i)\}, \tag{31.8}$$

$$\Delta v_i = \frac{+\gamma}{2\pi r^4} \{[(x_j - x_i)^2 - (y_j - y_i)^2](\Delta x_j - \Delta x_i)$$
$$+ 2(x_j - x_i)(y_j - y_i)(\Delta y_j - \Delta y_i)\}, \tag{31.9}$$

with

$$r^2 = (x_j - x_i)^2 + (y_j - y_i)^2. \tag{31.10}$$

The velocity fluctuations of the vortex identified by $i = 0$ can now be expressed as follows

$$\Delta u_0 = \frac{\gamma}{2\pi a^2} \left\{ \sum_m \frac{\Delta y_m - \Delta y_0}{m^2} - \sum_n \frac{[(n + \tfrac{1}{2})^2 - b^2](\Delta y_n - \Delta y_0) + 2(n + \tfrac{1}{2}) b(\Delta x_n - \Delta x_0)}{[(n + \tfrac{1}{2})^2 + b^2]^2} \right\},$$
$$\tag{31.11}$$

$$\Delta v_0 = \frac{\gamma}{2\pi a^2} \left\{ \sum_m \frac{\Delta x_m - \Delta x_0}{m^2} - \sum_n \frac{[(n + \tfrac{1}{2})^2 - b^2](\Delta x_n - \Delta x_0) - 2(n + \tfrac{1}{2}) b(\Delta y_n - \Delta y_0)}{[(n + \tfrac{1}{2})^2 + b^2]^2} \right\},$$

(31.12)

where the sums run from $-\infty$ to $+\infty$. Similar expressions formulated for every other vortex would result in an infinite system of equations in the unknowns Δx_j and Δy_j. In order to examine the oscillations that are periodic in space we introduce the notations

$$\Delta x_m = \xi e^{im\varphi + pt},$$ (31.13)

$$\Delta y_m = \eta e^{im\varphi + pt},$$ (31.14)

$$\Delta x_n = \lambda e^{in\varphi + pt},$$ (31.15)

$$\Delta y_n = \mu e^{in\varphi + pt},$$ (31.16)

where the angle φ is related to the wavelength as $\Lambda = 2\pi a/\varphi$. The time dependence is exponential, for all coefficients are independent of t.

Because of the symmetry between the effects of x_i and x_j and with relations such as $\Delta u_i = p \, \Delta x_i$, the equations for one vortex of the upper row (say $m = 0$) and one of the lower row (say $n = 0$) give

$$-p\xi - A\eta - B\lambda - C\mu = 0,$$ (31.17)

$$-A\xi - p\eta - C\lambda + B\mu = 0,$$ (31.18)

$$-B\xi + C\eta - p\lambda + A\mu = 0,$$ (31.19)

$$C\xi + B\eta + A\lambda - p\mu = 0,$$ (31.20)

where the coefficients A, B, and C are defined as

$$A = \frac{-\gamma}{2\pi a^2} \sum_{m=1}^{\infty} \frac{1 - \cos m\varphi}{m^2} + \frac{\gamma}{2\pi a^2} \sum_{n=-\infty}^{+\infty} \frac{(n + \tfrac{1}{2})^2 - b^2}{[(n + \tfrac{1}{2})^2 + b^2]^2},$$ (31.21)

$$B = \frac{-\gamma}{2\pi a^2} \sum_{n=0}^{\infty} \frac{2i(n + \tfrac{1}{2})b}{[(n + \tfrac{1}{2})^2 + b^2]^2} \sin (n + \tfrac{1}{2})\varphi,$$ (31.22)

$$C = \frac{-\gamma}{2\pi a^2} \sum_{n=0}^{\infty} \frac{(n + \tfrac{1}{2})^2 - b^2}{[(n + \tfrac{1}{2})^2 + b^2]^2}.$$ (31.23)

When the summations are extended to infinity, these coefficients can be expressed in terms of elementary functions (Lamb, 1932). This system of equations is homogeneous, hence has no solutions unless the determinant of the coefficient vanishes. The result leads to the eigenvalues p. To solve for p it is convenient to add the second row of the determinant to the first and then to subtract the first from the second. The operation is repeated for the third and fourth rows. The columns are treated in the same manner, and the result is a biquadratic form that leads to the four roots given by

$$p = \pm B \pm \sqrt{A^2 - C^2}. \tag{31.24}$$

Two roots correspond to $\xi = \lambda$, $\eta = -\mu$, and the other pair, to $\xi = -\lambda$, $\eta = \mu$. Because B is purely imaginary, the oscillation is unstable only if $A^2 - C^2$ is positive.

With an infinite number of vortices, all solutions have been found unstable, except for the staggered assembly in which $\varphi = \pi$. In this case $C = 0$ and the A vanishes if $\cosh \pi b = \sqrt{2}$. This leads to

$$b = 0.2805. \tag{31.25}$$

Thus the only stable configuration is the staggered double row with a spacing equal to the height of the cats eyes.

In reality, the number of vortices is limited by various factors. In the case $\varphi = \pi$ a calculation limited to a few terms of the expressions for A has shown that it is the nearest neighbors that are all-important. By considering only the effect of the nearest four vortices (two from each row) we find stability for $b = 0.24$. With eight vortices we find $b = 0.255$, and with 30 vortices b is within 0.001 of the limit value given in (31.25).

The double vortex row has been observed behind many bluff bodies in the range of Reynolds numbers of 40 to 300. It is sometimes accompanied by vibrations of the obstacle and by emission of sound waves. The singing of wires exposed to the wind is a good example. Behind a cylinder the frequency of shedding has been found experimentally near the value $f = \frac{1}{5}U_0/D$, where D is the diameter of the cylinder. This empirical result relates the diameter of the obstacle to the spacing between vortices. There is no theory to support this finding. The factor $\frac{1}{5}$ has a weak dependence on the Reynolds number, so that viscosity does not seem to be important. Perhaps we should examine oscillations that are purely periodic in time and growing in space. As discussed in Section 53, the eigenvalues have a singularity for values of the frequency of the order of $\omega = 0.55U_0/R$, where $\omega = 2\pi f$ and $R = \frac{1}{2}D$. However, these results stem from a linear analysis of a parallel mean flow.

An extensive treatment of the von Kármàn double row can be found in the text of Kochin *et al.* (1964), including a proof that the value $b = 0.2805$ is only a condition of minimum instability. It appears that even this configuration is unstable to higher order disturbances.

32. The Energy Method

This method was first proposed by Orr (1907) in flows that present parallel streamlines before oscillations occur.† Couette and Poiseuille flows have been treated by Orr, who obtained critical Reynolds numbers of about 88 in both cases. [His deductions have been confirmed by machine calculations (Conrad and Criminale, 1965a).] These results are in gross disagreement with experimental facts. The theory, however, is restricted to two-dimensional perturbations. Three-dimensional oscillations in parallel flow and flows with inflection points (shear layer, jet) have not yet been integrated. It is therefore not possible to give a complete assessment of the merits of the energy method, but it should be discussed because of its peculiarities and physical implications. Moreover, unexpected applications may perhaps be found in the study of anisotropic turbulent flows.

We start from the equations governing the perturbation $\tilde{u}_i$ of the mean flow U_i. The streamlines of the mean flow could be parallel or curved, and the governing equations are not linearized. We define the kinetic energy per unit mass of the perturbation as

$$e = \tfrac{1}{2}\tilde{u}_i\tilde{u}_i. \tag{32.1}$$

The rate of change of this energy of the perturbation can be determined and cast in the following form:

$$\frac{\partial e}{\partial t} = -\tilde{u}_i\tilde{u}_j\frac{\partial U_i}{\partial x_j} - \nu\frac{\partial \tilde{u}_i}{\partial x_j}\frac{\partial \tilde{u}_i}{\partial x_j}$$

$$+ \frac{\partial}{\partial x_j}\left[\frac{\tilde{u}_i\tilde{u}_i}{2}(U_j + \tilde{u}_j) + \frac{\nu}{2}\frac{\partial}{\partial x_j}\tilde{u}_i\tilde{u}_i - p\tilde{u}_j\right]. \tag{32.2}$$

We shall now assume that if (32.2) is integrated over a certain volume $\mathscr{V}$ the term contained in brackets in (32.2) will not contribute. Usually this volume is bounded by the walls and by the nodes of a periodic velocity field.

---

† It has been applied with some success to oscillating flows between rotating cylinders by Conrad and Criminale (1965b). Although the experiments of Taylor (1923) indicate a critical Reynolds number of 60, the energy method predicts 45.

This leads to the integrated energy equation

$$\frac{\partial E}{\partial t} = \frac{\partial}{\partial t} \iiint_{\mathscr{V}} e \, d\mathscr{V} = -\iiint_{\mathscr{V}} \tilde{u}_i \tilde{u}_j \frac{\partial U_i}{\partial x_j} \, d\mathscr{V} - \nu \iiint_{\mathscr{V}} \frac{\partial \tilde{u}_i}{\partial x_j} \frac{\partial \tilde{u}_i}{\partial x_j} \, d\mathscr{V}, \quad (32.3)$$

where E is the total perturbation energy in volume $\mathscr{V}$.

Note that the nonlinear terms have disappeared without linearization. They can only transport or shift energy from one velocity component to another and do not contribute to the total energy in the volume $\mathscr{V}$. The first term on the right-hand side of (32.3) could either increase or decrease the energy of the perturbation at the expense of the mean flow. The second term is always negative and represents the production of heat by friction.

If E is stationary, the left side vanishes and the production must match the dissipation. With the constant U_0 and L as references, this condition corresponds to

$$\mathscr{R} = \frac{U_0 L}{\nu} = \frac{U_0 L \iiint_{\mathscr{V}} \frac{\partial \tilde{u}_i}{\partial x_j} \frac{\partial \tilde{u}_i}{\partial x_j} \, d\mathscr{V}}{-\iiint_{\mathscr{V}} \tilde{u}_i \tilde{u}_j \frac{\partial U_i}{\partial x_j} \, d\mathscr{V}}. \quad (32.4)$$

Note that E is a function of time and that we are saying only that its time derivative is zero at one particular instant. If the $\tilde{u}_i$ correspond to some simple oscillation in time, this could indicate neutral stability. However, if the $\tilde{u}_i$ are some combination of various modes, the decay of some modes and the growth of others could momentarily render E stationary. It is obvious that for any given set of $\tilde{u}_i$ a value of ν can be found such that E is stationary and (32.4) is valid. The question raised by Orr is the following: "What functions $\tilde{u}_i$ make $\mathscr{R}$ minimum in (32.4)?" An answer can be found by applying the calculus of variations. If A is the numerator and B, the denominator, on the right-hand side of (32.4), a small variation of the functions $\tilde{u}_i$ will produce changes in A, B, and $\mathscr{R}$ such that

$$\mathscr{R} \, \delta B + B \, \delta \mathscr{R} = \delta A. \quad (32.5)$$

If $\mathscr{R}$ is extremum, $\delta \mathscr{R}$ vanishes, and standard variation calculus leads to the equations

$$\tfrac{1}{2} \tilde{u}_i \left(\frac{\partial U_i}{\partial x_j} + \frac{\partial U_j}{\partial x_i} \right) = \nu \nabla^2 \tilde{u}_j + \frac{1}{2} \frac{\partial \lambda}{\partial x_j}, \quad (32.6)$$

$$\frac{\partial \tilde{u}_i}{\partial x_i} = 0, \quad (32.7)$$

where $\lambda(x_i)$ is a Lagrange multiplier introduced to ensure that the continuity equation will be satisfied everywhere.

If the coefficients are functions of only one coordinate, the problem is reduced to a system of ordinary differential equations and a computer can be used. The extremum value of $\mathcal{R}$ can be determined from the solution.

Let us now make some terminal remarks. In a curved flow, in which this method is successful, the flow is unstable at large Reynolds numbers and viscosity is stabilizing. Thus we can surmise that the c_i's are relatively large, even near the neutral boundary. In general, the functions $\tilde{u}_i$ can be regarded as a superposition of modes in the sense of the initial value problem. If we are too far from the neutral line, the unstable mode will have large c_i's and the production of energy will exceed the dissipation. On the other hand, in the case of Poiseuille flow (and probably of boundary layers), the c_i's are small because the instability depends essentially on viscous effects. Thus, if the initial conditions contain some unstable mode, we may have to wait a long time before the growth of these modes becomes apparent against a background of decaying modes. The example discussed in Section 24 shows a marked decrease of the initial energy of the perturbation until the "winning" mode surges ahead.

Other uses of the energy method can be found in the work of Stuart (1958) and Serrin (1959). Critical reviews are also given by Shen (1964) and Stuart (1963).

Chapter VIII

Compressibility and Fluid Stratification Effects

33. Generalities

In this chapter we review the stability problem in which fluctuations of the density occur jointly with those of the velocity. To do so it is necessary to begin anew and to derive a new set of basic equations. These equations are applied to low supersonic flows, hypersonic flows, and, in a special section, to flows at very low Mach number at which the mean density is inhomogeneous and gravity forces cause convective instabilities.

The stability of compressible flows has been studied by Lees and Lin (1946), Lees (1947), and Dunn and Lin (1955); these early contributions were reviewed by Lin (1955). Several years later, further theoretical work was done by Reshotko (1960) and Lees and Reshotko (1962). During the same period Laufer and Vrebalovich (1957, 1958, 1960) and Demetriades (1958) gained experimental evidence of instabilities in compressible boundary layers. The integration of compressible stability equations by a numerical step-by-step method was first achieved by Brown (1961b). Shortly after a wealth of results were reported by Mack (1960, 1965a, 1965b, 1966). [Mack sometimes used double precision and complex arithmetic. A method similar to Kaplan (see Appendix III) should be advantageous for such tasks.]

In general, the stability of a supersonic boundary layer presents a complicated problem because of the significant increase in the number of parameters that must be considered. The mean velocity profile is affected by the thermal properties at the wall (insulated wall, cooling at the wall, etc.); the coefficients of viscosity and thermal conductivity of the fluid may vary with the temperature. Finally, the thermal inertia of the wall may become important as it affects the phase of the thermal fluctuations. Each of these points leads to various degrees of difficulty.

Unfortunately, there is not enough theoretical, numerical, or experimental evidence to allow a systematic discussion of this subject, and therefore we shall pursue only a few selected and essential topics. Our first goal is to examine the effect of compressibility when viscosity and thermal conductivity are set aside. Thus the first considerations deal with fluctuations at infinite Reynolds numbers. In Section 39 we restore both diffusive processes.

34. Thermodynamic Preliminaries

For notation we use capital letters for pressure P, temperature T, density Υ, and entropy S. From thermodynamics we have two independent relations for pressure, namely $P(\Upsilon, T)$ and $P(\Upsilon, S)$. We define S as the entropy per kilogram in place of the usual entropy per mole; this requires a slight modification of the common definitions of thermodynamics. For small fluctuations we have the relations

$$dP = \left(\frac{\partial P}{\partial \Upsilon}\right)_T d\Upsilon + \left(\frac{\partial P}{\partial T}\right)_\Upsilon dT, \tag{34.1}$$

$$dP = \left(\frac{\partial P}{\partial \Upsilon}\right)_S d\Upsilon + \left(\frac{\partial P}{\partial S}\right)_\Upsilon dS. \tag{34.2}$$

In these relations we identify the speed of sound in a compressible medium, $a = [(\partial P/\partial \Upsilon)_S]^{1/2}$ the specific heat at constant volume per kilogram, $c_v = T(\partial S/\partial T)_v$, the specific heat at constant pressure per kilogram, $c_p = T(\partial S/\partial T)_p$, and the ratio $\gamma = c_p/c_v$. It is also convenient to introduce the coefficient of thermal dilation $\lambda = -1/\Upsilon(\partial \Upsilon/\partial T)_P$. Having done this, we can now write (34.1) and (34.2) in the form

$$dP = \frac{a^2}{\gamma} d\Upsilon + \frac{\lambda \Upsilon a^2}{\gamma} dT, \tag{34.3}$$

$$dP = a^2 d\Upsilon + \frac{\lambda \Upsilon a^2 T}{c_p} dS. \tag{34.4}$$

We must now remember that these coefficients are related by the Maxwell equations of thermodynamics. Essentially, we have

$$\frac{\partial^2 E}{\partial V \partial S} = \left(\frac{\partial T}{\partial V}\right)_S = -\left(\frac{\partial P}{\partial S}\right)_V, \tag{34.5}$$

where E is the energy. This gives the following single restriction between the coefficients:

$$a^2 = \frac{(\gamma - 1)c_p}{\lambda^2 T}. \tag{34.6}$$

For a perfect gas we have $\lambda = 1/T$, $a^2 = \gamma RT$, and c_p and γ are assumed constants. This leads to the relation

$$c_v = \frac{R}{(\gamma - 1)}, \tag{34.7}$$

where R is the gas constant. As (34.3) to (34.5) stand, however, they are valid for any gas or liquid. This fact is referred to when we discuss the very low Mach number problem in liquids where gravity forces are important. The Mach number is to be based on the free stream velocity.

35. The Mean Flow of a Perfect Gas

We consider a parallel laminar flow with a constant mean pressure P in which the other mean quantities U, Υ, T, and S are functions of y only. It has been found that this system corresponds to practical cases (cf. Shen, 1952). With a constant pressure throughout the layer and a perfect gas, we are led to the following thermodynamic relations between the mean quantities:

$$\frac{\Upsilon'}{\Upsilon} + \frac{T'}{T} = 0, \tag{35.1}$$

$$\frac{\Upsilon'}{\Upsilon} + \frac{S'}{c_p} = 0. \tag{35.2}$$

To obtain a relation between U and T we must return to boundary layer theory and solve a problem similar to that leading to the Blasius profile. A detailed discussion has been given for this topic by Mack (1965a, 1965b). An example of a solution for a hypersonic boundary layer is shown in Figs. 35.1. Additional information is given for the same flow in Fig. 35.2. It can be seen that the curvature of U is concentrated near the outer edge. At the wall the temperature T tends to rise. This result can be understood if we note that the total enthalpy per kilogram $c_p T + \frac{1}{2}U^2$ stays almost constant across the layer. Of course, other boundary conditions, such as a cooled wall, would modify the curves shown in Figs. 35.1 and 35.2. Generally speaking, however, it is the temperature profile that exhibits the most notable change; the velocity is modified only slightly.

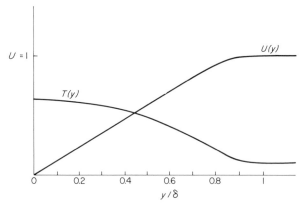

FIG. 35.1. Mean velocity and temperature profiles for a boundary layer at Mach number 5.6 (insulated wall). The inviscid oscillations have their critical layer at $y = 0.81$ where $d(U'/T)/dy = 0$. Then the sonic point $c - U = a$ occurs at $y = 0.46$.

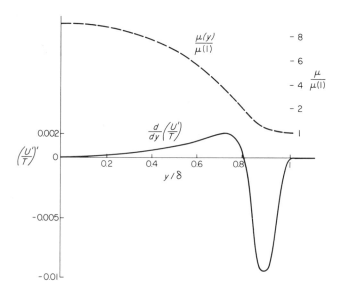

FIG. 35.2. Profiles of important functions in a boundary layer at Mach number 5.6.

36. Inviscid Fluctuations

By assuming that the fluid is nonviscous and in the absence of heat conductivity we can write the general equations for the velocity, pressure, density, and entropy as

$$\frac{\partial \rho}{\partial t} + \frac{\partial(\rho u)}{\partial x} + \frac{\partial(\rho v)}{\partial y} = 0, \tag{36.1}$$

$$\rho\left(\frac{\partial u}{\partial t} + u\frac{\partial u}{\partial x} + v\frac{\partial u}{\partial y}\right) + \frac{\partial p}{\partial x} = 0, \tag{36.2}$$

$$\rho\left(\frac{\partial v}{\partial t} + u\frac{\partial v}{\partial x} + v\frac{\partial v}{\partial y}\right) + \frac{\partial p}{\partial y} = 0, \tag{36.3}$$

$$\frac{\partial s}{\partial t} + u\frac{\partial s}{\partial x} + v\frac{\partial s}{\partial y} = 0. \tag{36.4}$$

For the moment we are limiting the investigation to that of a two-dimensional problem. The question of three-dimensionality is brought up in Section 43. This group of equations is standard, except for the use of the one for entropy. We prefer to use an entropy equation instead of the traditional one for energy, for, as we shall see, the fluctuations of entropy are simpler to analyze than those of temperature.

Introducing the symbols π, ρ, τ and σ as the fluctuating pressure, density, temperature, and entropy, respectively, we consider these quantities directly as complex amplitude functions. It is understood that the full solution to the fluctuating variable is the product of the complex amplitude and the factor $e^{i\alpha(x-ct)}$. Then we introduce two new quantities which are formed by combinations of the above. First, we define

$$\mathbf{m} = i\alpha\mathbf{u} + \mathbf{v}'. \tag{36.5}$$

In incompressible flows this quantity is zero and therefore serves as an indicator of the importance of compressibility effects. Second, we use a nondimensional density fluctuation given by

$$\theta = \frac{\rho}{\Upsilon}. \tag{36.6}$$

We retain the symbols U, P, T, S for the mean variables. The linearized equations for the fluctuations now read as

$$i\alpha(U - c)\theta + \frac{\Upsilon'}{\Upsilon}\mathbf{v} + \mathbf{m} = 0, \tag{36.7}$$

$$i\alpha(U - c)\mathbf{u} + U'\mathbf{v} + i\alpha \frac{\pi}{\Upsilon} = 0, \tag{36.8}$$

$$i\alpha(U - c)\mathbf{v} + \frac{\pi'}{\Upsilon} = 0, \tag{36.9}$$

$$i\alpha(U - c)\sigma + S'\mathbf{v} = 0. \tag{36.10}$$

In (36.7) we note that the term $\Upsilon'\mathbf{v}$ indicates that vertical oscillations in a stratified fluid will produce density fluctuations. In (36.10) a similar term in $S'\mathbf{v}$ produces entropy fluctuations which can be likened to the mechanism that produces vorticity fluctuations even if each fluid particle conserves its vorticity. This particular production of the entropy fluctuations represents a reversible effect in the thermodynamic sense.

Equations (36.7) to (36.10) can be supplemented by the thermodynamic relations

$$\pi = \frac{\Upsilon a^2}{\gamma} \left(\theta + \frac{\tau}{T} \right), \tag{36.11}$$

$$\pi = \Upsilon a^2 \left(\theta + \frac{\sigma}{c_p} \right). \tag{36.12}$$

Recalling our definition for the vorticity fluctuation $\mathbf{r} = \mathbf{u}' - i\alpha\mathbf{v}$, we can form an equation for the vorticity by multiplying (36.8) by $i\alpha$ and adding to (36.9) after taking the derivative with respect to y. We find this to be

$$i\alpha(U - c)\mathbf{r} + \left(U'' + U' \frac{\Upsilon'}{\Upsilon} \right)\mathbf{v} + U'\mathbf{m} + i\alpha(U - c)\frac{\Upsilon'}{\Upsilon}\mathbf{u} = 0. \tag{36.13}$$

The term $U'\mathbf{m}$ can be interpreted as a production of vorticity resulting from the change in cross section of a fluid particle having a mean vorticity U'. This is an obvious contribution when we think in terms of the preservation of angular momentum of a deformable body. The two terms proportional to Υ'/Υ are equivalent to $-\Upsilon'/\Upsilon(i\alpha\pi)$ by virtue of (36.8). This couple is a manifestation of the Bjørkness effect, which is well-known in meteorology. Indeed, in the general vorticity equation we find the term $(1/\rho^2)[(\partial\rho/\partial x_i)(\partial p/\partial x_j) - (\partial\rho/\partial x_j)(\partial p/\partial x_i)]$. It is easy to see that the fluctuation of this expression is precisely $-\Upsilon'/\Upsilon^2(i\alpha\pi)$.

Let us now eliminate $\mathbf{m}$ from the vorticity equation. Using the continuity equation (36.7), we can find $\mathbf{m}$ as a function of θ and $\mathbf{v}$. We then express θ in terms of π and σ by using (36.12). It remains to find σ in terms of v. We then obtain

$$\mathbf{m} = -i\alpha \frac{(U-c)\pi}{\Upsilon a^2} - \left(\frac{S'}{c_p} + \frac{\Upsilon'}{\Upsilon}\right)\mathbf{v}; \qquad (36.14)$$

but by (35.2) the second term in (36.14) vanishes. Finally, after substitution of (36.14) into (36.13), we are led to

$$i\alpha(U-c)\mathbf{r} + \left(U'' + U'\frac{\Upsilon'}{\Upsilon}\right)\mathbf{v} + i\alpha(U-c)\left(\frac{\Upsilon'}{\Upsilon}\mathbf{u} - \frac{U'}{a^2}\frac{\pi}{\Upsilon}\right) = 0. \quad (36.15)$$

Examination of this equation reveals an obvious analogy to that of the incompressible layer: to find an inviscid neutral solution the critical layer must be located at a point such that $U'' + U'\Upsilon'/\Upsilon = 0$. With reference to (36.1) we see that this is equivalent to requiring that $(d/dy)(U'/T) = 0$ at the critical layer. When the temperature profile is as shown in Fig. 35.1, U'/T is a maximum near $y = 0.81$.

Now, by the usual manipulation of (36.8) and (36.9) and by reuse of (36.9) to eliminate $\mathbf{v}$ and (36.14) for $\mathbf{m}$ we are able to obtain a relation for the pressure:

$$\pi'' - \left(\frac{2U'}{U-c} + \frac{\Upsilon'}{\Upsilon}\right)\pi' + \alpha^2\left[\left(\frac{U-c}{a}\right)^2 - 1\right]\pi = 0. \qquad (36.16)$$

If we replace π with $\mathbf{v}$ in (36.16), it becomes

$$\left[\frac{-U'\mathbf{v} + (U-c)\mathbf{v}}{\gamma RT - (U-c)^2}\right]' = \alpha^2 \frac{(U-c)}{\gamma RT}\mathbf{v}, \qquad (36.17)$$

a form often used by Lees and Lin (1964) and Lees and Reshotko (1962).

If the Mach number is sufficiently large, the denominator of the left-hand side of (36.17) vanishes for a particular value of y. This is the sonic point. An observer traveling with the phase velocity will find that, below the sonic point, the mean flow is supersonic. Thus the equation governing pressure fluctuations changes from elliptic to hyperbolic. The solution does not present an essential singularity at the sonic point, however. This is clear if, instead of a second-order equation, we write the equivalent system of two equations as used by Mack (1965a, 1965b)

$$(U-c)\mathbf{v}' = U'\mathbf{v} + \frac{i\alpha}{\Upsilon}\left[1 - \frac{(U-c)^2}{\gamma RT}\right]\pi \qquad (36.18)$$

$$\pi' = -i\alpha\Upsilon(U-c)\mathbf{v}. \qquad (36.19)$$

The only singularity occurs at the critical layer. These equations have the property that $\pi v^* + \pi^* v = K(U - c)$ where K takes one constant value below the critical layer and another one above. If the wall is rigid, $v(0) = 0$, and the first constant vanishes. If there is no accoustic radiation into the free stream, the second constant also vanishes. This function represents the transfer of energy by correlation between pressure and normal velocity fluctuations.

A few remarks on the effect of wall cooling are now in order. Cooling the wall modifies the functions $U(y)$ and $T(y)$. Mack noticed that cooling introduces a second maximum of the function U'/T, located near the wall. So far there is no evidence that this maximum allows new kinds of unstable waves. The phase velocity of such waves would be low, so that the pressure fluctuations would correspond to accoustic waves in the free stream. These would radiate energy away from the boundary layer and introduce substantial damping.

As the cooling of the wall increases, the first maximum of U'/T moves toward the wall. When this maximum occurs below the sonic point, Mack found that the instability disappears. With still more cooling, the two maxima merge into one, which eventually vanishes completely.

Stabilization by cooling has been established by Lees (1947) and reviewed by Lin (1955) and Shen (1964). It even occurs in the case of zero Mach number $(a \to \infty)$.

Some information on the stability of compressible wakes has been obtained by Gold (1963) and Lees and Gold (1964).

37. The Free Stream

In the free stream we can take $U = 1$, and all gradients of the mean quantities are zero. The governing equation for the pressure is a reduced form of (36.16), or

$$\pi'' + \alpha^2 \left[\left(\frac{1 - c}{a} \right)^2 - 1 \right] \pi = 0. \tag{37.1}$$

Solutions of this equation are readily obtained and are found to be

$$\pi = C \exp \left\{ \pm \alpha \left[1 - \left(\frac{1 - c}{a} \right)^2 \right]^{1/2} y \right\}. \tag{37.2}$$

Thus, if $1 - c < a$, we find the two familiar pressure waves from above and below. Conversely, if $1 - c > a$, the pressure oscillates with constant amplitude and there is the possibility of two sound waves.

Going back to (36.9) in the free stream and substituting the solution for the pressure, we obtain

$$\Upsilon(1 - c)\mathbf{v} \pm \left[\left(\frac{1 - c}{a} \right)^2 - 1 \right]^{1/2} \pi = 0. \tag{37.3}$$

Hence, if $|1 - c| > a$, the negative sign gives a $\mathbf{v}$ in phase with the pressure amplitude; with the positive sign it is out of phase. The flow of energy is proportional to the product $\pi\mathbf{v}^* + \pi^*\mathbf{v}$. From this fact we see that one of the acoustic waves takes energy away from the wall, whereas the other brings energy to the wall. The only possible relevant solution, therefore, is the one in which $\mathbf{v}$ and π are in phase and thus show a radiation of energy away from the wall. So far there is no experimental evidence to indicate that these radiating solutions actually exist. It would also be interesting to study the relation between an incoming and an outgoing wave. In hypersonic or supersonic flow between two walls this could lead to acoustic instability akin to resonance.

We can evaluate the amount of energy that would be radiated by an acoustic wave for the case $1 - c > a$. This power could be compared with the power released by the Reynolds stresses between the wall and the outer edge. Even if $\overline{\tilde{u}\tilde{v}}$ is of the order of $\overline{\tilde{v}^2}$, the fact that $\Upsilon(y)$ decreases near the wall suggests that a sound wave takes away more energy than the Reynolds stresses can release.

However, certain of Mack's numerical calculations have indicated the possibility of acoustic waves, as we report in the next section. Apparently, $\tilde{u}$ and $\tilde{v}$ are sometimes larger inside the layer than in the free stream, favoring the Reynolds stresses.

38. Inviscid Shear Flow

The inviscid pressure equation (36.16) has been integrated by Mack (1965a, 1965b) in laminar boundary layers along an insulated plate at various Mach numbers. He used the form given as (36.18) and (36.19). At the wall we have $\mathbf{v} = 0$, corresponding to $\pi' = 0$. The critical layer is located at the point at which $(U'/T)' = 0$.

As shown in Fig. 38.1, there is only one mode at low Mach numbers. If the Mach number is higher than 3, other modes appear, and at Mach 10 the proliferation of modes bears an obvious resemblance to acoustic processes.

Indeed, at such high Mach numbers the term in π' of (36.16) plays a

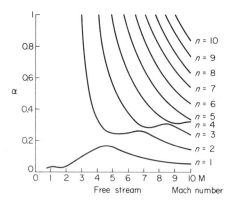

Fɪɢ. 38.1. Wave numbers of inviscid neutral solutions (insulated wall-boundary layer). (From Mack, 1965a, 1965b.) (At critical layer $d(U'/T)/dy = 0$.)

minor role, and essentially we have a wave equation. In Fig. 38.2 we show some pressure profiles for Mach 7 for the first four modes. Notice that the number of zeros of π is equal to the mode number minus one. The critical layer is located at $y = h$ and the sonic point at $y = s$.

The behavior of π for the lowest mode at different Mach numbers is shown in Fig. 38.3. It shows the transition between the almost constant pressure at Mach 2.2 to the marked variation at Mach 7.

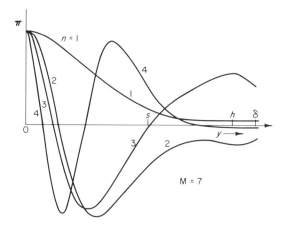

Fɪɢ. 38.2. Pressure fluctuations through a hypersonic boundary layer for the first four modes. The wall is insulating. (From Mack, 1965a, 1965b.)

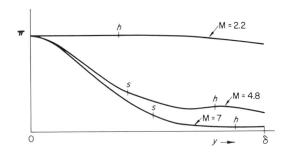

Fig. 38.3. Pressure fluctuation through a compressible boundary layer, at various Mach numbers, for the lowest mode. The wall is insulating. (From Mack, 1965a, 1965b.)

In addition to the results that correspond to the curves of Fig. 38.1, the calculations of Mack have revealed some entirely unexpected waves, which seem to form two classes: waves with a phase velocity of unity and waves with outgoing acoustic power.

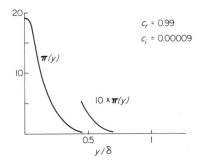

FIG. 38.4. The pressure fluctuation for an oscillation of the type confined to the hypersonic boundary layer. There is strong wall cooling. Note that the oscillation does not extend into the free stream. (From Mack, 1965a, 1965b.)

In Fig. 38.4 the pressure distribution of a wave with $c_r = 1$ is shown. Here the wall is highly cooled, but this condition is not necessary for the existence of such waves. Furthermore, great numerical difficulties are encountered when c_i is exactly zero, so that we must use a limiting process to locate the neutral eigenvalue. It appears in Fig. 38.4 that π does not penetrate the free stream and all other fluctuations have the same character. These oscillations, which are similar to acoustic oscillations, occur between the wall and the outer edge, where it is a fact that $(U'/T)'$ has another zero. The wavenumbers of the two lowest modes are shown as dotted lines in Fig. 38.1.

The other type of unexpected wave has a phase velocity so low that an observer moving with the wave would see a supersonic free stream. These waves have been found only in the calculations of cooled boundary layers.

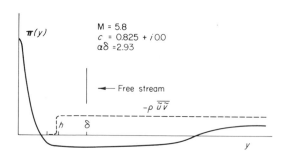

Fig. 38.5. The pressure fluctuations of a neutral supersonic oscillation. A sound wave is emitted. The behavior of the Reynolds stress is indicated by the dotted line. This wave was found in the case of substantial wall cooling, but there still is a maximum of (U'/T). (From Mack, 1965a, 1965b.)

An example is given in Fig. 38.5 in which we show the profile of π from the wall up to a point located far inside the free stream. A single acoustic wave in a constant flow has a constant Reynolds stress $-\rho\overline{\tilde{u}\tilde{v}}$. This can be seen from (36.8) and (36.9) with $U' = 0$ and π' taken as proportional to $i\pi$.

The value of $-\rho\overline{\tilde{u}\tilde{v}}$ inside the boundary layer has been calculated by Mack and indicated by a dotted line in Fig. 38.5. Apparently the shear stress exists only above the critical layer. Between the critical layer and the outer edge of the laminar layer U' is not zero, which means that energy is transferred from the mean flow to the oscillation, according to $e_1 = -\int \rho\overline{\tilde{u}\tilde{v}}\, U'\, dy$.

This energy is radiated as an acoustic power given by $e_2 = -\rho\overline{\tilde{u}\tilde{v}}a$.

39. Effects of Viscosity and Heat Conduction

If we consider a viscous, heat-conducting fluid, the right-hand side of (36.2), (36.3), and (36.4) must be modified to include the full complexity of all diffusion terms.

First, in a perfect gas the dynamic viscosity $\mu = \rho v$ is insensitive to the pressure, but it is sufficiently dependent on the temperature to warrant caution. An example of a viscosity distribution is given in Fig. 35.2. At Mach numbers smaller than two or three these variations of the dynamic viscosity with temperature can be neglected. Otherwise, the viscosity is a variable, and consequently a fluctuation $\tilde{\mu}$ is introduced with a relation such as

$$\tilde{\mu} = \frac{d\mu}{dT}\tau. \tag{39.1}$$

In addition, when the volume of a fluid particle is subject to change and not merely the shape, the second coefficient of viscosity μ_2 must be introduced. Finally, we assume that the heat conduction coefficient κ is proportional to μ in such a way that the Prandtl number $\mathscr{P}\imath$ is constant. We define the Prandtl number as

$$\mathscr{P}\imath = \frac{\mu c_p}{\kappa}. \tag{39.2}$$

Beginning with the basic expression for the viscous stresses,

$$\tau_{ij} = \mu\left(\frac{\partial u_i}{\partial x_j} + \frac{\partial u_j}{\partial x_i}\right) + \frac{2}{3}(\mu_2 - \mu)\frac{\partial u_k}{\partial x_k}\delta_{ij}, \tag{39.3}$$

the complete equation in terms of the perturbation quantities for conservation of momentum can now be written as follows:

$$\Upsilon[i\alpha(U - c)\mathbf{u} + U'\mathbf{v}] + i\alpha\pi = \mu(\mathbf{u}'' - \alpha^2\mathbf{u}) + i\alpha\mu\mathbf{m} + \tfrac{2}{3}i\alpha(\mu_2 - \mu)\mathbf{m}$$
$$+ \left(\frac{d\mu}{dT}U'\tau\right)' + \mu'(\mathbf{u}' + i\alpha\mathbf{v}), \tag{39.4}$$

$$\Upsilon i\alpha(U - c)\mathbf{v} + \pi' = \mu(\mathbf{v}'' - \alpha^2\mathbf{v}) + \mu\mathbf{m}' + \tfrac{2}{3}(\mu_2 - \mu)\mathbf{m}$$
$$+ i\alpha\frac{d\mu}{dT}U'\tau + 2\mu'\mathbf{v}' + \tfrac{2}{3}(\mu_2' - \mu')\mathbf{m}. \tag{39.5}$$

The general relation for the release of heat Q per unit volume and unit time is

$$Q = \Upsilon c_v\left(\frac{\partial T}{\partial t} + u_j\frac{\partial T}{\partial x_j}\right) = \frac{\partial}{\partial x_j}\left(\kappa\frac{\partial T}{\partial x_j}\right) + \mu\left(\frac{\partial u_i}{\partial x_j}\frac{\partial u_i}{\partial x_j} + \frac{\partial u_j}{\partial x_i}\frac{\partial u_i}{\partial x_j}\right)$$
$$+ \tfrac{2}{3}(\mu_2 - \mu)\frac{\partial u_k}{\partial x_k}\frac{\partial u_j}{\partial x_j}. \tag{39.6}$$

Introducing the fluctuations, we further obtain

$$\mathbf{q} = \Upsilon c_v[i\alpha(U - c)\tau + T'\mathbf{v} + (\gamma - 1)T\mathbf{m}]$$

$$= \frac{c_p}{\mathscr{P}_\imath}\left[\mu(\tau'' - \alpha^2\tau) + \left(\frac{d\mu}{dT}\,T'\tau\right)' + \mu'\tau'\right]$$

$$+ 2\mu U'(\mathbf{u}' + i\alpha\mathbf{v}) + \frac{d\mu}{dT}\,U'^2\tau. \tag{39.7}$$

These equations were first obtained by Lees and Lin (1946), who defined nondimensional fluctuations by referring to the free-stream velocity, temperature, pressure, and viscosity. In these particular notations (39.7) takes the following form:

$$\Upsilon[i\alpha(U - c)\tau + T'\mathbf{v} + (\gamma - 1)T\mathbf{m}]$$

$$= \frac{\gamma}{\mathscr{R}\mathscr{P}_\imath}\left[\mu(\tau'' - \alpha^2\tau) + \left(\frac{d\mu}{dT}\,T'\tau\right)' + \mu'\tau'\right]$$

$$+ \frac{\gamma(\gamma - 1)}{\mathscr{R}}\,M^2\left[2\mu U'(\mathbf{u}' + i\alpha\mathbf{v}) + \frac{d\mu}{dT}\,U'^2\tau\right], \tag{39.8}$$

where $\mathscr{R}$ is the free-stream Reynolds number and M is the free-stream Mach number.

From (39.7) we can write the entropy relation in the form

$$i\alpha(U - c)\sigma + S'\mathbf{v} = \frac{\mathbf{q}}{\Upsilon T}. \tag{39.9}$$

The factor $1/\Upsilon$ is necessary because σ is an entropy per kilogram. Note that ΥT is constant, for the mean pressure is assumed to be independent of y.

40. Free-Stream Modes with Viscosity and Heat Conduction

In the free stream we take all mean quantities to be constants. This being true, all coefficients in the set of differential equations are constants and solutions that are proportional to $e^{\beta y}$ can be found. Making this substitution in our equations, we obtain the following set of algebraic equations for the velocity, pressure, and the entropy fluctuations:

$$i\alpha(U - c)\theta + \mathbf{m} = 0, \tag{40.1}$$

$$\Upsilon i\alpha(U - c)\mathbf{u} + i\alpha\pi = \mu(\beta^2 - \alpha^2)\mathbf{u} + i\frac{\alpha\mu}{3}\,\mathbf{m} + i\frac{2\alpha}{3}\,\mu_2\mathbf{m}, \tag{40.2}$$

$$\Upsilon i\alpha(U - c)\mathbf{v} + \beta\pi = \mu(\beta^2 - \alpha^2)\mathbf{v} + \frac{\beta\mu}{3}\mathbf{m} + \frac{2\beta}{3}\mu_2\mathbf{m}, \qquad (40.3)$$

$$\Upsilon T i\alpha(U - c)\sigma = \frac{c_p}{\mathscr{P}_\imath}\mu(\beta^2 - \alpha^2)\tau. \qquad (40.4)$$

Multiplying (40.2) by $i\alpha$ and (40.3) by β, we are able to form an equation for the pressure:

$$(\beta^2 - \alpha^2)\pi + \Upsilon i\alpha(U - c)\mathbf{m} = \tfrac{4}{3}\mu(\beta^2 - \alpha^2)\mathbf{m} + \tfrac{2}{3}\mu_2(\beta^2 - \alpha^2)\mathbf{m}. \qquad (40.5)$$

Using equation (40.1) for $\mathbf{m}$ and one of our thermodynamic relations, we can change (40.5) to contain the entropy:

$$\left[(\beta^2 - \alpha^2) + \alpha^2\left(\frac{U - c}{a}\right)^2 \right]\pi - \alpha^2 \Upsilon(U - c)^2 \frac{\sigma}{c_p}$$

$$= \left(\frac{4}{3}\mu + \frac{2}{3}\mu_2\right) i\alpha(U - c)(\beta^2 - \alpha^2)\frac{\sigma}{c_p}$$

$$- \left(\frac{4}{3}\mu + \frac{2}{3}\mu_2\right) i\alpha(U - c)(\beta^2 - \alpha^2)\frac{\pi}{\gamma R\Upsilon T}. \qquad (40.6)$$

The first term on the left-hand side of (40.6) is nothing more than the wave equation operator operating on the fluctuating pressure. The two terms on the right-hand side are both operated on by the Laplace operator, hence form the diffusion of the respective quantities. The form of the second term on the left-hand side might be deceiving, but it is also a diffusion term, as can be seen from (40.4). Let us substitute for this term and use again one of the thermodynamic relations. This leaves us with the final form

$$\left[\beta^2 - \alpha^2 + \alpha^2\left(\frac{U - c}{a}\right)^2 \right]\pi$$

$$= -\left(\frac{4}{3} + \frac{2}{3}\frac{\mu_2}{\mu} + \frac{(\gamma - 1)}{\mathscr{P}_\imath}\right)\mu i\alpha(U - c)(\beta^2 - \alpha^2)\frac{\pi}{\Upsilon a^2}$$

$$+ \left(\frac{4}{3} + \frac{2}{3}\frac{\mu_2}{\mu} - \frac{1}{\mathscr{P}_\imath}\right)\mu i\alpha(U - c)(\beta^2 - \alpha^2)\frac{\sigma}{c_p}. \qquad (40.7)$$

We now note that the pressure fluctuations are coupled to the entropy only by a dissipative term. This fact was first pointed out by Kovásznay (1953) who made one of the first systematic studies of modes in fluid mechanics. A subsequent work was also made by Chu and Kovásznay (1958), in which they treated the nonlinear interactions. In the original work Kovásznay assumed that $\mu_2 = 0$ and a Prandtl number of $\frac{3}{4}$ which makes π independent of σ in the free stream. According to Stokes and subsequent rigorous theories for a monatomic gas, $\mu_2 = 0$. This assumption is supported by measurements of shock thicknesses in real gases in which the estimate that $\mathscr{P}\imath = \frac{3}{4}$ is acceptable.

It can easily be verified that the vorticity fluctuations in the free stream obey the usual diffusion equation.

The entropy relation has the form

$$i\alpha(U - c)\sigma = \frac{1}{\mathscr{P}\imath}\,(\beta^2 - \alpha^2)\left(\frac{\mu}{\Upsilon}\,\sigma + \frac{\mu}{\Upsilon}\,\frac{\pi}{\Upsilon T}\right). \tag{40.8}$$

Consequently, we now see that there are basically three modes for the compressible problem. Hence, in treating the complete system, the computer could start at the outer edge of the layer with each solution and integrate through the layer. Once the result of the three passes is known the boundary conditions at the wall can be satisfied. For an insulated wall this would mean that $\mathbf{u} = \mathbf{v} = 0$, $\tau = 0$. The physical significance of the pressure and the vorticity fluctuations is the same as in the incompressible case. The presence of entropy fluctuations is simply due to the fact that the layer produces temperature fluctuations that diffuse into the free stream.

Kaplan's method could be generalized by storing two badly divergent solutions corresponding to the vorticity and the entropy free-stream modes. Then at each station the third solution is amended by solving a system of two equations in two unknowns. These amplitudes represent the additional amounts of the two stored functions necessary to diminish the oscillations.

41. The Two-Dimensional Supersonic Boundary Layer

Some numerical results obtained by Mack (1965a, 1965b) for a supersonic boundary layer are shown in Fig. 41.1. The data correspond to a Mach number of 2.2 and an insulated wall. Comparison has also been made with experimental measurements of Laufer and Vrebalovich (1957, 1958), indicating good agreement. The trend is quite like that of the normal stability curves obtained in the incompressible problem. At this Mach number and values above, one should consider the effects of obliquity of the wave and of the V component of the mean flow. This will be done in Section 43.

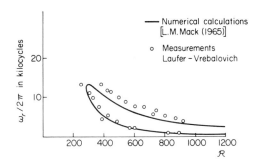

Fɪɢ. 41.1. Neutral stability curve of a compressible boundary layer at Mach 2.2 (insulated wall). (From Mack, 1965a, 1965b.)

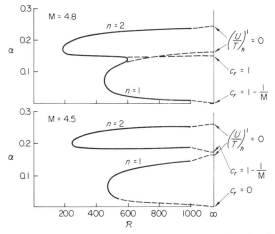

Fɪɢ. 41.2. Neutral stability curves for boundary layer along insulated wall. The dotted lines indicate presumed behavior up to $\mathcal{R} = \infty$. (From Mack, 1965a, 1965b.)

When the Mach number is increased, Mack finds curves like those shown in Fig. 41.2. Of course, there are multiple modes, as we have already mentioned in the inviscid theory; but, comparing this figure with the single mode of Mach 2.2, we see that the nature of the curve has changed. There is no longer the pronounced characteristic hairpin shape where there is an increased wavenumber (or frequency) as the Reynolds number is decreased. This changeover occurs at about a Mach number of 3. Thus below this value we recognize the instability as the resistivity variety that we have discussed throughout. Above this Mach number, however, the instability is related to a form of acoustic resonance in a shear flow and the effect of viscosity is primarily stabilizing.

In these considerations the wall is always assumed to be completely rigid. Other modes may be uncovered when the elasticity or ablation of the wall is taken into consideration with the supersonic flow.

42. Subsonic Oscillations in Stratified Layers

We now consider oscillations in which density fluctuations are important because of the presence of gravity forces. Generally, when this occurs, the flow is significantly subsonic, that is, when the speed of sound is much larger than the free-stream velocity. This fact enables us to make a number of simplifications to the problem.

Going back to (36.11) and (36.12) for a general fluid, we have

$$\pi = \frac{\Upsilon a^2}{\gamma} (\theta + \lambda \tau), \tag{42.1}$$

$$\pi = \Upsilon a^2 \left(\theta + \frac{\lambda T}{c_p} \sigma \right). \tag{42.2}$$

Thus we see that as $a \to \infty$ it is necessary that the terms multiplying a^2 be very small. In the limit we can combine (42.1) and (42.2) to eliminate θ and have

$$c_p \tau = T \sigma. \tag{42.3}$$

For the mean flow we still require the relations

$$-\frac{\Upsilon'}{\Upsilon} = \lambda T' = \lambda T \frac{S'}{c_p} \tag{42.4}$$

to be true. It is clear from this that the mean pressure and the fluctuating pressure remain finite in the limit as $a \to \infty$.

Now, by using (36.7) (continuity), (39.8) (complete energy), and the fact that $\theta = -(\lambda T/c_p)\sigma$ must be true from (42.2) we find that

$$\mathbf{m} = \frac{\lambda}{\Upsilon c_p} \mathbf{q}, \tag{42.5}$$

where $\mathbf{q}$ is given by (39.6) and indicates the fluctuation in heat input.

Our previous equations do not contain buoyancy forces, and, in view of this, we now consider two different kinds of problem.

(a) Buoyancy across the Mean Flow

The buoyancy adds a force $-\rho g$ parallel to the y-axis which is balanced by a special hydrostatic pressure. The fluctuations of a density add a term $-\Upsilon g\theta$ on the right-hand side of (39.5).

An equation for the vorticity can be derived in the usual way. We find this to be

$$i\alpha(U - c)\mathbf{r} + U'\mathbf{m} + U''\mathbf{v} - i\alpha \frac{\Upsilon'}{\Upsilon} \frac{\pi}{\Upsilon} = i\alpha g\theta + \Phi, \qquad (42.6)$$

where Φ stands for the entire group of viscous terms. Now, by use of continuity and the definition of vorticity we can replace $\mathbf{m}$, θ, and $\mathbf{r}$ with $\mathbf{v}$. This gives

$$(U - c)(\mathbf{v}'' - \alpha^2\mathbf{v}) - U''\mathbf{v} + i\alpha \frac{\Upsilon'}{\Upsilon} \frac{\pi}{\Upsilon} - g \frac{\Upsilon'}{\Upsilon} \frac{\mathbf{v}}{(U - c)} = \Phi. \qquad (42.7)$$

Suppose we now neglect the viscous and heat-conduction terms. From (42.5) it follows that $\mathbf{m} = 0$. Moreover, for this case we can eliminate π quite easily from the reduced form of (42.7) by using (36.8); the result is the equation

$$(U - c)(\mathbf{v}'' - \alpha^2\mathbf{v}) - U''\mathbf{v} - \frac{\Upsilon'}{\Upsilon}\left[\frac{g\mathbf{v}}{(U - c)} - (U - c)\mathbf{v}' + U'\mathbf{v}\right] = 0. \qquad (42.8)$$

One last reduction can be made. We note that if $U'(U - c) \ll g$ the terms that resulted from the fluctuations of the pressure can be neglected altogether. Consequently, the final form becomes

$$(U - c)(\mathbf{v}'' - \alpha^2\mathbf{v}) - U''\mathbf{v} - \frac{g}{(U - c)} \frac{\Upsilon'}{\Upsilon} \mathbf{v} = 0. \qquad (42.9)$$

This equation was solved by Drazin (1958); the results were also reviewed by Chandrasekhar (1961). The basis used by Drazin for obtaining (42.9), however, was the Boussinesq approximation. Essentially, this approximation is the combination of $\mathbf{m} = 0$, zero viscosity, and the neglect of the pressure compared with the gravity force. The heat conductivity remains finite. It can be noted that this equation produces no difficulties, except possibly a singularity at the critical layer. The particular solution, as examined by Drazin, is such that $\mathbf{v} = 0$ at the critical layer. It is not unique since there is a symmetric solution if $g = 0$.

Drazin computed the stability for the problem in which $\Upsilon'/\Upsilon = -\beta = \text{constant}$, $U = U_0 \tanh(y/L)$, and found that the system is unstable if $g\beta L^2/U^2 < \frac{1}{4}$ with $0 < \alpha < 1$.

Let us now return to this problem from the outset and retain the diffusion terms. We see that it is not unduly complicated, and, of course, the would-be singularity no longer exists.

Examining the energy equation in terms of the temperature fluctuations, we have at our disposal a wealth of experience from the incompressible problem. For example, let us suppose that the viscosity is small, but not zero, and that the Prandtl number is constant. We can then neglect the viscous heat production compared with the diffusion of the temperature and write the equation for the temperature as

$$i\alpha(U - c)\tau + T'\mathbf{v} = \frac{\mathbf{q}}{\Upsilon c_p}.$$ (42.10)

Substituting the value for $\mathbf{q}$ in this case, we may make a direct estimate of the width of the "thermal critical layer." Calling this width $2l$, we find that

$$l \doteq \left(\frac{\kappa}{\Upsilon c_v \, \mathbf{v}(h) \, U'(h)}\right)^{1/3},$$ (42.11)

where $\mathbf{v}(h)$ and $U'(h)$ are the values of these quantities at the critical layer or $y = h$. This answer bears a strong analogy to the viscous critical layer obtained from the vorticity relation of the incompressible case. From (42.11) we can find the maximum value of the temperature at the critical layer:

$$\tau_{\max} \doteq T'(h)\left(\frac{\Upsilon c_v}{\kappa \alpha^2 U'^2(h)}\right)^{1/3} \mathbf{v}(h).$$ (42.12)

Finally, it is possible to evaluate the change in τ as we cross the critical layer. We obtain

$$\int_{-l}^{+l} \tau \, dy \doteq \frac{T'(h) \, \mathbf{v}(h)}{\alpha U'(h)}.$$ (42.13)

These results also follow the pattern of the vorticity and the viscous critical layer.

At the critical layer $U = c$ and from (42.10) we simply have

$$T'\mathbf{v} = \frac{\mathbf{q}}{\Upsilon c_p}.$$ (42.14)

Thus at this location the diffusion is balanced by the production of temperature. We note that $\mathbf{v}$ is no longer required to vanish at the critical layer. Finally, by knowing τ it is possible to find θ, which is now finite at the critical layer, and, in turn, all of the perturbation quantities. Practically speaking, this is best done on the computer when viscosity is taken into account. For this combined problem we see that there are two superimposed interacting critical layers.

(b) Buoyancy along the Mean Flow: Free Convection

We shall now review a particular class of problem in which a body force proportional to temperature difference acts along the direction of mean flow. In fact it often is the primary cause of the mean flow. Such flows occur along vertical heated plates at which the pressure is very nearly constant and the temperature produces density differences. The gravity forces set the fluid in motion. Similar flows occur in rotating machinery, in which the centrifugal forces play the usual role of gravity. The stability of such flows is related to many important problems of heat transfer.

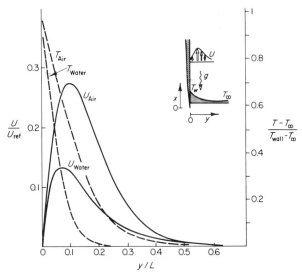

FIG. 42.1. Mean temperature and mean velocity profiles in free convection flow. Air: $\mathscr{P}_i = 0.733$ (from Ostrach, 1953). Water: $\mathscr{P}_i = 6.7$ (from Nachtsheim, 1963).

We now examine the case of the heated wall and use the coordinates shown in Fig. 42.1. The acceleration is parallel to the x-axis and y is the distance from the wall. The wall is at a constant temperature T_w, and the temperature of the fluid, at $y = \infty$, is T_∞. The lower edge of the wall is at $x = 0$, and a laminar profile of temperature and velocity can be found by the well-known boundary-layer approximation, shown by Ostrach (1953).

For a given Prandtl number, $\mathscr{P}\imath = c_p\mu/\kappa$, the effects of the acceleration g and the viscosity ν can be absorbed by a proper choice of a reference length L_0 and a reference velocity U_0 defined as

$$L_0 = \left(\frac{4\nu^2}{g\lambda(T_w - T_\infty)}\right)^{1/3}, \tag{42.15}$$

$$U_0 = (16\nu g\lambda(T_w - T_\infty))^{1/3}. \tag{42.16}$$

We take for ν and λ the values at temperature T_∞. Thus $\lambda = 1/T_\infty$ for a perfect gas, but λT_∞ varies between 0.2 and 0.05 for ordinary liquids. The laminar flow can be described by the expressions

$$U(x, y) = U_0\left(\frac{x}{L_0}\right)^{1/2} F'(\eta), \tag{42.17}$$

$$T(x, y) = T_\infty + (T_w - T_\infty) H(\eta), \tag{42.18}$$

where $\eta = y/L_0^{3/4} x^{1/4}$. The functions F and H are solutions of

$$F''' + 3FF'' - 2(F')^2 + H = 0, \tag{42.19}$$

$$H'' + 3\mathscr{P}\imath F''H' = 0, \tag{42.20}$$

with boundary conditions $F(0) = F'(0) = F''(\infty) = H(\infty) = 0$ and $H(0) = 1$. Note that $U_0 L_0/\nu = 4$ and that there is only the parameter $\mathscr{P}\imath$.

Tables of F and H have been supplied by Ostrach (1953) and Nachtsheim (1963). When $\mathscr{P}\imath$ is small, F' and H decay. When $\mathscr{P}\imath$ is large, H vanishes rapidly, but F' extends further out. For experimental results see Schmidt and Beckmann (1930), and Eichhorn (1962).

The profiles for air and water at normal conditions are shown in Fig. 42.1.

These results are valid if $(x/L_0)^{3/2}$ is large. It is customary to define a Grasshof number $\mathscr{G}$ as

$$\mathscr{G} = \frac{gx^3}{\nu^2} \lambda(T_w - T_\infty), \tag{42.21}$$

and (42.15) shows that $(x/L_0)^{3/2} = \frac{1}{2}\mathscr{G}^{1/2}$.

The stability of these free convection flows has been studied by Szewczyk (1962) and Nachtsheim (1963). At some distance x_1 from the leading edge the mean flow is considered parallel. It is now convenient to use a new reference length and a new reference velocity defined as

$$L_1 = L_0 \left(\frac{x_1}{L_0} \right)^{1/4}, \tag{42.22}$$

$$U_1 = U_0 \left(\frac{x_1}{L_0} \right)^{1/2}. \tag{42.23}$$

In addition to the number $\mathscr{P}\imath$, we have a new parameter, varying with x_1, which can be taken either as $\mathscr{G}$ or as $\mathscr{R} = (U_1 L_1/\nu) = 2\sqrt{2}\mathscr{G}^{1/4}$. The temperature fluctuations are expressed in units of $T_w - T_\infty$.

The equations for the small fluctuations can be derived from the general equations (36.11), (36.12), (39.4), (39.5), and (39.7), with the addition of a buoyancy term in (39.4) and in the limit $a \to \infty$. The resulting equations for vorticity and temperature read as follows:

$$i\alpha(F' - c)\mathbf{r} + F'''\mathbf{v} = \frac{1}{\mathscr{R}} (\mathbf{r}'' - \alpha^2 \mathbf{r}) + \frac{\tau}{\mathscr{R}}, \tag{42.24}$$

$$i\alpha(F' - c)\tau + H'\mathbf{v} = \frac{1}{\mathscr{P}\imath\mathscr{R}} (\tau'' - \alpha^2 \tau). \tag{42.25}$$

To integrate this sixth-order complex system Nachtsheim used a computer, starting from the wall and matching the analytic solution at the outer edge, where F' and H are nearly zero. For a given set of values of α, $\mathscr{P}\imath$, and $\mathscr{R}$ we must find the values of c, $\mathbf{v}''(0)$, $\mathbf{v}'''(0)$, and $\tau'(0)$, such that $\mathbf{v}$, $\mathbf{v}'$, and τ will vanish as $y \to \infty$.

At the wall we have $\mathbf{v} = \mathbf{v}' = \tau = 0$, and one of the quantities $\mathbf{v}''$, $\mathbf{v}'''$, or τ' can be arbitrary, thereby fixing the amplitude and phase of the oscillation. This leads to three equations in the three unknowns (say c, $\mathbf{v}'''(0)$, and $\tau'(0)$ if $\mathbf{v}''(0) = 1$) whose determinant must vanish. The search in a complex three-dimensional space is greatly simplified by application of the Cauchy relations, but it still involves some intricacies.

The neutral curve for air is shown in Fig. 42.2, in which the solid line corresponds to the sixth-order system of (42.24) and (42.25); the dotted line is for the fourth-order system given by (42.24), with $\tau = 0$ and complete neglect of (42.25). These results suggest that there are two modes of oscillation: one similar to the ordinary oscillation of isothermal parallel flows and the other essentially depending on temperature fluctuations. Note that for $\alpha < 0.13$ the phase velocity c_r is larger than the maximum value of F'. There is no critical layer for the lower branch of the neutral curve.

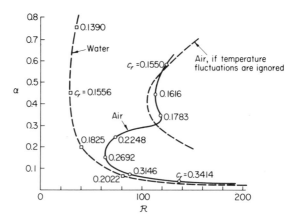

Fig. 42.2. Neutral curves for the free convection flows of Fig. 42.1. (From Nachtsheim, 1963.)

For $\alpha > 0.13$ there are two critical layers, as in the case of the jets. Some experimental confirmation has been reported by Čolak-Antić (1964a, 1964b).

In water Nachtsheim finds a minimum Reynolds number of about 30 for αL_1 between 0.1 and 0.8. Without the thermal fluctuations, the critical Reynolds number occurs at $\mathscr{R} = 380$ with $\alpha L_1 = 0.4$.

43. Three-Dimensional Compressible Effects

The possibility of using the Squire transformations for the compressible problem was originally investigated by Dunn and Lin (1955). Because of the discarding of certain viscous dissipation terms, however, their results are incomplete. Indeed, it is only in the true inviscid problem that the conclusions drawn in their work can be considered strictly correct.

From the bases developed in Section 18 and from the examination of the full three-dimensional equations for the compressible problem we are led to the following transformations. All mean quantities and thermodynamic variables transform identically. They include U, P, Υ, T, S, c_v, c_p, and a. In the transformed notation we write $\overline{U}$, $\overline{P}$, $\overline{\Upsilon}$ $\overline{T}$, $\overline{S}$, $\overline{c_v}$, $\overline{c_p}$, $\overline{a}$. In addition, we have

$$\bar{\alpha}^2 = \alpha^2 + \beta^2, \tag{43.1}$$

$$\bar{\alpha}\,\bar{\mathbf{u}} = \alpha\mathbf{u} + \beta\mathbf{w}, \tag{43.2}$$

$$\bar{\mathbf{v}} = \mathbf{v}, \tag{43.3}$$

$$\bar{c} = c, \tag{43.4}$$

where the meaning of $\mathbf{w}$ is obvious and all fluctuating quantities are proportional to $e^{i(\alpha x + \beta z)}e^{-i\alpha c t}$. Going further, we have

$$\bar{\alpha}\,\bar{\pi} = \alpha\pi, \tag{43.5}$$

$$\bar{\alpha}\,\bar{\tau} = \alpha\tau, \tag{43.6}$$

$$\bar{\alpha}\,\bar{\theta} = \alpha\theta, \tag{43.7}$$

$$\bar{\alpha}\,\bar{\sigma} = \alpha\sigma. \tag{43.8}$$

The group of relations (43.6) to (43.8) follows directly from that for pressure, or (43.5), because of the necessity of having compatible thermodynamic relations, such as (36.1) and (36.2). It should be noted by comparison with (18.5), that the pressure transformation is not the same as that in the incompressible problem.

Finally, the viscosity transforms exactly as it does in the incompressible case, or

$$\alpha\bar{\mu} = \bar{\alpha}\mu. \tag{43.9}$$

For the Prandtl number we take

$$\overline{\mathscr{P}\imath} = \mathscr{P}\imath \tag{43.10}$$

Substitution of these relations into the governing equations produces the desired results. First, in place of having $\bar{\pi}$ occur with a coefficient of unity, we find $(\bar{\alpha}/\alpha)^2\bar{\pi}$. This would correspond exactly to Dunn and Lin's problem if we had used a Mach number in the equations. Second, in the energy equation there is a viscous dissipation term that cannot be put in terms of the transformed variables at all. Specifically, in the expression for $\mathbf{q}$ we have the quantity

$$2\mu U'(\mathbf{u}' + i\alpha\mathbf{v}). \tag{43.11}$$

It is impossible to put (43.11) in terms of $\bar{\mathbf{u}}$ and $\bar{\mathbf{v}}$. This last point could not be noted by Dunn and Lin, for they omitted this term from the outset in their equations.

If we consider only the inviscid problem, we have a two-dimensional problem—in principle. The only difference is the factor that multiplies the pressure. The three-dimensional equation corresponding to (36.16) (which is valid for the inviscid two-dimensional problem) is

$$\bar{\pi}'' - \left(\frac{2\overline{U}'}{\overline{U} - \bar{c}} + \frac{\overline{\Upsilon}'}{\overline{\Upsilon}}\right)\bar{\pi}' + \bar{\alpha}^2\left[\left(\frac{\overline{U} - \bar{c}}{\bar{a}}\right)^2\cos^2\theta - 1\right]\bar{\pi} = 0. \tag{43.12}$$

Here we have used the factor $\cos^2 \theta$ instead of $(\bar{\alpha}/\alpha)^2$, which is an identity, because

$$\cos^2 \theta = \frac{\alpha^2}{\bar{\alpha}^2}. \qquad (43.13)$$

Thus by choosing a particular θ or β we can solve the problem.

In general, the reduction of a three-dimensional inviscid problem to a similar two-dimensional problem lowers the order of the system of equations and saves one integration through the shear layer whenever an eigenvalue is sought. Once the eigenvalue is known, however, the determination of the eigenfunctions requires as many integrations as there are components of the velocity fluctuations.

The transformation is therefore of some help, but we are far from having the advantages provided by the Squire theorem for viscous incompressible flows. There a table of eigenvalues for $\beta = 0$ could readily give the eigenvalues for any other value of β. In other words, transformations are possible with limitations, but the theorem is no longer valid.

Let us now return to compressible and viscous flows with three-dimensional disturbances. If the complete energy equation is used, there is no convenient transformation and the full eight-order system must be integrated. This was done by Brown (1965). His program starts from the wall, finds four particular solutions, and applies the conditions at the outer edge so that the oscillations vanish at infinity. His first results with two-dimensional waves did not agree with those of the experiments of Laufer and Vrebalovich (1960) and Demetriades (1958). Brown examined the effect of the obliquity and found that at Mach 5 the lowest critical Reynolds number occurs near $\theta = 56°$ (see Fig. 43.1). He also realized that the V-component of

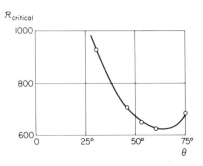

Fig. 43.1. Variation of the critical Reynolds number with the obliquity of the wave at Mach $= 5$. (From Brown, 1965.)

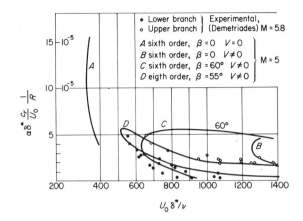

FIG. 43.2. Theoretical and experimental neutral stability curves. (From Brown, 1965.)

the mean flow has an important effect and added the necessary terms to the Lees and Lin equations (see Cheng, 1953). With the obliquity of 56° and the effects of V, Brown obtained good agreement with the experiments (see Fig. 43.2). Recently, Mack (1966) found that the obliquity destabilizes the first mode but stabilizes the higher ones. Simultaneously, Kendall (1966) measured the rate of growth along a flat plate at Mach 4.5 with $\mathscr{R} = 1400$. The perturbations were introduced by a controlled glow discharge produced between electrodes imbedded in the plate at angles of obliquity equal to 0°, 30°, and 55°. The discharge oscillates with a frequency that can be varied between 1 and 30 kilocycles. The growth of the oscillation is detected with a hot wire anemometer.

At low frequencies, the observed rate of amplification increases with obliquity and agrees with theoretical results obtained for the first mode and the angle of 55°. At high frequencies, the most unstable waves have an angle of 0°, and their amplification agrees with the viscous theory for the second mode.

Measurements of the phase velocity are also in agreement with theory.

Additional information on the effect of obliquity on the stability of jets and wakes can be found in the work of Lessen *et al.* (1965). For inviscid two dimensional jets at Mach numbers varying between 1 and 5 they found substantial variations of the rate of growth between 0 and 75°. These authors used an algebraic method to obtain solutions.

Magnetohydrodynamic Effects

44. General Equations

Magnetohydrodynamics (MHD) is the name given to an area of study in which fluid dynamics and electrodynamics overlap. It deals with situations in which the motion of a fluid is affected by Lorentz forces—that is, the forces produced by a magnetic field on a current carrying conductor. Thus the fluid must be able to conduct current. Mercury, salt water, ionized gases all have this property to various degrees. In solar phenomena the Lorentz force assumes a dominant role. If the magnetic field is externally produced, the motion of a homogeneous fluid has no effect on the current. Consequently, the Lorentz forces have the character of external or body forces. It is true that there is the possibility of a fluid with a variable conductivity that would modify the electric current. On the other hand, the most interesting part of magnetohydrodynamic work is concerned with the motion of the fluid modifying the magnetic field. Therefore we limit our discussion to this central problem.

If the fluid carried unequal densities of positive and negative electric charges, it would be subject to electrostatic forces. This falls outside the range of magnetohydrodynamics as formulated and, in practice, does not commonly occur, for it would require a fluid with a high resistivity to prevent the leaking of the charges. Finally, magnetohydrodynamics is restricted to systems in which electromagnetic waves do not occur. This means that the frequencies are sufficiently small and that the electric field has negligible effect on the magnetic field. Thus the displacement current of the Maxwell equations is neglected.

Let us start from the general equations given by Maxwell for the evolution of the magnetic field, $\vec{H}$, and the electric field, $\vec{E}$

$$\frac{\partial \vec{H}}{\partial t} = -\frac{1}{\mu}\,\text{curl}\,\vec{E}, \tag{44.1}$$

$$\frac{\partial \vec{E}}{\partial t} = \frac{1}{\varepsilon}(\text{curl}\,\vec{H} - \vec{J}). \tag{44.2}$$

The vector $\vec{J}$ is the electric current per unit area. In the MKS system, the constants are $\mu = 4\pi \times 10^{-7}$ and $\varepsilon = 8.8 \times 10^{-12}$.

If there are no free charges, we have the relation

$$\text{div}\,\vec{E} = 0. \tag{44.3}$$

The equation

$$\text{div}\,\vec{H} = 0 \tag{44.4}$$

must also be satisfied at all times. Moreover it follows that, if it is true at one instant, it must be valid at all other times, past or future. From (44.2) and (44.3) we can form

$$\text{div}\,\vec{J} = 0. \tag{44.5}$$

If the current is zero everywhere, the solutions are electromagnetic waves propagating at the velocity $c = 1/\sqrt{\mu\varepsilon}$. Except for very large wavelengths, these waves have such high frequencies that there is no coupling with fluid motions. At a wavelength of one meter, for example, the frequency is 300 megacycles. Since the constant ε is very small, this means that the right-hand side of (44.2) must always be approximately zero. This consequence gives us the relation

$$\text{curl}\,\vec{H} = \vec{J}. \tag{44.6}$$

If the magnetic field violates this condition, rapid changes of $\vec{E}$ are set up which adjust $\vec{H}$ and produce waves which radiate energy away. If the current is constant in time, but still a function of the space coordinates, it is possible to integrate (44.6) by the Biot-Savart formula. This gives a stationary $\vec{H}$, and it follows from (44.1) that $\vec{E}$ is the gradient of a potential Φ. By (44.3) this potential must obey Laplace's equation. This potential is generally determined by boundary conditions imposed by electrodes or insulating surfaces. So far we have treated $\vec{E}$ and $\vec{J}$ as two independent fields, but in fluid media they are related by microscopic processes. If the fluid is at rest and has simple electric properties, we can postulate Ohm's law in the form

$$\vec{J} = \sigma\vec{E}, \tag{44.7}$$

where σ is the conductivity in ohm^{-1} meter^{-1}.

Now let us turn to the possibility of currents which are not constant in time. In order to make such a probe we shall consider a particular case. Let us place on a table a system of batteries, coils, electrodes, and stationary fluids which do not vary with time. The instruments fixed to the table register the vectors $\vec{E}'$, $\vec{H}'$, $\vec{J}'$, and the potential Φ'.

The table is now set in uniform motion along the x-axis say with a velocity U. The uniform motion cannot modify the readings of the instruments fixed on the table for this would be contrary to the second postulate of the theory of relativity. On the other hand, the instruments fixed in the laboratory will now read time-varying signals. It should be clear, however, that these signals will be functions of $(x - Ut)$, y, and z. For this reason, (44.1) can be rewritten as

$$\mu U \frac{\partial}{\partial x} \vec{H} = \text{curl } \vec{E}, \tag{44.8}$$

where $\vec{E}$, $\vec{H}$, $\vec{J}$ and Φ now refer to laboratory observables. (An engineer will say that the variable magnetic field induces an electric field.)

For the currents, we will have $\vec{J} = \vec{J}'$. Indeed, in the coordinates moving with the table let us say that the ions are at rest and that the electrons have some motion, v. In the laboratory coordinates, the same ions have a velocity U, while the electrons have the velocities of $U + v$. Thus the net transport of charges is the same in both systems of coordinates. By (44.6) this leads to $\vec{H} = \vec{H}'$. Similar considerations of electrostatic forces show that $\Phi(x - ct, y, z) = \Phi'$. But the electric fields must differ. Integration of (44.8) gives

$$-\mu(\vec{U} \times \vec{H}) = \vec{E} - \text{grad } \Phi. \tag{44.9}$$

Thus the two fields are related by

$$\vec{E}' = \vec{E} + \mu(\vec{U} \times \vec{H}). \tag{44.10}$$

This relation is quite general and valid as long as $\vec{U}$ is small in comparison with the speed of light. It is clear, then, that Ohm's law in the form (44.7) is valid only for fixed media. In the laboratory coordinates, it becomes

$$J = \sigma[\vec{E} + \mu(\vec{U} \times \vec{H})]. \tag{44.11}$$

This relation can be used to eliminate $\vec{E}$ from (44.1) and $\vec{J}$ can be expressed by means of (44.6). The result is

$$\frac{\partial \vec{H}}{\partial t} = -\lambda \text{ curl curl } \vec{H} + \text{curl}(\vec{U} \times \vec{H}),$$

where $\lambda = 1/\mu\sigma$ is the magnetic diffusivity. In Cartesian tensor notation, and after some further reductions, (44.10) takes the classical form used in magnetohydrodynamics:

$$\frac{\partial h_i}{dt} + u_j \frac{\partial h_i}{\partial x_j} - h_j \frac{\partial u_i}{\partial x_j} + \frac{\partial u_j}{\partial x_j} h_i = \lambda \frac{\partial^2 h_i}{\partial x_j \partial x_j} \qquad (44.12)$$

where h_i is a component of the magnetic field.

These are the basic equations of magnetohydrodynamics. They show how the combined presence of u_i and h_j can alter a magnetic field. All other electric quantities can be derived from the magnetic field by (44.6) and (44.11). If λ is zero and u_i is constant, h_j is transported with the fluid. The equations for h_i have a certain analogy with the vorticity equations, especially in the form given in (27.4). In particular, the analogy can be used to demonstrate that the square of $\vec{H}$ or $h_i h_i$, that is, the density of magnetic energy, can be increased by stretching of the magnetic field lines. The coupling between the magnetic field and the velocity is however very different from that between the vorticity and the velocity.

The Lorentz force per unit volume is simply $\mu(\vec{J} \times \vec{H})$. The Navier–Stokes equations, then, must be written as follows:

$$\frac{\partial u_i}{\partial t} + u_j \frac{\partial u_i}{\partial x_j} + \frac{1}{\rho} \frac{\partial p}{\partial x_i} = \nu \frac{\partial^2 u_i}{\partial x_j \partial x_j} + \frac{\mu}{\rho} \left(h_j \frac{\partial h_i}{\partial x_j} - h_j \frac{\partial h_j}{\partial x_i} \right). \qquad (44.13)$$

Let us now choose our electrical units so that the ratio μ/ρ becomes unity and is eliminated from these equations. Because μh^2 is an energy density, our new magnetic field will have the physical dimensions of a velocity. The new magnetic field is the classical field multiplied by $(\mu/\rho)^{1/2}$. Thus in mercury, with a density of 1.3×10^4 kg/m^3, a field of 10^4 gauss corresponds to $\mu h = 1$ in MKS units and $h = 8$ m/sec in the new units. With another fluid, we would find a different result.

In addition to (44.13) we have the relation

$$\frac{\partial u_j}{\partial x_j} = 0. \qquad (44.14)$$

Let us now consider the magnetic boundary conditions. Essentially, the Maxwell equations must be satisfied inside the walls and the magnetic field in the flow must be joined to the field in the wall. Inside the wall each component of the magnetic field must obey a diffusion equation, as seen in

(44.12), when the velocity is set to zero and λ is replaced by its appropriate value. In addition, (44.4) must also be satisfied. If the wall is perfectly insulating, its resistivity is infinite and the magnetic field fluctuations must obey Laplace's equation. If the wall is perfectly conducting, the magnetic fluctuations must be zero. In this case the magnetic field component normal to the wall must be zero, whereas the component tangent to the wall, but already on the side of the flow, can have an arbitrary value.

45. The Two-Dimensional Problem

We now investigate a two-dimensional problem with $V = 0$. The perturbation equations are obtained by linearization of (44.12) and (44.13). We denote the components of the magnetic fluctuations as **g** and **h** and obtain (**p** is the kinematic pressure):

$$i\alpha\left(U - \frac{\omega}{\alpha}\right)\mathbf{u} + U'\mathbf{v} + i\alpha\mathbf{p} = i\alpha G\mathbf{g} + H\mathbf{g}' + G'\mathbf{h} + v\nabla^2\mathbf{u} - i\alpha(G\mathbf{g} + H\mathbf{h}),$$
$$(45.1)$$

$$i\alpha\left(U - \frac{\omega}{\alpha}\right)\mathbf{v} + \mathbf{p}' = i\alpha G\mathbf{h} + H\mathbf{h}' + v\nabla^2\mathbf{v} - (G\mathbf{g}' + H\mathbf{h}'), \qquad (45.2)$$

$$i\alpha\left(U - \frac{\omega}{\alpha}\right)\mathbf{g} - i\alpha G\mathbf{u} - H\mathbf{u}' - U'\mathbf{h} + G'\mathbf{v} = \lambda\nabla^2\mathbf{g}, \qquad (45.3)$$

$$i\alpha\left(U - \frac{\omega}{\alpha}\right)\mathbf{h} - i\alpha G\mathbf{v} - H\mathbf{v}' = \lambda\nabla^2\mathbf{h}, \qquad (45.4)$$

$$i\alpha\mathbf{u} + \mathbf{v}' = 0, \qquad (45.5)$$

where

$$\nabla^2 = \frac{d^2}{dy^2} - \alpha^2 \equiv \text{Laplace operator.}$$

It is possible to obtain an equation for the pressure by multiplying (45.1) by $i\alpha$ and summing the result with the derivative of (45.2) with respect to y. This leads to

$$\mathbf{p}'' - \alpha^2\mathbf{p} + 2i\alpha U'\mathbf{v} = -(G\mathbf{g}'' + H\mathbf{h}'') + \alpha^2(G\mathbf{g} + H\mathbf{h}) + 2i\alpha G'\mathbf{h} - G'\mathbf{g}.$$
$$(45.6)$$

If we define a magnetic pressure as

$$\pi = \tfrac{1}{2}h_i h_i, \qquad (45.7)$$

the complex amplitude of the fluctuating pressure is

$$\pi = G\mathbf{g} + H\mathbf{h}. \tag{45.8}$$

We then have for the pressure equation the relation

$$\nabla^2\mathbf{p} + 2i\alpha U'\mathbf{v} = -\nabla^2\pi + 2i\alpha G'\mathbf{h} + G'\mathbf{g} \tag{45.9}$$

between the fluid and magnetic pressures. An equation for the complex amplitude of the vorticity $\mathbf{r}$ can easily be extracted from (45.1) and (45.2). We then find

$$i\alpha\left(U - \frac{\omega}{\alpha}\right)\mathbf{r} + U''\mathbf{v} = i\alpha G\eta + H\eta' + G''\mathbf{h} + v\,\nabla^2\mathbf{r}, \tag{45.10}$$

where η is defined as

$$\eta = \mathbf{g}' - i\alpha\mathbf{h} = \frac{i}{\alpha}(\mathbf{h}'' - \alpha^2\mathbf{h}) = \frac{i}{\alpha}\nabla^2\mathbf{h}. \tag{45.11}$$

Thus η is simply the curl of the magnetic field and corresponds to the current fluctuations. For this two-dimensional system the current has only one component perpendicular to the xy-plane. The same is true for the fluctuating vorticity. The corollary to the vorticity equation for the magnetic field is therefore one for the current. From (45.3) and (45.4) we obtain

$$i\alpha\left(U - \frac{\omega}{\alpha}\right)\eta - i\alpha G\mathbf{r} - H\mathbf{r}' - U''\mathbf{h} - 2U'\mathbf{h}' = \lambda\,\nabla^2\eta. \tag{45.12}$$

46. Magnetohydrodynamic Squire Theorem

Let us consider the two-dimensional mean flow $U(y)$ in the presence of an oblique mean magnetic field described by the components $G(y)$, H, $K(y)$, where H must be constant in order to satisfy the magnetic divergence equation. The density is constant. In Section 44 we gave the general equations of magnetohydrodynamics in Cartesian coordinates. From these equations the linearized relations between the fluctuations $\mathbf{u}$, $\mathbf{v}$, and $\mathbf{w}$ of the velocity and $\mathbf{g}$, $\mathbf{h}$, and $\mathbf{k}$ of the magnetic field can be obtained. It is convenient to define the total pressure π_T as the sum of the ordinary pressure and the magnetic pressure:

$$\pi_T = \mathbf{p} + (G\mathbf{g} + H\mathbf{h} + K\mathbf{k}). \tag{46.1}$$

All fluctuations vary as $e^{i(\alpha x + \beta z)}e^{-i\alpha ct}$. These equations can now be transformed into a simpler system by introducing the following quantities which are similar to those used in the incompressible case.

$$\bar{\alpha}^2 = \alpha^2 + \beta^2, \tag{46.2}$$

$$\bar{\alpha}\,\bar{\mathbf{u}} = \alpha\mathbf{u} + \beta\mathbf{w}, \tag{46.3}$$

$$\bar{\mathbf{v}} = \mathbf{v}, \tag{46.4}$$

$$\bar{\alpha}\,\bar{\mathbf{g}} = \alpha\mathbf{g} + \beta\mathbf{k}, \tag{46.5}$$

$$\bar{\mathbf{h}} = \mathbf{h}, \tag{46.6}$$

$$\bar{U} = U, \tag{46.7}$$

$$\bar{c} = c, \tag{46.8}$$

$$\bar{G} = G + \frac{\beta}{\alpha}K, \tag{46.9}$$

$$\bar{H} = \frac{\bar{\alpha}}{\alpha}H, \tag{46.10}$$

$$\frac{\bar{v}}{v} = \frac{\bar{\lambda}}{\lambda} = \frac{\alpha}{\bar{\alpha}}, \tag{46.11}$$

$$\bar{\pi}_T = \frac{\bar{\alpha}}{\alpha}\bar{\pi}_T. \tag{46.12}$$

Note the similarity between the transformation for G, H, K and that for U, V, W in the general incompressible case. The pressure transformation is also similar to that of the incompressible case.

It is found that the transformed system corresponds exactly to a two-dimensional problem. Thus the transformation is helpful when $K(y)$ is present. Furthermore, if $K = 0$ and the eigenvalues are known for all values of H, no further integrations are needed to know the eigenvalues of the oblique waves.

In that sense we have an extension of the Squire theorem. In a less general context the validity of the theorem in magnetohydrodynamics was first pointed out by Michael (1953) and later by Stuart (1954).

47. The Free Stream

We now examine the case in which the mean flow and the mean magnetic field are constants. All of the coefficients in (45.1) to (45.4) then become constants, and we can seek solutions for the complex amplitude functions in the form $e^{\beta y}$. Making this substitution, we obtain the following set of algebraic equations that relates the five complex constants **u**, **v**, **h**, **g**, **p**:

$$(U - c)\mathbf{u} + \mathbf{p} + \pi = \left(G - \frac{i\beta}{\alpha}H\right)\mathbf{g} - \frac{iv}{\alpha}(\beta^2 - \alpha^2)\mathbf{u}, \qquad (47.1)$$

$$(U - c)\mathbf{v} - \frac{i\beta}{\alpha}(\mathbf{p} + \pi) = \left(G - \frac{i\beta}{\alpha}H\right)\mathbf{h} - \frac{iv}{\alpha}(\beta^2 - \alpha^2)\mathbf{v}, \qquad (47.2)$$

$$(U - c)\mathbf{g} = \left(G - i\frac{\beta}{\alpha}H\right)\mathbf{u} - \frac{i\lambda}{\alpha}(\beta^2 - \alpha^2)\mathbf{g}, \qquad (47.3)$$

$$(U - c)\mathbf{h} = \left(G - i\frac{\beta}{\alpha}H\right)\mathbf{v} - \frac{i\lambda}{\alpha}(\beta^2 - \alpha^2)\mathbf{h}, \qquad (47.4)$$

$$\mathbf{u} = \frac{i\beta\mathbf{v}}{\alpha}. \qquad (47.5)$$

The definition of π is recalled from (45.7); this quantity is not a sixth unknown. We have used the relation $\omega = \alpha c$.

This system of equations is homogeneous, and therefore a nontrivial solution is obtained when the determinant of the coefficients vanishes. In the expression of this determinant we find that the highest power of β is eight. Thus we can expect eight different solutions which correspond to eight modes in the free stream. A systematic analysis that results from the expansion of the determinant shows that the values of β form three separate categories.

(a) Pressure Waves

This group contains two solutions in which $\beta = \pm\alpha$ and with no restrictions on c. Taking **p** as an arbitrary complex constant, we find that

$$\mathbf{u} = -\frac{\mathbf{p}}{(U - c)}, \qquad (47.6)$$

$$\mathbf{v} = \mp\frac{i\mathbf{p}}{(U - c)}, \qquad (47.7)$$

$$\mathbf{r} = 0, \qquad (47.8)$$

$$\mathbf{g} = (G \mp iH)\frac{-\mathbf{p}}{(U - c)^2}, \qquad (47.9)$$

$$\mathbf{h} = \mp(G \mp iH)\frac{i\mathbf{p}}{(U - c)^2}, \tag{47.10}$$

$$\boldsymbol{\eta} = 0, \tag{47.11}$$

$$\pi = -\frac{(G^2 + H^2)}{(U - c)^2}\,\mathbf{p}. \tag{47.12}$$

Hence the velocity field is exactly the same as in the absence of Lorentz forces. Indeed, the Lorentz forces are zero and the velocity field is decoupled from the magnetic field. The velocity fluctuations, however, induce magnetic fluctuations and it is not surprising to find π different from zero. Both the velocity and magnetic fields are potential, and as a result v and λ are ineffective. These solutions can be designated as pressure waves from above and from below.

(b) MAGNETOHYDRODYNAMIC WAVES

For this category $\beta^2 \neq \alpha^2$, and therefore, by referring to (45.9), we see that $\mathbf{p} + \pi = 0$ must follow. The only other possibilities correspond to the roots of the following fourth-order equation:

$$\left[U - c + \frac{iv}{\alpha}(\beta^2 - \alpha^2)\right]\left[U - c + \frac{i\lambda}{\alpha}(\beta^2 - \alpha^2)\right] = \left(G - i\frac{\beta}{\alpha}H\right)^2. \tag{47.13}$$

In general it is difficult to solve this equation explicitly as long as both G and H occur simultaneously. On the other hand, we can discuss some particular cases. If $G = H = 0$, we find that the vorticity and current are transported and diffused without interacting; v controls the diffusion of $\mathbf{r}$ and λ controls that of $\boldsymbol{\eta}$.

Let us now consider $v = \lambda = 0$. Here we can rewrite (47.13) in the form

$$c = U \pm \left(G - i\frac{\beta}{\alpha}H\right). \tag{47.14}$$

If the variation with y is periodic, β is purely imaginary and c is purely real, according to (47.14). This is the classic form of Alfvén waves (cf. Thompson, 1962, or Cowling, 1957). If β is complex, it is possible for c to have a positive imaginary part. This does not represent an instability in the ordinary sense. Physically, this case corresponds to an initial distribution of energy that is uneven in space so that the propagation of waves tends to increase the amplitude in the regions originally depleted. The same situation is encountered in acoustics in which the pressure fluctuation obeys the equation

$$\frac{\partial^2 p}{\partial t^2} = a^2 \nabla^2 p.$$ (47.15)

This leads to an algebraic relation for the propagation speed:

$$c = \pm a\left(1 - \frac{\beta^2}{\alpha^2}\right)^{1/2}.$$ (47.16)

A complex β gives a complex c; yet there is no instability. Note that if $\beta = -i\alpha H/G$, (47.1) to (47.4) are greatly simplified. The magnetic fluctuations exert only potential forces on the flow through the term π, and the magnetic fluctuations are no longer produced by induction.

Let us now restore the diffusive effects and consider the sums and differences of (45.10) and (45.11). In the free stream we obtain the equation

$$\left[i\alpha(U - c) \mp (i\alpha G + \beta H) - \frac{(\lambda + \nu)}{2}(\beta^2 - \alpha^2)\right](\mathbf{r} \pm \mathbf{\eta})$$

$$= -\frac{(\nu - \lambda)}{2}(\beta^2 - \alpha^2)(\mathbf{r} \mp \mathbf{\eta}).$$ (47.17)

When ν and λ are neglected, we recognize that (47.14) for the Alfvén wave speeds implies that $\mathbf{r} + \mathbf{\eta} = 0$ for one wave and $\mathbf{r} - \mathbf{\eta} = 0$ for the other. If $\nu = \lambda$, we see from this equation that the two Alfvén waves are damped but do not interact (see Hasimoto, 1960, for further discussion of this possibility).

If λ is different from ν, it is clear from (47.17) that the two Alfvén waves will interact. Going back to (47.13), we can solve for the complex phase velocity rather than the wave number and find that

$$c = U - i\frac{(\lambda + \nu)}{2\alpha}(\alpha^2 - \beta^2) \mp \left[\left(G - i\frac{\beta}{\alpha}H\right)^2 - \frac{(\lambda - \nu)^2}{4\alpha^2}(\alpha^2 - \beta^2)\right]^{1/2}.$$ (47.18)

If β is mostly imaginary, it can be seen that the sum of λ and ν will introduce a damping but that their difference tends to decrease the magnetic effect on the Alfvén wave speed.

Thus magnetohydrodynamic waves can finally be recognized as the appearance of a pair of waves from above and a pair from below. Each pair can be described as two Alfvén waves ($\mathbf{r} + \mathbf{\eta}$ and $\mathbf{r} - \mathbf{\eta}$) or as one vorticity $\mathbf{r}$ and one current wave $\mathbf{\eta}$. A very general and elucidating treatment of magnetohydrodynamic modes in a constant field has been given by Clauser (1963), who pointed out that wakes and boundary layers tend to be swept away from their normal positions and laid out along the lines of force of the mean magnetic field.

(c) Frozen Magnetic Waves

The last two solutions are of a peculiar type. They occur at $\beta = \pm \alpha$ and $c = U$. In addition, $\mathbf{u} = \mathbf{v} = \mathbf{p} = \mathbf{r} = 0$. With π as the arbitrary complex constant, we find that

$$\mathbf{g} = \frac{\pi}{(G \mp iH)},\tag{47.19}$$

$$\mathbf{h} = \mp \frac{i\pi}{(G \mp iH)},\tag{47.20}$$

$$\mathbf{\eta} = 0.\tag{47.21}$$

Physically speaking, these fluctuations describe the field produced by a periodic array of magnetic poles. These poles must move with the free stream or currents will be produced that will in turn result in Lorentz forces and velocity fluctuations. With respect to stability, however, these solutions have no special significance. Even more, this point shows that the stability analysis needs to be considered only with a sixth-order system rather than the full eighth and the procedure examined in Appendix III might become useful for calculations.

48. Shear Flows and Magnetic Gradients

We now discuss the solutions in which at least one of the mean quantities U or G varies with y; H must always be a constant by the continuity equation. We can expect pressure fluctuations according to (45.9). We can also expect damped Alfvén waves but with the understanding that in the limit of large λ one of the Alfvén waves will degenerate into the ordinary oscillation of the vorticity.

By forming the sums and differences of (45.10) and (45.11) we obtain the equation

$$\left[i\alpha(U - c) \mp \left(i\alpha G + H \frac{d}{dy} \right) - \frac{(\lambda + v)}{2} \left(\frac{d^2}{dy^2} - \alpha^2 \right) \right](\mathbf{r} \pm \mathbf{\eta})$$

$$+ U''(\mathbf{v} \mp \mathbf{h}) - \left(G'' \pm 2U' \frac{d}{dy} \right)\mathbf{h} = \frac{(v - \lambda)}{2} \left(\frac{d^2}{dy^2} - \alpha^2 \right)(\mathbf{r} \mp \mathbf{\eta}). \quad (48.1)$$

Note that if the terms in U'', G'', U', λ, and v are all zero in (48.1) we will have the true Alfvén waves discussed in Section 47. The term in U'' couples one Alfvén with the other, provided that $\mathbf{v}$ and $\mathbf{h}$ are simply the integrals of $\mathbf{r}$ and $\mathbf{\eta}$ without any additional curl-free contribution. The role of the terms in G'' and U' have no simple interpretations.

A close inspection of (48.1) indicates that the question of a critical layer must be reconsidered from the beginning. There is no obvious singularity at the point $U - c = 0$ in the absence of the v or λ terms. Therefore, if some terms of the equation are zero, the remaining terms tend to cancel each other. To facilitate the discussion of a magnetohydrodynamic critical layer we introduce further simplifications in Section 49.

49. The Low Magnetic Reynolds Number Approximation

Recalling (45.4), we note that the terms in G and H act as a forcing function for the fluctuation $\mathbf{h}$. In most laboratory applications of magnetohydrodynamics the resistivity λ is sufficiently high and the term $U - c$ can be neglected. This corresponds to the assumption that the magnetic Reynolds number $\mathcal{R}_m = U_0 L/\lambda \ll 1$ and the nondimensional wave number α is not much larger than one. The simplified form of (45.4) can be combined with the definition of the current to get

$$\mathbf{\eta} = (\alpha G \mathbf{v} - iH\mathbf{v}')/\lambda\alpha. \quad (49.1)$$

We now take the vorticity equation (45.10) and assume that the mean magnetic field is everywhere constant. This results in the loss of the single term G'' only. By substituting (49.1) into this new equation we obtain

$$i\alpha\left(U - c - i \frac{H^2}{\lambda\alpha} \right)\mathbf{r} - 2 \frac{GH}{\lambda} \mathbf{v}' + \left[U'' - i \frac{\alpha}{\lambda}(G^2 - H^2) \right]\mathbf{v} = v(\mathbf{r}'' - \alpha^2\mathbf{r}).$$

$$(49.2)$$

Now we can see that the equivalent of a critical layer for the magnetohydrodynamic system occurs when $U - c_r = 0$ and $c_i = -H^2/\lambda\alpha$. We further recognize that there are no longer any singularities when the flow is inviscid at this point because of the presence of the term GH which stems from the fact that the magnetic field may be oblique to the direction of the flow.

Stuart (1954) investigated the stability of plane Poiseuille flow in the realm of the approximations used to obtain (49.2). In particular, he treated this flow with $H = 0$ and G constant. It can be verified that the mean flow still has a parabolic distribution, for U and G are parallel. From (49.2) we find the well-known Orr-Sommerfeld equation with only one new term. For this case, however, there is a critical layer, but the phase of the vorticity production is different from the classical problem. The results of Stuart's calculations (which are also reproduced in Cowling, 1957) indicate that there is a stabilizing effect, and in the limit $G \to \infty$ the hairpin tends to disappear altogether. There is some question of the validity of this last point, however, because of the approximations that have been made along the way. Rossow (1958a) examined the very same problem that was done by Stuart and, starting from the same equations, found results that are in slight disagreement. Rossow (1958b) also considered the case of the Blasius boundary layer for $H = 0$ and G constant and found that the magnetic field is stabilizing. [An additional fact of note is that the G-field affects the mean flow along a wall in the boundary layer approximation, as shown by Greenspan and Carrier (1959).] Both authors used power series for the inviscid solutions and a Tietjens function at the wall.

Recently the problem of Poiseuille flow was treated by Nachtsheim and Reshotko (1965). Using a computer, they found that the critical Reynolds number was about double that found by Stuart. The magnetic Reynolds number was of the order of 10^{-7}, so that the approximation leading to (49.1) was excellent. As the magnetic Reynolds number increases, (49.2) loses its validity, and it becomes necessary to integrate a sixth-order system. Indeed, if $H = G' = 0$, (45.10) is of order four in $\mathbf{v}$ and contains a term in $\mathbf{h}$. To determine $\mathbf{h}$ we must integrate the second-order equation (45.4), giving $\mathbf{h}$ as a function of $\mathbf{v}$.

Nachtsheim and Reshotko found that when the magnetic Reynolds number raises above unity, while the parameters σ, G^2, and α are kept constant, the wave is destabilized. With a magnetic Reynolds number of 100, c_i has become roughly 60 times larger and seems to have reached a ceiling. These changes are made with constant mean flow and constant viscosity ($\nu = 10^{-4}$).

For the case in which $G = 0$ and H is constant, Lock (1955) treated the mean plane Poiseuille flow. The mean solution is better known as a Hartmann flow, and the profile is generally flatter than the simple parabola. The curvature U'' occurs near the wall. Lock examined the stability of these particular profiles and used the ordinary form of the Orr-Sommerfeld equation instead of (49.2). This is acceptable as long as $H^2 L / U_0 \lambda \ll 1$.

As can be expected from the conclusions of Section 11, the concentration of the curvature of the mean profile toward the wall has a stabilizing effect.

Let us now consider a boundary layer flow along a wall in the presence of a strong magnetic field H perpendicular to the wall. It does not matter whether the permanent magnets or coils used to produce this steady and homogeneous field move with the free-stream velocity, for the magnetic field is uniquely specified. The observer whose coordinates are fixed to the wall will note that in (44.11) the cross term between $U(y)$ and H will result in the appearance of an mean electric current parallel to the z-axis, to an electric field, or to both. The electric field parallel to the z-axis cannot vary with y, lest its curl demand some variation of the magnetic field according to (44.1). Thus the transverse electric field must be constant. If the flow is limited laterally by insulating walls, the net current $\int_0^\infty J\,dy$ must vanish. This requirement defines the constant electric field. From a physical point of view, we can say that negative ions are deflected one way by the magnetic field, positive ions the other way. The result is deposition of charges on the insulating sidewalls. This creates the electric field. The observer moving with the free stream, or any other constant speed, will observe a different electric field and a different mean flow, but these vectors will satisfy his version of (44.3).

The sidewalls can be made to conduct current and can even be divided into thin strips connected to assorted current generators. As long as E stays constant, a family of solutions $J(y)$ can be found that will differ only by a constant current. Let us now consider the mean Lorentz forces, proportional to $\vec{J} \times \vec{H}$. In parallel laminar flows they must be balanced by viscous stresses or pressure gradients. This gives us two limiting cases: (a) the Lorentz forces vanish in the free stream; (b) the net current vanishes (insulated sidewalls).

When the current vanishes in the free stream, the velocity takes the exponential profile given by Hartmann (1937). The curvature is concentrated near the wall and the critical Reynolds number is increased.

If the net current is made to vanish, a special velocity profile is found (Rossow, 1958a) with a curvature U'' located away from the wall. Again we are not surprised to find that the critical Reynolds number is low (Rossow, 1958b).

50. The Magnetic Critical Layer

When the mean magnetic field is parallel to the mean velocity, a new phenomenon is encountered if the phase velocity matches the Alfvén wave velocity. This process, first pointed out by Velikhov (1959), is particularly evident if viscosity and resistivity are neglected.

Referring to the general equations (45.1) to (45.5), with the limitations $H = 0$, $v = \lambda = 0$, and $G = \text{constant}$, we find that $\mathbf{v}$ is proportional to $\mathbf{h}$. Thus we can express the vorticity equation (45.10) in the form

$$[(U - c)^2 - G^2](\mathbf{h}'' - \alpha^2\mathbf{h}) + 2(U - c)U'\mathbf{h} = 0. \qquad (50.1)$$

This equation presents the same difficulty as the vorticity equation of the incompressible ordinary flow: if c coincides somewhere with the value $U \pm G$, the coefficient of the highest derivative vanishes and the solution has a singularity. Note that the critical speeds are those of the two possible Alfvén waves. Thus we can have two magnetic critical layer. If λ or v or both are retained, a term of higher order will prevent the solution from diverging because it contains derivatives of $\mathbf{h}$ higher than the second.

It is interesting to note that a small degree of obliquity of the mean magnetic field can also control the magnetic critical layer. This can be understood by examining the general vorticity equation (45.10). It contains a term in $H\eta'$ that is of higher order than the term in $G\eta$. Thus any small amount of H will have an effect similar to the dissipative processes. In fact, the Alfvén wave no longer propagates exactly along the streamlines, so that the perturbations can be carried away from the critical layer.

Velikhov examined $H = 0$ with finite v and λ between two parallel and perfectly conducting walls, using asymptotic series and the Tietjens function. He also used a parabolic mean flow and found the neutral curves shown in Fig. 50.1. Note that as the conductivity of the fluid approaches

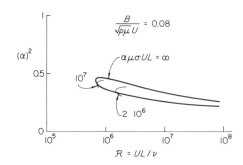

Fig. 50.1. The stability of Poiseuille flow in the presence of a longitudinal magnetic field. The resistivity σ is stabilizing. (After Velikhov, 1959.)

zero the results do not seem to tend toward the ordinary neutral curve of the channel flow (see Fig. 15.1). This suggests the existence of different modes and invites further investigation.

51. The Oblique Magnetic Field

The presence of a term proportional to GH in (49.2) has prompted us to a limited investigation of oblique magnetic fields. The presence of a new term in v' in this generalized form of the Orr-Sommerfeld equation raises immediately the question of the critical layer. There may be solutions in which the viscous corrections are no longer necessary because the term in GHv' balances the term in $U''v$.

As an example we assume that the mean velocity corresponds to $U = \tanh y$ and that it is not affected by the mean magnetic field. This may not comply with the mean flow equations, but it does allow us to isolate the effects and clarify a few points. We take a constant free stream for $y > 3$ and place a rigid wall at $y = -3$. In the absence of any magnetohydro-dynamic effects we know from Section 5 that the unstable oscillations are limited to the range $0 < \alpha < \alpha_{max}$ and the numerical calculations have suggested $\alpha_{max} = 1$. For $\alpha > 1$ we must consider the viscous effects, as in Section 13; the calculations have shown damped solutions as for the free shear layer.

The viscous terms are neglected, and the appropriate equation is obtained from (49.2):

$$\left(U - c - i\frac{H^2}{\lambda\alpha}\right)(v'' - \alpha^2 v) + 2\frac{GH}{\lambda}v' - (U'' + i\alpha H^2 - i\alpha G^2)v = 0. \quad (51.1)$$

In the free stream the roots of the characteristic equation are always complex, so that one solution vanishes as $y \to \infty$ and the other diverges. It is easy therefore to specify $v'(3)$, with the arbitrary choice of $v(3) = 1$. The computer then proceeds to the wall and indicates $v(-3)$. The quantity varies with α, c, G, and H, and we examine the quantities $[\partial v(-3)]/\partial c$ and $[\partial v(-3)]/\partial G$ or $[\partial v(-3)]/\partial H$ for various c.

If $H = 0$, we find that $\partial v/\partial G$ is such that c_i must decrease to return $v(-3)$ closer to zero. Thus G has a stabilizing influence, as determined by Stuart (1954) in a boundary layer profile without inflection.

If $G = 0$, we find that c_i must again decrease, as H is increased, at a rate higher than in $H = 0$, $G \neq 0$. This seems surprising, but if the term $iH^2/\lambda\alpha$ is removed from the equation or we examine $c_i + H^2/\lambda\alpha$ instead of c_i, we find that H^2 has an effect opposite that of G^2, as it should be.

Thus G is stabilizing because of its effect proportional to $\mathbf{v}$; H is also stabilizing because of its effect proportional to $\mathbf{v}''$.

When both G and H exist, the values of c_i, which are negative, tend to increase when the product GH is negative. Thus the obliquity can render the flow less damped.

All of these observations are pertinent for $\alpha < 1$. Equation (51.1) suggests that when α is very large the term proportional to $H^2\mathbf{v}$ could lead to a new instability, for the term in $H^2\mathbf{v}''$ decreases with α.

However, the viscous nonmagnetic solutions are strongly damped, so that the effects of H^2 are likely to be small. At the same time this does not rule out the existence of a distinct magnetic mode which would not be connected to an inviscid solution.

52. Magnetohydrodynamic Energy Equation

With our special choice of the units of mass and of magnetic field, the density of energy of an oscillation is given by

$$e = \tfrac{1}{2}(\overline{\tilde{u}_i\tilde{u}_i + \tilde{h}_i\tilde{h}_i}), \qquad (52.1)$$

where it is assumed that the terms with a tilde have a zero average. The rate of change of this energy can be expressed by cross multiplication of the basic equations (44.12) and (44.13). Before averaging, we obtain the following relation, in which certain terms have already been grouped conveniently

$$\frac{1}{2}\frac{\partial}{\partial t}(\tilde{u}_i\tilde{u}_i + \tilde{h}_i\tilde{h}_i) + \frac{1}{2}\frac{\partial}{\partial x_k}[U_k\tilde{u}_i\tilde{u}_i + U_k\tilde{h}_i\tilde{h}_i] + \frac{\partial U_i}{\partial x_k}\tilde{u}_i\tilde{u}_k$$

$$+ \frac{\partial}{\partial x_i}(\tilde{u}_i\tilde{p}) - \frac{\partial U_i}{\partial x_k}\tilde{h}_i\tilde{h}_k + \frac{\partial H_i}{\partial x_k}(\tilde{h}_i\tilde{u}_k - \tilde{h}_k\tilde{u}_i)$$

$$= \frac{\partial}{\partial x_k}[H_k\tilde{u}_i\tilde{h}_i - H_i\tilde{h}_i\tilde{u}_k] - v\tilde{u}_i\frac{\partial^2\tilde{u}_i}{\partial x_k\,\partial x_k} - \lambda\tilde{h}_i\frac{\partial^2 h_i}{\partial x_k\,\partial x_k}. \qquad (52.2)$$

Let us now apply an average in space in a region confined by horizontal walls along which we have $U_i = \tilde{u}_i = 0$ and by vertical control surfaces placed one wavelength apart. A horizontal wall can also be moved to infinity.

All the terms grouped in brackets in (52.2) average to zero, and the final equation reads

$$\frac{\partial e}{\partial t} = \mu_1 + \mu_2 + \mu_3 + \mu_4 + \mu_5, \qquad (52.3)$$

with the definitions

$$\mu_1 = -\overline{\tilde{u}_i \tilde{u}_k} \frac{\partial U_i}{\partial x_k}, \tag{52.4}$$

$$\mu_2 = \overline{\tilde{h}_i \tilde{h}_k} \frac{\partial U_i}{\partial x_k}, \tag{52.5}$$

$$\mu_3 = \overline{\tilde{u}_i \tilde{h}_j} \left(\frac{\partial H_i}{\partial x_j} - \frac{\partial H_j}{\partial x_i} \right). \tag{52.6}$$

The term μ_1 comes from the work of the Reynolds stresses, the term μ_2, from the work of the Maxwell stresses, and the term μ_3 represents the power of the Lorentz forces. The terms μ_4 and μ_5 are always negative; they give the dissipation by viscous and ohmic processes according to

$$\mu_4 = -\nu \overline{\frac{\partial \tilde{u}_i}{\partial x_j} \frac{\partial \tilde{u}_i}{\partial x_j}}, \tag{52.7}$$

$$\mu_5 = -\lambda \overline{\frac{\partial \tilde{h}_i}{\partial x_j} \frac{\partial \tilde{h}_i}{\partial x_j}}. \tag{52.8}$$

Thus energy can be extracted from the mean flow by the terms μ_1 and μ_2 or from the mean magnetic field, if there is a mean electric current, as implied by the term (52.6) corresponding to curl $\vec{H}$.

Convince yourself that the pressure wave described approximately by (17.12) could not interact with a mean shear flow or a mean electric current. For Alfvén waves we find from (45.1) to (45.5) that $\mu_1 + \mu_2 = 0$ and that $\mu_3 = 0$. Thus, if we approximate an exact solution of the magnetohydrodynamic equations by *one* simple wave, in a region of mean shear flow or mean electric current, we find no energy exchange. It is only when mean shear or electric currents have distorted the solution that energy exchange is possible.

Of course, we can also have the interaction between two waves, depending on an unspecified phase factor.

Chapter X

Additional Topics and Complexities

53. Spatially Growing Oscillations

(a) Growth in Time Versus Growth in Space

So far we have dealt only with situations in which the oscillation is periodic in space and proportional in time to $e^{-i\alpha ct}$. In reality, most fluid oscillations have an amplitude that is constant with time but growing in some spatial direction. Hence, this situation corresponds to equations in which α is complex and the frequency ω is real, with the definition

$$\omega = \alpha c. \tag{53.1}$$

It should be noted that if α is complex then ω cannot be real unless c is also complex. From the point of view of mathematical analysis it is much more convenient to have a complex ω than a complex α. This becomes even more significant when viscous effects are considered (e.g., Tietjens function or the Squire theorem). Furthermore, from a fundamental point of view, we note from the governing equations proper, namely (1.1), (1.2), (1.3), that the temporal derivatives are not so involved as those of the spatial variety; the time derivative is always first-order and has the coefficient of unity. On the other hand, when machine computations are utilized, the transition to complex α adds only a few more operations to the program. In this respect it is often good policy to plan for a complex α at the beginning. We could modify the searching procedure in such a way that, given a particular ω, the computer will search the complex α that will satisfy the boundary conditions. Thus it is α that becomes an eigenvalue in place of ω.

It is important to note that, whether the growth occurs in time or in space, the neutral line is always the same. Indeed, if $\alpha_i = c_i = 0$, we have $\omega_i = 0$.

Some consideration has been given to this topic by Gaster (1965a, 1965b). In the first paper a general discussion is put forth together with a specific example. The second work specifically demonstrates that the spatial problem corresponds more correctly to the experiments of Schubauer and Skramstad (1943).

(b) ANALYTIC RELATIONS

In general, we can select a complex value of α and find a value of c that is consistent with the boundary conditions which lead to the notion that α is a function of c that can be mapped in the complex c-plane. In symmetric oscillations of the jet, given by $U = 1/\cosh^2 y$, the systematic

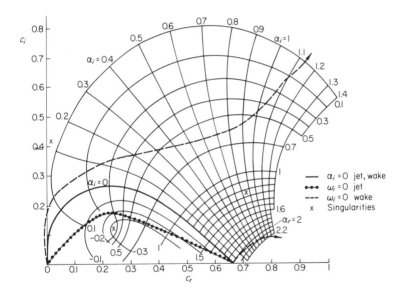

FIG. 53.1. Spatial amplification for the inviscid jet, displayed in the plane of c. Results for Mode I only.

search of eigenvalues leads to the map in Fig. 53.1. The curve for $\alpha_i = 0$ corresponds to the neutral line in Fig. 6.2. Note that the lines at constant α_r are orthogonal to the lines at constant α_i. Hence we can see that α is an analytic function of c. At some special points the function exhibits remarkable behavior and becomes multivalued. These singularities become the object of a special discussion which is given later.

If we know the values of c_r and c_i as a function of α_r along the neutral line, we can obtain the derivatives $\partial c_r/\partial\alpha_r$ and $\partial c_i/\partial\alpha_r$. Then, according to the Cauchy-Rieman equations, we have

$$\left(\frac{\partial c_i}{\partial\alpha_i}\right)_{\alpha_r} = \left(\frac{\partial c_r}{\partial\alpha_r}\right)_{\alpha_i}, \qquad (53.2)$$

$$\left(\frac{\partial c_r}{\partial\alpha_i}\right)_{\alpha_r} = -\left(\frac{\partial c_i}{\partial\alpha_r}\right)_{\alpha_i}. \qquad (53.3)$$

With these relations we can evaluate the distance between the neutral line and a line with a small and constant α_i. It is quite clear that this process can be carried on in principle and that the whole map could be drawn from a single line. However, it would require extreme accuracy and the singularities might hinder the procedure.

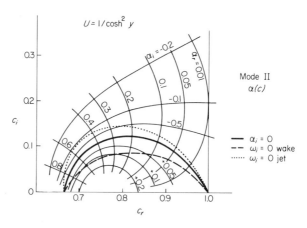

FIG. 53.2. Spatial amplification for the inviscid jet, displayed in the plane of c. Results for Mode II only.

(c) Applications to Inviscid Jet

The map in Fig. 53.1 was obtained with a computer by Betchov and Criminale (1966), and for every point they also determined the complex quantity ω according to (53.1).

The solutions in which $\omega_i = 0$ correspond to the physically significant cases. By interpolation between the computed results the locus of points such that $\omega_i = 0$ can be determined; it is shown in Fig. 53.1 as a dotted line. Similar results for the asymmetric solutions are given in Fig. 53.2 and

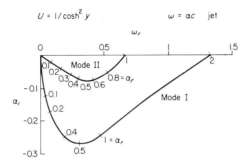

Fig. 53.3. Spatial growth as function of frequency for a symmetric jet.

indicated as Mode II. These curves correspond to negative values of α_i and therefore to spatial growth in x proportion to $e^{-\alpha_i x}$. These rates of growth appear in Fig. 53.3 as a function of ω_r for both modes.

Judging from the maps in Figs. 53.1 and 53.2, it seems impractical to use the technique of analytic continuation to locate the curve $\omega_i = 0$ from the neutral curve $\alpha_i = 0$.

(d) Application to Inviscid Wake

The maps in Figs. 53.1 and 53.2 contain the information necessary to compute the spatial amplification for an entire family of wake flows. We define the mean flow

$$U(y) = -U_0 + \frac{1}{\cosh^2 y}, \tag{53.4}$$

where U_0 is a positive constant. Note that the central velocity $U(0)$ is zero if $U_0 = 1$.

Let us momentarily use the coordinates x', y', t', which are fixed to the body creating the wake. In this frame of reference the body is at rest and the mean flow proceeds to the left with velocity $-U_0$. A perturbation is proportional to $e^{i\alpha'(x'-c't')}$.

An observer moving with the mean flow will find that the mean profile is exactly that of the jet and that his stability analysis will lead to the map in Figs. 53.1 and 53.2.

Thus we label his coordinates x, y, t, with the relations $x = x' + U_0 t$, $y = y'$, $t = t'$. Substitutions in the exponential function show that the two observers use the same wavenumbers, $\alpha' = \alpha$. However, the phase velocities are related by $c' = c - U_0$ and the frequencies, by $\omega' = \alpha(c - U_0)$.

The condition for a real frequency in the wake is clearly different from the corresponding condition in the jet because of the presence of the factor $-\alpha U_0$.

Let us now take $U_0 = 1$. In symmetric disturbances of **v** the locus of points at which ω' is real forms the two separate branches indicated by broken lines in Fig. 53.1.

This surprising result is related to the fact that the function $\omega'(\alpha)$ has a singularity at $\omega' = 0.55 - i0.05$. A map of α as a function of ω' is shown in Fig. 53.4.

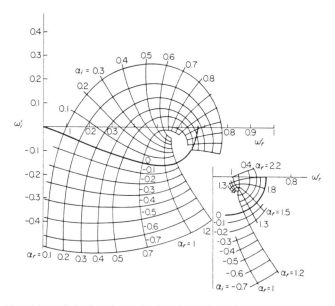

FIG. 53.4. Map of the function α in the plane of ω', for a wake flow, Mode I.

This singularity leads to multiple values of α, and it becomes necessary to use a special insert to map one of the branches of α. The neutral curve is shown without interruption from one neutral point at $\alpha = 0$ to the other point at $\alpha = 2$. The values of α_i and α_r along the line $\omega_i' = 0$ are shown in Fig. 53.5, as functions of ω_r'.

The second mode presents no difficulties; the results are also shown in Fig. 53.5. The rates of growth of the second mode are smaller than those of the first.

Let us now consider other values of U_0. A calculation along the line $\alpha_r = 1.25$ from the neutral line $\alpha_i = 0$ to the singularity in Fig. 53.1 showed

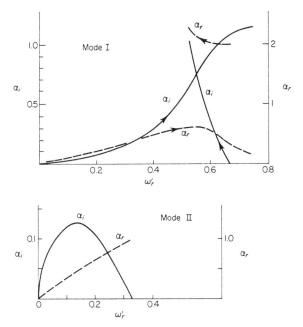

Fig. 53.5. Spatial growth as function of frequency for a symmetric wake. Note the change in scale for α_i, in Mode II.

that as U_0 increases from 0 to 0.94 the line $\omega'_i = 0$ still passes between the neutral line and the singularity.

This indicates that for lower values of U_0 the singularity in Fig. 53.4 probably will remain above the line $\omega'_i = 0$. In this eventuality the rates of growth form a single curve.

Physically, a wake with a central velocity of zero would be extremely unstable, and we can expect that the nonlinear effects will rapidly "fill" the mean profile and create a less unstable situation. The entire matter, however, deserves further consideration. Moreover, viscosity and even the growth of the wake downstream should be included before final conclusions are made. A comparison between theory and experiment for a wake has been made by Sato and Kuriki (1961). These authors, however, used a wake given by a Gaussian profile rather than the one used in (53.4). Their central velocity was of the order of $U(0) > 0.4$, and spatial amplification can be obtained from the curve $\alpha_i = 0$ by the linear approximation. They examined the phase velocity, frequency of oscillations, rate of amplification, and even the amplitude of the amplification, and obtained good agreement. They also studied the nonlinear region experimentally.

(e) ON THE SINGULARITIES OF A COMPLEX FUNCTION

To understand the peculiar behavior pointed out in Fig. 53.1 in the vicinity of $c = 0.23 + i0.1$ we consider the complex function given as

$$F(z) = \varphi(z) + i\psi(z), \tag{53.5}$$

where $z = x + iy$ is the independent variable. If the function can be expanded in power series in the vicinity of a point z_0, we have

$$F = F_0 + F_0'(z - z_0) + \tfrac{1}{2}F_0''(z - z_0)^2 + \cdots. \tag{53.6}$$

If F_0' is different from zero, the mapping of F near z_0 will be regular. The Cauchy–Riemann relations will guarantee the orthogonality of the lines at constant φ or ψ. However, if F_0' is zero, the variations of F will be quadratic. In the vicinity of z_0 we have

$$\Delta F = \frac{F_0''}{2}\left[(\Delta x)^2 - (\Delta y)^2 + 2i\,\Delta x\,\Delta y\right]. \tag{53.7}$$

Let us take $F_0'' = 2$ as an example. The real part of ΔF is constant along a hyperbola defined by $\Delta\varphi = (\Delta x)^2 - (\Delta y)^2$; the imaginary part is constant along the hyperbola $\Delta\psi = 2\,\Delta x\,\Delta y$. If other complex values of F_0'' are encountered, the results will be the same except for a rotation of the system of hyperbolas.

Let us now plot z in the F-plane. The curves at constant Δx will correspond to the parabolas $\Delta\varphi = (\Delta x)^2 - (\Delta\psi)^2/4(\Delta x)^2$. Those at constant Δy lead to $\Delta\varphi = (\Delta\psi)^2/4(\Delta y)^2 - (\Delta y)^2$. These sets of curves are sketched in Fig. 53.6.

A comparison with the singularity noted in the map of $\alpha(c)$ (see Fig. 53.1) shows that we have a singularity of the type $c = c_0 + Q(\alpha - \alpha_0)^2$. Thus $\partial\alpha/\partial c = \infty$ at the point of interest. Note that the curves of positive Δx rotate clockwise as Δy increases, whereas those of negative Δx rotate counterclockwise.

Similar conclusions apply to the singularity observed in Fig. 53.4; thus we have $\omega' = \omega_0 + Q'(\alpha - \alpha_0)^2$. Analogous results have also been found in the study of plasma instabilities. (See Briggs, 1964.)

(f) BOUNDARY LAYERS

In the Blasius boundary layer numerical calculations with complex α have been carried out independently by Kaplan (1964), Raetz (1964), and

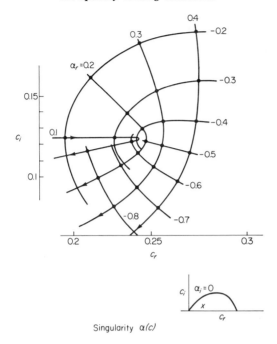

Singularity $\alpha(c)$

FIG. 53.6. Details of the singularity of the function α in the plane of c. This singularity is already visible in Fig. 53.1.

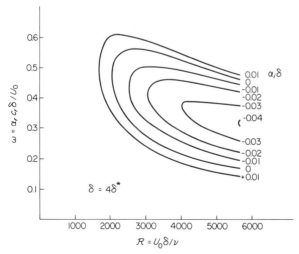

FIG. 53.7. The spatial rate of growth of an oscillation for a Blasius boundary layer. (After Kaplan, 1964.)

Wazzan, Okamura and Smith (1966). As far as the unstable solutions are concerned, the values of α_i are small. The results obtained by Kaplan are shown in Fig. 53.7.

Let us now examine the relation between the values of α when $\omega_i = 0$ and the values of c when $\alpha_i = 0$. The first set gives the rate of growth in space in a laboratory experiment, whereas the second set can be read from the usual plot of c_r and c_i in the plane showing α as a function of the Reynolds number.

If the curve corresponding to constant amplitude in time and growth in space (i.e., $\omega_i = 0$) is not too far from the original curve obtained for α purely real, we can use a linear approximation. The condition $\omega_i = 0$ implies

$$\omega_i = \alpha_i c_r + \alpha_r c_i = 0. \tag{53.8}$$

In general, a change in α corresponds to the following change in c:

$$\Delta c = \frac{\partial c_r}{\partial \alpha_r} \Delta \alpha_r + \frac{\partial c_r}{\partial \alpha_i} \Delta \alpha_i + i\left(\frac{\partial c_i}{\partial \alpha_r} \Delta \alpha_r + \frac{\partial c_i}{\partial \alpha_i} \Delta \alpha_i\right). \tag{53.9}$$

Now, from a point at $\alpha_i = 0$ we shall move at constant α_r and locate the line where $\omega_i = 0$. Thus we take $\Delta \alpha_r = 0$ and, because of the smallness of α_i, we use the approximation $\Delta \alpha_i = -\alpha_r(c_i/c_r)$. After use of the Cauchy-Riemann relations (53.2) and (53.3) we obtain from (53.9)

$$\Delta \alpha_r = \alpha_r \frac{c_i}{c_r} \frac{\partial c_i}{\partial \alpha_r}, \tag{53.10}$$

$$\Delta c_i = -\alpha_r \frac{c_i}{c_r} \frac{\partial c_r}{\partial \alpha_r}. \tag{53.11}$$

We can also consider ω as an analytic function of α. This aspect has been investigated by Gaster (1963).

However, certain results of Wazzan, Okamura, and Smith, which were obtained for Blasius boundary layers, indicate that the linear approximation of (53.9) may not be sufficiently accurate for the purpose of forecasting transition.

54. Quasi-parallel Mean Flows

The only mean flows that are exactly parallel are those constrained by two parallel walls. All actual jets, wakes, and boundary layers, however, have a mean flow in which both U and V vary with x as well as with y.

For incompressible flows V is smaller than U and proportional to $U\mathscr{R}^{-n}$. We denote such flows as quasi-parallel and review the findings of the few authors who have examined their stability. This subject is indeed an important link between the Orr-Sommerfeld equation and reality. (See also Alekseyev and Korotkin, 1966.)

According to Cheng (1953), the terms proportional to V are more important than the terms involving the derivatives of U with respect to x. Thus we can freeze the coordinate x and examine the stability problem as if V and its derivatives $d^n V/dy^n$ were functions of y alone. The continuity equation for the mean flow establishes a relation between V and U.

In a Blasius boundary layer we have $V \sim U/\mathscr{R}$, but Cheng points out that in a hypersonic boundary layer V becomes proportional to $UM^4/\mathscr{R}$, where M is the Mach number. Thus the lack of parallelism can have serious consequences. Cheng derived the full set of linearized equations for two-dimensional fluctuations in a compressible flow. They form a sixth-order system and contain a great many more terms proportional to the x-derivatives of U, V, or T.

These equations have been extended by Brown (1965) to include three-dimensional fluctuations. We have seen that even in a parallel mean flow these equations cannot be transformed to appear like those for two-dimensional perturbations with unusual coefficients. In particular, the viscous term in the entropy equation cannot be expressed in terms of the new variables. Thus Brown obtained an eight-order system that could not be reduced. In the free stream his equations indicate four pairs of modes: pressure, vorticity perpendicular to the xy-plane, vorticity along the x-axis, and entropy.

Each pair has one solution vanishing as $y \to \infty$. Using a computer and performing four straightforward integrations from the wall to the edge of the boundary layer, Brown obtained the results reproduced in Fig. 43.2. The abscissa is the Reynolds number based on the displacement thickness and the ordinate is essentially the frequency. The calculations have been made with a real α. The points indicate experimental results of Demetriades (1958) at a Mach number of 5.8. The curve which stays below $\mathscr{R}_{\delta*} = 400$ is obtained from the equations of Lees and Lin (1946) [see (39.4), (39.5), and (39.7)].

With the quasi-parallel approximation, Brown finds the curve staying above $\mathscr{R}_{\delta*} = 1200$. Thus the mean flow perpendicular to the wall seems to have a stabilizing influence. Still with a sixth-order system, Brown examined the effects of obliquity by approximating the troublesome term in the entropy equation.

The angle of the most unstable wave is determined from a plot similar to Fig. 43.1, and it is found that the neutral curve begins around $\mathscr{R}_{\delta*} = 630$, which agrees with the measurements of the lower branch but diverges from those of the upper branch. Finally, Brown used the full eight-order system, thus including quasi-parallel terms, obliquity, and the exact entropy equation. This leads to the cuve beginning at $\mathscr{R}_{\delta*} = 520$ and to good agreement with experiment.

It is not possible to evaluate the quality of the experimental results, and the successive steps of the theoretical studies have clearly shown that they are sensitive to a plurality of factors. This should not invite us to criticism but to a continuation of the effort.

The stability of incompressible quasi-parallel flows has also been studied recently by Lanchon and Eckhaus (1964) and discussed by Reid (1965). They considered the fact that the boundary layer thickness δ grows as $\sqrt{x}$ and define nondimensional variables $\eta = y/\delta$ and $\xi = x/\delta$. This new pair of coordinates contains the parabolic growth of boundary layers and therefore leads to mathematical simplicity. The linearized equation for vorticity fluctuations is derived and its solution is expressed by a power series in terms of the viscosity. The "inviscid" approximation has the exact form of the Rayleigh equation, although the fluctuations are proportional to $\mathbf{r}(\eta)e^{i\alpha(\xi - c\tau)}$, where $\tau = Ut/\delta$. Thus the special geometry of the growing layer can be encompassed without difficulty. At the critical layer the effects of a very small viscosity are the same as in the theory of parallel flows.

The other viscous effects, corresponding to the diffusion of vorticity from the wall, are modified in the special geometry by the terms proportional to V. In boundary layers and in the limit of large Reynolds numbers Lanchon and Eckhaus find that the parallel flow approximation is valid in the first approximation only.

The limitation in jets is more severe: only the "inviscid" approximation, which includes the special geometry and neglects the viscous stresses and the terms in V, is permissible.

These conclusions of a rigorous analysis can be compared with the properties of the flow endowed with a constant V_0, which was the object of our attention in Section 18. Referring to (18.13), we note that if we consider V_0 as proportional to $\mathscr{R}^{-1}$ it contributes to the inviscid equation in the usual Cartesian coordinates. At the critical layer we know that the term in $\mathbf{v}''''$ plays the leading role and it can be shown that V_0 is ineffective in the limit $v \to 0$. Its effect on the transport of vorticity, however, cannot be disregarded.

In boundary layers with suction through the wall the magnitude of V_0 is not related a priori to that of v and the inviscid equation is already affected. The order of this equation is raised from second to third; when the oscillation is neutral the vertical transport of vorticity prevents the accumulation of vorticity at the critical layer. Thus there may be inviscid instabilities when V_0 is constant. This is a moot point however, for suction is used to control the resistive oscillations of boundary layers.

55. Forecasting Transition

One of the most promising applications of linear stability theory was found by Smith (1957). A review is given in Schlichting (1960). Let us consider a Blasius layer along a rigid plate. If a perturbation is introduced during a short time and at a place at which the Reynolds number is less than critical, its effects will be carried downstream. These effects will decay in the course of the time until they reach the distance x_C, where the Reynolds number becomes larger than critical. The process of amplification, characterized by c_i, then begins.

Because c_i varies with the Reynolds number, the total amplification between the onset of instability and a point x can be approximated as a factor

$$\varphi = \exp\left(\int_{x_C}^{x} \frac{\alpha c_i}{U_0} \, dx\right). \tag{55.1}$$

In a real flow the laminar layer becomes oscillatory and at some distance x_T, turbulent transition takes place.

Using theoretical values of c_i and experimentally determined values of x_T, Smith found that the numerical values of the amplification agreed with the relation

$$\varphi_T = \exp\left(\int_{x_C}^{x_T} \frac{\alpha c_i}{U_0} \, dx\right) = e^9 \doteq 10^4. \tag{55.2}$$

He extended his work to a variety of flows, including the entrance region of pipe flows and boundary layer with pressure gradients. Apparently the location of transition always occurs when the amplification has the same value e^9. We could say that the laminar layer behaves as an amplifier with an energy gain of 80 dB. If the perturbations are continuously introduced by roughness at the wall, it is clear that the most important region is located near x_C, where the Reynolds number is critical. The effect of a unique perturbation concentrated at x_C is given by (55.2), but the effect of a distribution of perturbations would formally involve the solution of the inhomogeneous Orr-Sommerfeld equation. Because the size of the sensitive region near x_C is not precisely known, the effect of distributed perturbation can only be evaluated. These studies also lead to the idea that thermal motion, often referred to as Brownian motion, might be the trigger of boundary layer oscillations under sufficiently quiet conditions. In this case it would be useless to improve the smoothness of the walls or the uniformity of the free stream (Betchov, 1961).

56. Axisymmetric Flows

(a) Flows in Circular Tubes

In this section we consider another class of parallel flow which has a simple geometry in cylindrical-polar coordinates. The flow is incompressible, viscous, and three-dimensional without body forces. The mean flow proceeds along the x-axis and U is a function of the radius ρ only. The fluctuations are functions of x, ρ, θ, and t. Because the fluctuations must be periodic in θ, which is the only possibility by the boundary conditions, we can consider one Fourier component at a time and take every fluctuation as complex and proportional to $\exp[i\alpha(x - ct) + in\theta]$ in which n is an integer. Thus for $n = 0$ we have a rotationally symmetric flow; with $n = 1$, the perturbations are constant along a simple helix; with $n = 2$, the properties are constant along two intertwined helices, and so on. With $\mathbf{u}(\rho)$ for the complex amplitude of the fluctuation parallel to the x-axis, $\mathbf{v}(\rho)$ for the radial velocity, and $\mathbf{w}(\rho)$ for the azimuthal velocity, the linearized equations of Navier and Stokes read as follows:

$$i\alpha\mathbf{u} + \mathbf{v}' + \frac{1}{\rho}\mathbf{v} + \frac{in}{\rho}\mathbf{w} = 0, \tag{56.1}$$

$$i\alpha(U - c)\mathbf{u} + U'\mathbf{v} + i\alpha\mathbf{p} = v\left[\mathbf{u}'' + \frac{1}{\rho}\mathbf{u}' - \left(\alpha^2 + \frac{n^2}{\rho^2}\right)\mathbf{u}\right], \tag{56.2}$$

$$i\alpha(U - c)\mathbf{v} + \mathbf{p}' = v\left[\mathbf{v}'' + \frac{1}{\rho}\mathbf{v}' - \left(\alpha^2 + \frac{n^2 + 1}{\rho^2}\right)\mathbf{v} - i\frac{2n}{\rho^2}\mathbf{w}\right], \tag{56.3}$$

$$i\alpha(U - c)\mathbf{w} + \frac{in}{\rho}\mathbf{p} = v\left[\mathbf{w}'' + \frac{1}{\rho}\mathbf{w}' - \left(\alpha^2 + \frac{n^2 + 1}{\rho^2}\right)\mathbf{w} + i\frac{2n}{\rho^2}\mathbf{v}\right]. \tag{56.4}$$

The prime indicates the differential operator $d/d\rho$.

If $n = 0$ the function **w** is completely independent of the others. Within the context of the stability theory there is therefore no production of **w**, and we can directly assume that it vanishes under the effects of friction.

Going further for the case $n = 0$, it becomes possible to use a stream function so defined that

$$\mathbf{u} = \frac{1}{\rho} \psi', \qquad \mathbf{v} = -\frac{i\alpha}{\rho} \psi. \tag{56.5}$$

Then, in terms of **v**, the vorticity equation reads

$$i\alpha(U - c)\left(\mathbf{v}'' + \frac{1}{\rho}\mathbf{v}' - \frac{1}{\rho^2}\mathbf{v} - \alpha^2\mathbf{v}\right) + i\alpha\rho\left(\frac{U'}{\rho}\right)' \mathbf{v}$$

$$= \nu\left(\mathbf{v}'''' + \frac{2}{\rho}\mathbf{v}''' - \frac{3}{\rho^2}\mathbf{v}'' + \frac{3}{\rho^3}\mathbf{v}' - \frac{3}{\rho^4}\mathbf{v} - 2\alpha^2\mathbf{v}'' - \frac{\alpha^2}{\rho}\mathbf{v}' + \alpha^4\mathbf{v}\right).$$

$$\tag{56.6}$$

Note the presence of the term in $(d/d\rho)(U'/\rho)$. It represents the production of vorticity in the cylindrical geometry and its presence can easily be justified. Consider an annulus of fluid of radius ρ and a rectangular cross section $\Delta\rho \, \Delta x$. The vorticity r is tangential to the annulus, and therefore the flux through the cross section is $r\rho \, \Delta\rho \, \Delta x$. The volume of fluid contained in the annulus is $2\pi\rho \, \Delta\rho \, \Delta x$. Let us now imagine that the annulus changes from radius ρ_1 to radius ρ_2. The condition of incompressibility gives the relation $\rho_1 \, \Delta\rho_1 = \rho_2 \, \Delta\rho_2$. It is also well known that in the absence of external or friction forces the flux of the vorticity remains constant. Therefore we must have $r_1 \, \Delta\rho_1 = r_2 \, \Delta\rho_2$, and, in view of the previous relation, this implies that $r_1/\rho_1 = r_2/\rho_2$. Thus, as a fluid particle changes its distance ρ from the axis as part of an axisymmetric motion, it is not its vorticity r that remains constant but the ratio r/ρ. The production of vorticity fluctuations therefore must be proportional to $(d/d\rho)(U'/\rho)$ in place of the usual term U'', as found in two-dimensional flows in Cartesian coordinates.

The well-known Poiseuille flow in a single circular pipe has $U = U_0(1 - \rho^2)$, if we take the radius as unity. As a consequence, the vorticity production term vanishes identically. It is not surprising to learn that, as far as is known, this flow is stable for axisymmetric disturbances ($n = 0$). Many workers have attempted this particular problem: Sexl (1927a, 1927b),

Pretsch (1941), Perkis (1948), and, more recently, Corcos and Sellars (1959). In each case the results were negative. It is important to remember, however, that all of this work is limited to linear analysis.

We have encountered another flow that is stable in the same sense as pipe flow—namely, plane Couette flow. In particular, in the analysis of this flow we noted that it was precisely because there was no production of vorticity that the flow was stable. Indeed, the analogy between the two cases has already been given by Pretsch (1941). When, however, there are other possible modes present (where n takes on higher integral values) for pipe flow, the investigation has not been made—even though this is one of the oldest problems. This is an interesting case, for when n is different from zero the production term no longer disappears from the governing equation. This can be verified easily and could mean that the answer to the stability of this flow is to be found here.

For flow between two concentric circular tubes parallel to the x-axis (Hagen-Poiseuille flow) the profile is $U = U_0(1 + K \log \rho - \rho^2)$. For this case the term $(d/d\rho)(U'/\rho)$ does not vanish when $n = 0$, and thus it is possible to have vorticity production. To date, however, we have not been able to obtain results for this profile, even for axisymmetric disturbances.

Experimentally, the theoretical findings of stability to axisymmetric disturbances have been confirmed for Poiseuille flow. Specifically, the work of Leite (1956, 1959), and Kuethe and Raman (1959) should be cited. A later experiment has been reported by Lessen et al. (1964), in which disturbances corresponding to the mode $n = 1$ have been explored. According to their results, this mode is unstable. Thus there is support for the points in the theory mentioned above. In any case, this is but a first step needed in an extensive investigation of this aspect of the problem.

One last treatment of this problem should be mentioned. Gill (1965) considered the stability from the point of view of disturbances growing in the downstream x-direction (cf. Section 53). The results here are again for axisymmetric disturbances and they are stable, but a new avenue has been opened in the general field. For example, in the inviscid limit there are no solutions to the problem if we look only for temporal instability. On the other hand, when allowance is made for spatial growth, a solution can be found (Bessel functions of order one). It indicates that the eigenvalues all present damping and that the disturbance decays in the downstream direction. Gill further checks his results with the experimental work of Leite and the agreement is good.

(b) ROUND JET

It is reasonable to enquire if there are analogues in this geometry to the theorems on stability that have been used in Cartesian geometries, such as those due to Rayleigh (cf. Section 20). This proves to be true, and Rayleigh himself showed that a necessary (but not sufficient) condition for instability for axisymmetric disturbances in the inviscid limit is the vanishing of $d/d\rho(U'/\rho)$ somewhere within the region of the flow. Clearly, this corresponds to the condition of the inflection point in the velocity profile of two-dimensional flows. Also, if a neutral disturbance exists (also for $n = 0$), then $(U - c)$ must vanish somewhere in the flow. This is tantamount to the result in the other system which states that the phase velocity lies between the limits of that of the mean flow. Finally, the condition for the jump of the perturbation across the critical layer in the limit of zero viscosity has been given by Tollmien (1935) and follows the general results given for the two-dimensional system. The applicability of these theorems is best found in problems in which flows away from walls are considered, such as the round jet or wake. This fact has also already been revealed in the discussion of two-dimensional cases and reference is made to Sections 5, 6, 15, and 21 for further elaboration.

Essentially all of the features to be found in an axisymmetric stability problem in which there are no walls can be illustrated by considering a round jet. Batchelor and Gill (1962) have made a thorough analysis of this system and their work is outlined here. Indeed, the example carries all of the generalities of any axisymmetric flow of this kind, subject to inviscid stability considerations.

The mean profile for the axisymmetric jet is given by the stream function (cf. Landau and Lifschitz, 1959)

$$\Psi = \frac{2\nu\rho \sin \theta}{\text{sech } \theta_0 - \cos \theta}, \tag{56.7}$$

where $\tan \theta = \rho/x$ and the coordinates are as already defined. Batchelor and Gill utilize (56.7), together with the approximation corresponding to far downstream conditions, and express the mean flow by

$$U = \frac{U_0}{[1 + (\rho/\rho_0)^2]}. \tag{56.8}$$

Perturbations are introduced and follow the governing equations (56.1) to (56.4) when $\nu = 0$. Boundary conditions require all perturbations to vanish at infinity. Batchelor and Gill then go on to show that the other condition at $\rho = 0$ must be such that multiple values of the pressure and velocity can be avoided. Thus **u** and **p** must vanish if $n = 0$ and **v** and **w** must vanish if $n = 1$. In the latter case, of course, the combination $\mathbf{v} + i\mathbf{w}$ must be equal to zero at the origin by the flow geometry.

Considering the inviscid set of equations, it is possible to eliminate **u**, **w**, and **p** and to obtain the Rayleigh equation valid for any n, namely,

$$(U - c)\frac{d}{d\rho}\left[\frac{\rho(d/d\rho)(\rho\mathbf{v})}{n^2 + \alpha^2\rho^2}\right] - (U - c)\mathbf{v} - \rho\mathbf{v}\frac{d}{d\rho}\left(\frac{\rho U'}{n^2 + \alpha^2\rho^2}\right) = 0. \quad (56.9)$$

This question is the counterpart to (56.6) when $v = 0$.

Now, several general results follow from an inspection of (56.9). The generalization of the inflection-point theorem, that is, when $n \neq 0$, is expressed by the requirement that

$$\frac{d}{d\rho}\left(\frac{\rho U'}{n^2 + \alpha^2\rho^2}\right) = 0, \quad (56.10)$$

must occur somewhere. Batchelor and Gill use this result to show that when n is large complete stability occurs, regardless of the jet velocity profile.

Detailed computations were made by Batchelor and Gill for the stability of a jet with a tophat profile. In order to do this, of course, it was necessary to match the solutions at the edge of the jet in addition to meeting the boundary conditions. This is easy enough to do, for the solutions to (56.9) in this case are Bessel functions on either side of the profile edge. The behavior of **v** for small ρ is as ρ^{n-1}. For this profile only the solution for $n = 1$ has an inviscid instability. This conclusion becomes even stronger, for Batchelor and Gill find that the $n = 1$ is the only solution leading to instability in the inviscid limit. It is further shown that the results apply equally well when the jet profile (56.8) is considered.

The periodicity of the fluctuations in x and θ allows a special transformation (proposed by Batchelor and Gill) which plays a role similar to the Squire transformation. At an arbitrary point we consider the helix defined by $\rho = $ constant and $\alpha x + n\theta = $ constant. We now take three orthogonal axes directed along the radius, the tangent to the helix, and the direction per endicular to both the radius and the tangent. Anywhere in the fluid we can define the velocity fluctuations along these three local axes:

$$\bar{\alpha}\bar{\mathbf{u}} = \alpha\mathbf{u} + \frac{n\mathbf{w}}{\rho}, \qquad \perp \text{ to radius}, \quad \perp \text{ to helix}, \quad (56.11)$$

$$\bar{\mathbf{v}} = \mathbf{v}, \qquad \parallel \text{radius}, \tag{56.12}$$

$$\bar{\alpha}\mathbf{w} = \alpha\mathbf{w} - \frac{n\mathbf{u}}{\rho}, \qquad \parallel \text{tangent}, \tag{56.13}$$

$$\frac{\bar{\mathbf{p}}}{\bar{\alpha}} = \frac{\mathbf{p}}{\alpha}, \tag{56.14}$$

where the total wavenumber $\bar{\alpha}$ is defined as

$$\bar{\alpha} = \left(\alpha^2 + \frac{n^2}{\rho^2}\right)^{1/2}. \tag{56.15}$$

In terms of these new variables, the equations of continuity and conservation of momentum are

$$i\bar{\alpha}\bar{\mathbf{u}} + \bar{\mathbf{v}}' + \frac{\bar{\mathbf{v}}}{\rho} = 0, \tag{56.16}$$

$$i\bar{\alpha}(U - c)\bar{\mathbf{u}} + U'\bar{\mathbf{v}} + i\bar{\alpha}\bar{\mathbf{p}} = 0, \tag{56.17}$$

$$i\bar{\alpha}(U - c)\bar{\mathbf{v}} + \bar{\mathbf{p}}' = 0, \tag{56.18}$$

$$i\bar{\alpha}(U - c)\bar{\mathbf{w}} - \frac{n}{\bar{\alpha}\rho} U'\bar{\mathbf{v}} = 0. \tag{56.19}$$

The component $\bar{\mathbf{w}}$ appears only in the last equation and can therefore be integrated last. The similarity of the other variables with axisymmetric flows is then obvious.

(c) COMPRESSIBLE FLOWS

Gold (1963) and Lees and Gold (1964) extended the works of Batchelor and Gill (1962) for the jet and wake when the fluid is compressible. For the inviscid case it is found that the $n = 1$ perturbation is the most unstable but with possible instabilities for other integral values of n. Additional comments are made on the role of the temperature of the core of a wake flow. Apparently an increase in temperature leads to a destabilizing situation.

57. Flexible Boundaries

(a) INTRODUCTION

In other sections of this book the boundaries confining the flow have been taken as rigid or infinitely far away. Such boundaries promote mathematical simplicity, but important physical effects are introduced when a displacement of the wall accompanies the fluid oscillation. In practice, the flexibility of boundaries may be significant in aeronautical engineering (panel fluttering), naval engineering, and oceanography and in cardiovascular and even zoological studies. A review of this vast array of possibilities opens a Pandora's box of boundary conditions.

The wall could not only respond to pressure fluctuations but could also be affected by the fluctuating shear stresses. For example, in a seal skin a change in the inclination of the hairs would alter the location of the effective "surface." A wall could also be coated with a viscoelastic film or backed by a similar fluid.

To demonstrate the importance of wall effects in hydrodynamical stability we limit the scope of this section to a general category of walls.

Let us consider a boundary layer along a flexible wall located at rest at $y = 0$. Let $\tilde{a}(x, t)$ be the modification in the ordinate of the wall and $\tilde{p}(x, 0, t)$ be the pressure at the wall. Because we always assume that $\tilde{a}$ is small and that $\tilde{p}$ varies slowly with y, this approximation is sufficient. Moreover, for small displacements we have the approximation

$$\frac{\partial \tilde{a}}{\partial t} = \tilde{v}(x, 0, t). \tag{57.1}$$

We now assume that the wall responds to the pressure and not the shear stress. In addition, we neglect the normal component of the viscous stress. A simple wall is characterized by a spring constant K, an equivalent mass m per unit area, and a coefficient of resistance r such that the equation of motion of the wall can be expressed as

$$-\tilde{p} = K\tilde{a} + r \frac{\partial \tilde{a}}{\partial t} + m \frac{\partial^2 \tilde{a}}{\partial t^2}. \tag{57.2}$$

Note that the fluid has unit density so that the parameters K, r, and m are expressed in a special system of units. We could include a term $-T(\partial^2 \tilde{a}/\partial x^2)$ to represent the surface tension of the wall on the right-hand side of (57.2). Its effect would simply modify K.

Because all fluctuations are proportional to the real part of $e^{i\alpha(x-ct)}$, we can introduce the complex amplitude $\mathbf{a}$ and define a complex stiffness Z:

$$Z = -\frac{\mathbf{p}}{\mathbf{a}}. \tag{57.3}$$

In the simple wall described by (57.2) the function Z becomes

$$Z = K - i\alpha rc - m\alpha^2 c^2. \tag{57.4}$$

Surface tension would simply add a term $T\alpha^2$, which could be amalgamated with K.

The oscillations of the flow are governed by the Orr-Sommerfeld equation and the integration requires two passes, with a tentative value of c. The two solutions are combined linearly so that $\mathbf{u}(0) = 0$, which is the correct boundary condition as long as the wall is not free to move in the x-direction. We shall later recognize that boundary conditions should apply at $y = \tilde{a}$, instead of $y = 0$.

The solution now yields a value of $\mathbf{v}(0)$ and the values of $\mathbf{p}(0)$ must be determined. In principle, we could obtain $\mathbf{p}$ from the equation of momentum in the x-direction but local errors are likely to be annoying. In practice, it is preferable to obtain $\mathbf{p}$ by integration of the equation of momentum in the y-direction. This procedure tends to average out the small errors.

Knowing $\mathbf{p}$ we can ascertain the displacement $\mathbf{a}$ by considering the mechanical properties of the wall (57.3) and (57.4). The flow will accompany the wall if (57.1) is satisfied. In practice, it is convenient to express this relation in the following equivalent form:

$$-\frac{i}{\alpha c}\frac{\mathbf{v}(0)}{\mathbf{p}(0)} = \frac{1}{Z}, \tag{57.5}$$

where the left-hand side is a property of the oscillating flow and the right-hand side is a property of the wall. If these two quantities are not equal, other values of c must be fed to the computer until an eigenvalue is located. Note that Z varies also with c.

If the wall is allowed to move in the x-direction, another mechanical relation must be given to relate horizontal displacement to shear stresses. In general, a matrix of four complex transfer coefficients can be defined as relating the two displacements to the two components of the force. This leads to two boundary conditions and (57.5) is replaced by the vanishing of a determinant. This procedure has been discussed by Kaplan (1964) and Landahl and Kaplan (1965). Some agreement between theory and experiment is reported by Korotkin (1966).

(b) SIMPLIFIED ANALYSIS

Let us now explore the effects of a flexible wall in the context of the simplified analysis of boundary layer oscillations given in Section 7. In this analysis we used a relation between **p** and **r'** at the wall. Along a rigid wall both **u** and **v** vanish, and **p** is simply proportional to **r'**. With a flexible wall we must replace (7.4) with

$$U'\mathbf{v}(0) + i\alpha\mathbf{p}_1(0) = v\mathbf{r}_1'(0). \tag{57.6}$$

Because **v**(0) is related to **a**, which in turn is tied to **p** by the wall properties, we can substitute from (57.1) and (57.3) and obtain (57.6) in the form

$$i\alpha\mathbf{p}_1(0)\left[1 + \frac{cU'(0)}{Z}\right] = v\mathbf{r}_1'(0). \tag{57.7}$$

The simplified analysis can now proceed without significant change through Sections 8 and 9, carrying the new term along. The viscous component given by (9.16) contains a special factor and now reads as follows

$$\mathbf{u}_2(0) = -\frac{\mathbf{p}(0)}{c}(1 + \mathscr{F}(\eta))\left(1 + \frac{cU'}{Z}\right). \tag{57.8}$$

The final equation (11.2), in which **u** vanishes, appears in the form

$$\frac{2}{\alpha}\frac{hU'}{(1 - hU')}\left[1 + i\pi\left(\frac{-hU''(h)}{U'(0)}\right)\right] = (1 + \mathscr{F}(\eta))\left(1 + \frac{cU'}{Z}\right). \tag{57.9}$$

We are now ready to discuss separately the role of the amplitude and phase of Z. First we note from (57.4) that the real part of Z is related to reversible effects such as inertia or a restoring force. If the term K is dominant, Z is real positive and the function $1 + \mathscr{F}$ is multiplied by a constant factor larger than unity. In an examination of Fig. 11.1 we see that the first contact between the left-hand side and the right-hand side of (57.9) now occurs for a lower value of α compared with that in the rigid wall. In effect, the right-hand side is moved away from the origin. For a given h the change in α is matched by a change in v. Thus the Reynolds number increases all along the neutral curve. This means that the flow is stabilized.

If the inertia is sufficiently large, Z is real negative and the boundary layer is destabilized by the reverse effect.

If the wall resistance r is zero, Z vanishes when c is equal to the velocity c_1 characteristic of the propagation of waves in the wall. From (57.4) we have

$$c_1 = \left(\frac{K}{m\alpha^2}\right)^{1/2}. \tag{57.10}$$

This very special situation corresponds to a marked departure from the ordinary behavior.

If K and m are negligible, the term in r makes Z imaginary negative. Thus the right-hand side of (57.9) is rotated in the counterclockwise direction, around the origin. The first contact occurs for a higher wave-number and for a larger value of v, for η remains essentially unchanged in Fig. 11.1. Thus the resistance of the wall is destabilizing and the effect is more pronounced than that of m. Indeed, it has the same effect as a rotation of the function $1 + \mathscr{F}$ in place of a mere contraction toward the origin.

There is little doubt that r will be positive for all ordinary walls. The work performed by the oscillating fluid in massaging the wall is given by

$$- \overline{\tilde{p}\tilde{v}} = - \frac{\mathbf{p}\mathbf{v}^* + \mathbf{p}^*\mathbf{v}}{4} = \frac{\alpha c}{4}\,\mathbf{p}\mathbf{p}^*\!\left(\frac{i}{Z^* - Z}\right). \qquad (57.11)$$

If r is positive, energy is transferred to the wall. Any negative value of r would imply that the wall contains a source of energy such as muscles or pistons (Betchov, 1960b). We are therefore faced with a form of resistive instability and clarification of this point would be helpful.

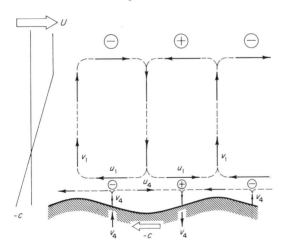

FIG. 57.1. The Reynolds stresses created by a resistive wall, leading to instability.

Consider the oscillating flow of Fig. 57.1, as seen by the observer moving with the phase velocity c. The wall moves to the left, and we assume that it has a purely resistive response. The main components of the oscillation are indicated as $\tilde{v}_1$ and $\tilde{u}_1$ and shown as long arrows. The pressure fluctuations are generated in the free stream and reach the wall through the layer. An excess of pressure causes a downward velocity of the wall, indicated as $\tilde{v}_4$. Since the wall moves to the left, its lowest position is to the left of each high pressure region. In the fluid this displacement of the wall introduces a new component to the flow, indicated as $\tilde{v}_4$ and $\tilde{u}_4$. Note that $\overline{\tilde{u}_4 \tilde{v}_4}$ is zero but that the Reynolds stresses produced by the interaction of $\tilde{u}_4$ and $\tilde{v}_4$ with the inviscid solution $\tilde{u}_1, \tilde{v}_1$ do not vanish. In fact $\overline{\tilde{u}_1 \tilde{v}_4}$ and $\overline{\tilde{u}_4 \tilde{v}_1}$ are negative all along the lower portion of the boundary layer. Thus the motion of the wall produces a special contribution to the Reynolds stresses, which increases the production term in the energy equation [see (2.13) and (2.14)]. This is the destabilizing mechanism.

An animal such as the dolphin swims with a Reynolds number of the order of 10^3 and could benefit from stabilization of the boundary layer. Although the term in K could be of some advantage, the term in r would be catastrophic. A negative resistance might be helpful, but an examination of the skin of the animal fails to reveal any adequate muscular structure, thus leaving no evidence of a natural use of the idea.

(c) PANEL FLUTTERING

After discussing the modifications introduced in boundary layer oscillations by the response of the wall, it is appropriate to look through the other end of the telescope and to ask how a flow affects the vibrations of a wall? This will show that a certain special kind of wave can be expected and that a general analysis is necessary. The bases can be found in the papers of Landahl (1962) and Benjamin (1963).

Let us consider a situation in which a parallel flow of constant velocity U_0 proceeds along an elastic wall of stiffness Z [see definition (57.3)] with a boundary layer so thin that it can be disregarded entirely. In fact, we consider that the displacement $\tilde{a}$ of the wall is equal to that of the outer edge of the boundary layer and leads to the relation

$$i\alpha(U_0 - c)\mathbf{a} = \mathbf{v}(1). \tag{57.12}$$

Note that this result differs by the presence of the term U_0 from the preceding one, in which $\mathbf{a}$ was defined at the base of the layer. If $U_0 > c$, a difference is introduced in the sign of the displacement. Thus (57.12) is valid if the displacement of the wall is matched by that of the entire boundary layer. The oscillations of a free stream of large Reynolds number can be approximated by a simple pressure wave, and we have

$$\mathbf{p} = i\mathbf{v}(U_0 - c). \tag{57.13}$$

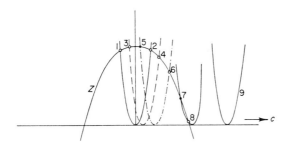

FIG. 57.2. Characteristic functions for panel fluttering.

Elimination between (57.13) and (57.4) produces the condition

$$\alpha(U_0 - c)^2 = K - i\alpha c r - m\alpha^2 c^2. \tag{57.14}$$

This relation can be solved for c to give the "dispersion equation." It is more informative to consider the case $r = 0$ first and to plot the left-hand and right-hand sides of (57.14) as functions of c, as shown in Fig. 57.2. For a given panel and a given α the right-hand side corresponds to a parabola. The left-hand side is a family of parabolas. Let us begin with the frictionless wall in contact with the immobile fluid. With $U_0 = 0$, we find two waves of equal and opposite phase velocities. See points 1 and 2, Fig. 57.2.

If we restore the density ρ of the fluid, we find $c = \pm c_2$, where

$$c_2 = \left(\frac{K/\alpha\rho}{1 + m\alpha^2/\rho}\right)^{1/2}. \tag{57.15}$$

In this expression m is the mass of the wall per unit area. This relation gives one of the parabolas which is the response of the wall loaded with immobile fluid.

Let us now set the fluid in motion. As U_0 increases, the parabola moves and the wave to the right becomes faster than c_1, whereas the wave to the left slows down (see point 3). If $U_0^2 > (K/\alpha)^{1/2}$, both waves proceed to the right and we can say that one wave has been forced back by the wind (see points 5 and 6). As shown in Fig. 57.2, some waves are still possible if $U_0 > c_1$, provided that the two parabolas still intersect (points 7 and 8). A brief calculation shows that the double root occurs at $U_0 = c_1\sqrt{1 + \alpha m}$. For larger values of U_0, the two roots of (57.14) are complex conjugates. Thus the wave proceeds with $c_r = U_0/(1 + \alpha m)$ and one of the two solutions grows with $c_i = [\alpha m U_0^2 - (1 + \alpha m)K]^{1/2}/(1 + \alpha m)$. This situation is analogous to the case studied in Section 5 and is known as the Helmholtz instability. Essentially, the wall plays the role of the lower fluid layer. Note that the basic equations do not contain dissipative terms, either in the fluid or in the wall. Each unstable solution has a decaying counterpart so that the motion can be reversed and still conform to the equations. Note also that the instability would not occur if the term in m did not appear. Thus the inertia of the wall is essential. In fact, it introduces a necessary delay. Such oscillations between two fluid layers or between one fluid and one wall are called "dynamically unstable." They can also be qualified as "reversible."

The general effect of resistance in the wall can be examined most easily by solving (57.14) for r and forming the partial derivative $\partial r/\partial c$, in which all other parameters remain constants. We find

$$\frac{\partial r}{\partial c} = i\left(1 + \frac{K}{\alpha c^2} + \alpha m - \frac{U_0^2}{c^2}\right). \tag{57.16}$$

Thus, if a small amount of resistance is introduced, c will differ slightly from the real values indicated by the intersection of two parabolas in Fig. 57.2.

If $U_0 = 0$, we find that both waves are damped by Δr. As U_0 becomes greater than $c_1\sqrt{1 + \alpha m}$, we also find that the wave which has been turned back (points 5 or 7) is destabilized by the resistance of the wall.

The solutions which have this property of resistive instability are indicated in Fig. 57.2 by full circles, whereas those that are resistively stable are indicated by empty circles.

A rapid way of assessing the stability of the wave is the following. Let $Z(c)$ be the stiffness of the wall, or the right-hand side of (57.14). Let $F(c)$ be the "stiffness" of the flow, defined as the left-hand side of (57.14). It is regarded as an arbitrary function of c. Then (57.14) with $r \to 0$ can be written

$$\Delta r = \frac{Z' - F'}{c_r} \Delta c_i, \tag{57.17}$$

where $Z' = dZ/dc$ and $F' = dF/dc$.

Thus, if c_r is positive, the resistive instability will occur if $F' < Z'$. Considering an intersection of the two functions F and Z, we can say that as c increases the function F will pass below the function Z. If c is negative, the resistive instability will occur if F passes above Z.

If the parabolas do not intersect, the effect of r is not so simple and has lesser control of the stability of the wave.

(d) Viscous Shear Flow Oscillations

Although the panel fluttering approximation has some merits, it does not account for the behavior of the boundary layer which separates the free stream from the wall. To supplement the results of paragraph (a) of this section, we now discuss the effects of the wall in greater detail, much along the lines developed by Benjamin (1960, 1963) and Landahl (1962).

Let us consider a neutral solution of the Orr-Sommerfeld equation along a rigid wall. The functions u and v vanish at the wall for the particular eigenvalue $c = c_e$. If c differs slightly from c_e and we modify v and α in such a way that $u(0)$ remains zero, there will be some changes in $v(0)$. From the general aspect of the inviscid approximation discussed in Section 8 and Fig. 8.2, we can expect a relation in the form of

$$v(0) = -A(c - c_e) \, v(1), \qquad (57.18)$$

where A is a positive number. The inviscid treatment indicates that A is real, but the viscous effects might add a relatively small imaginary part. Because $v(1)$ is related to $p(1)$ and $p(y)$ is almost constant through the layer, we have, approximately,

$$v(0) = iA \, \frac{c - c_e}{1 - c} \, p(0). \qquad (57.19)$$

The displacement of the wall is related to $v(0)$ by (57.1), and the "dispersion relation" takes the form

$$F(c) = \alpha \, \frac{1 - c}{A} \, \frac{c}{c - c_e} = Z(c). \qquad (57.20)$$

If r is zero, we can plot F and Z as functions of a real c. This means that the neutral oscillations will be found at the intersections of the hyperbola F with the parabola Z. The principal cases are illustrated in Fig. 57.3, in which it is assumed that F is a valid approximation over a certain range of values of c.

If K is large and m is small, we will have, essentially, the ordinary boundary layer oscillation indicated as A_1 in Fig. 57.3. A discussion of the role of small wall resistance leads to the same conclusions as found in panel fluttering. Resistive instability occurs if $F' < Z'$, assuming that c_e is positive.

If m is large enough, another wave, which is resistively stable, will appear, as shown by B_1, Fig. 57.3.

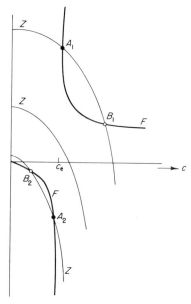

FIG. 57.3. Stability of a boundary layer along a deformable wall.

If the stiffness K is decreased, the two roots A_1 and B_1 will merge and the intersection will eventually disappear.

Because (57.20) is cubic, the discussion is slightly more complicated, but we can establish the existence of two complex conjugate roots. Thus once again we find that dynamical instability is possible.

If we further reduce K or increase m, the functions F and Z may intersect again. This gives the resistively unstable solution A_2 and its counterpart B_2.

With the help of a computer or by using such approximations as the Tietjens functions it is possible to specify the imaginary part of A, which is usually small.

If the imaginary part of A is positive and $c > c_e$, the effect will be similar to a negative wall resistance. If $c < c_e$, the effect will be similar to a positive wall resistance.

Thus the imaginary part of A can be stabilizing or destabilizing. We can therefore find a situation in which the flow along a dissipative wall is stable and in which the instability begins only when the resistance of the wall exceeds a certain level.

A classification has been proposed by Benjamin which can be summarized as follows: Class A—waves that are neutral if the wall has no resistance and the fluid is inviscid, yet are destabilized if r appears; Class B —similar waves that are stabilized if r appears; Class C—waves that are dynamically unstable and therefore correspond to cases in which the equation $F = Z$ has no real roots, even with $r = 0$.

(e) Inviscid Shear Flow Oscillations

When the viscosity is negligible, the imaginary part of A can be specified without major difficulties. Indeed, the effect of the critical layer can be approximated and the effects of the viscous layer at the wall neglected as far as pressure and normal velocity are concerned. This situation has been studied by Miles (1957) in relation to the formation of ocean waves by wind. In this problem the flow of air is turbulent and the molecular viscosity must be replaced by the eddy viscosity giving a special velocity profile together with appropriate boundary conditions.

Let us consider the ordinary kind of laminar boundary layer. We start from a phase velocity c_e such that the real part of $\mathbf{v}$, the inviscid solution, vanishes at the wall, as in Fig. 8.2, which corresponds to a rigid wall. In the inviscid limit we have seen that a single row of vortices is concentrated at the critical layer and that below the critical layer it induces the velocity fluctuations $\mathbf{u}_3 = \pi\, U''(h)\, \mathbf{v}_1(h)/\alpha U'(h)$. By virtue of the equation of continuity this requires the presence at the wall of the following normal velocity fluctuation:

$$\mathbf{v}_3 = i\pi \frac{h\, U''(h)}{U'(h)} \mathbf{v}_1(h). \tag{57.21}$$

In viscous flows along rigid walls this term is compensated by viscous terms.

In the present situation we must consider both the inviscid wall component and the effect of the critical layer. The combined result is

$$\mathbf{v}(0) = -A(c - c_e)\, \mathbf{v}_1(1) + \frac{i\pi h\, U''(h)}{U'(h)} \mathbf{v}_1(h). \tag{57.22}$$

For $\mathbf{v}_1(h)$ we use a linear approximation

$$\mathbf{v}_1(h) = \mathbf{v}_1'(0)h + \mathbf{v}(0). \tag{57.23}$$

For $\mathbf{v}_1'(0)$ we use the continuity equation and further define the ratio g such that

$$\mathbf{u}_1(0) = \frac{g}{\alpha} \frac{\mathbf{p}(0)}{1 - c}. \tag{57.24}$$

We can express $\mathbf{v}(0)$ in terms of $\mathbf{v}(1)$ and $\mathbf{u}(1)$. These quantities are related to $\mathbf{p}(1)$, which we regard as almost identical to $\mathbf{p}(0)$. Thus we obtain the relation

$$\mathbf{v}(0) = \frac{\left[iA \dfrac{c - c_e}{1 - c} + \pi \dfrac{h\, U''(h)}{U'(h)} \dfrac{g}{1 - c} \right] \mathbf{p}(0)}{1 - (i\pi h\, U''(h))/U'(h)} \tag{57.25}$$

The "dispersion relation" can now be established by referring to (57.1) and (57.3) and we find

$$\frac{\alpha c(1 - c)}{A(c - c_e) - \dfrac{i\pi h\, U''(h)\, g(\alpha, c)}{U'(h)}} \left(1 - i\pi \frac{h\, U''(h)}{U'(h)} \right) = Z. \tag{57.26}$$

Because the term containing U'' in the denominator of the left-hand side is more influential than the term in U'' appearing in the numerator, it follows from (57.26) that the critical layer has a stabilizing influence. This occurs whether the real part of A is negative or positive.

Let us now remark on viscous flow along a rigid wall. We can perhaps retain the idea that the viscous critical layer has the effect of a very thin row of vortices and introduce the idea that the effects of viscosity along the wall can be compared with those of a flexible wall. Thus we regard the layer of viscous fluid near the wall as equivalent to a flexible wall. Because the term in $U''(h)$ is stabilizing, it becomes quite clear that the resistance of this layer has a destabilizing influence and that it must exceed a certain level to permit instability.

This reasoning suggests that if the viscosity immediately near the wall is reduced while the viscosity of the rest of the layer is kept constant, the flow will be stabilized. This remark has special significance in the study of turbulent flows in which the eddy viscosity falls in the vicinity of the wall.

Further complications may arise if the eddy viscosity near the wall has the nature of a tensor rather than that of a scalar. Such an event could be possible if the big eddies were anisotropic.

(f) BOUNDARY CONDITIONS AT A WAVY WALL

In the preceding sections, we have applied the boundary conditions at $y = 0$ instead of $y = \tilde{a}$. This corresponds to the mean surface approximation. In the linearized approach, this is acceptable except for an interaction with the mean flow. At an altitude $\tilde{a}$ above a rigid wall, the mean flow has the velocity $U'\tilde{a}$. If the wall rises to this altitude, the condition of no flow along the wall must be

$$\tilde{u} + U'\tilde{a} = 0, \tag{57.27}$$

as pointed out by Landahl (1962) and Benjamin (1963).

It follows that (7.4) includes a term $-i\alpha c\mathbf{u}$ in addition to the term $U'\mathbf{v}$ already mentioned in (57.6). These two new terms cancel each other so that the balance of horizontal forces is exactly as along a rigid wall. However, the term $U'\tilde{a}$ must appear in the left-hand side of (7.1) so that, after some substitutions, (57.9) is replaced by

$$\frac{2}{\alpha}\frac{hU'}{(1 - hU')}\left\{1 + i\pi\left[\frac{-hU''(-h)}{U'(0)}\right]\right\} = 1 + \mathscr{F}(\eta) + \frac{cU'}{Z}. \qquad (57.28)$$

The difference between (57.9) and (57.28) is only quantitative for most cases. If the wall is purely resistive, it becomes clear that the effects of viscosity along the wall, which produce the term $\mathscr{F}$, are analogous to those of a resistive wall (Landahl, 1962).

The comparison between (57.9) and (57.28) also shows that the boundary conditions can be applied at $y = 0$ without major error. For accurate results though, one should use (57.27), together with a relation between the stiffness of the wall and that of the flow. Thus the interaction between the flow and the wall becomes more intricate. Numerical computations could include the possibility of a tangential displacement of the wall in response to pressure fluctuations, as well as the effects of viscous shear stresses. According to Benjamin (1963) these are small and of the order $(\alpha\mathscr{R})^{-1/2}$.

58. The Effect of Centrifugal Forces

In this section we make a brief excursion outside the field of parallel flows. Many flows, for example along slightly curved walls or between large rotating cylinders, are almost but not quite parallel and, as a consequence, there is the question of any new effect in the problem.

The fact that centrifugal forces have a destabilizing effect was pointed out by Rayleigh (1917). The flow between rotating cylinders, coaxial but not necessarily rotating at the same rate, can be unstable, as discovered by Taylor (1923), and a great amount of work has been done on the problem, reported in Lin (1955) and in Chandrasekhar's treatise (1961). The flow along a single wall with a free stream was first studied by Görtler (1940a, 1940b), who reported similar instabilities caused by centrifugal forces. The mathematical difficulties are considerable. This problem was reconsidered by Smith (1955), who used the Galerkin method and a card-programmed calculator. There is little doubt that the problem could be treated with even greater accuracy by using a high-speed computer. Several restrictions and simplifications could be omitted.

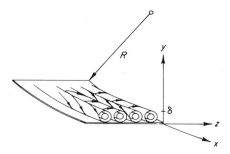

Fig. 58.1. Instability along a curved wall in a boundary layer of thickness δ and curvature R. (From Görtler, 1940a, 1940b.)

(a) The Basic Mechanism of Instability in a Rotating Flow

Let us consider a two-dimensional flow proceeding in the direction of the curvilinear x-coordinate along a concave wall, as shown in Fig. 58.1.

The constant radius of curvature R is larger than the boundary layer thickness δ. The main flow $U(y)$ takes the constant value U_0 in the curved free stream.

In the boundary layer an increase Δy corresponds to an increase ΔU of the mean velocity with approximately

$$\frac{\Delta U}{\Delta y} \approx \frac{U_0}{\delta}. \qquad (58.1)$$

The mean pressure is not constant. Because the centrifugal forces tend to press the fluid against the wall, a pressure gradient appears, which balances the centrifugal forces. Thus we must have

$$\frac{\Delta P}{\Delta y} = -\frac{U^2}{R}. \qquad (58.2)$$

Let us now imagine that a fluid particle is displaced from altitude y to altitude $y + \Delta y$ by some disturbance perhaps like one of the eddies sketched in Fig. 58.1. Let us assume that the energy of the particle stays constant. Neglecting a quadratic term in v, we have

$$\Delta P + U \, \Delta u = 0. \qquad (58.3)$$

Referring to (58.2) and equating the pressure changes, we find that the velocity of the fluid particle will be below the local mean velocity. Indeed, we have

$$\Delta u - \Delta U = \left(\frac{U}{R} - \frac{U_0}{\delta}\right) \Delta y, \qquad (58.4)$$

where R is perhaps 10 times larger than δ.

As the velocity of the particle drops, so does the centrifugal force, and the pressure gradient accelerates the particle to larger values of y. Thus along a concave wall, we have an instability of a peculiar kind. Along convex walls the effect is reversed, and the particle is returned to its original position.

A general discussion of the vorticity equations in a rotating flow is due to Görtler (1940a, 1940b). It is found that the stability of the z-component of the vorticity is controlled by the function $U'' + U'/R$. A dynamical instability occurs if this quantity vanishes somewhere between the walls or inside the layer. This instability does not occur unless R is rather short or unless the profile is already close to possessing a regular inflection point.

Other instabilities which involve both the y- and x-components of the vorticity are possible, and we now present a simplified theory of the case treated first by Görtler. He considered a flow homogeneous in x and growing in time. Later Smith (1955) introduced a growth in x and faced greater difficulties.

(b) THE GENERAL EQUATIONS

The exact form of the Navier-Stokes equations for an incompressible flow along a curved wall has been described in detail by Smith (1955).

In the approximation $R \gg \delta$ a certain number of terms drop out and the equations read as follows:

$$\frac{\partial v}{\partial y} - \left[\frac{v}{R}\right] + \frac{\partial w}{\partial z} = 0, \tag{58.5}$$

$$\frac{\partial u}{\partial t} + v\frac{\partial u}{\partial y} - \frac{vu}{R} + w\frac{\partial u}{\partial z} = \nu\left(\frac{\partial^2 u}{\partial y^2} + \frac{\partial^2 u}{\partial z^2} - \left[\frac{1}{R}\frac{\partial u}{\partial y}\right]\right), \tag{58.6}$$

$$\frac{\partial v}{\partial t} + v\frac{\partial v}{\partial y} + \frac{u^2}{R} + w\frac{\partial v}{\partial z} + \frac{\partial p}{\partial y} = \nu\left(\frac{\partial^2 v}{\partial y^2} + \frac{\partial^2 v}{\partial z^2} - \left[\frac{1}{R}\frac{\partial v}{\partial y}\right]\right), \tag{58.7}$$

$$\frac{\partial w}{dt} + v\frac{\partial w}{\partial y} + w\frac{\partial w}{\partial z} + \frac{\partial p}{\partial z} = \nu\left(\frac{\partial^2 w}{\partial y^2} + \frac{\partial^2 w}{\partial z^2} - \left[\frac{1}{R}\frac{\partial w}{\partial y}\right]\right), \tag{58.8}$$

where u, v, and w are the classical Eulerian velocities, along the curvilinear axes, as shown in Fig. 58.1.

The term $-vu/R$ of (58.6) corresponds to the Coriolis force, whereas the term u^2/R of (58.7) stems from the centrifugal force. The other terms in $1/R$, contained in brackets, are introduced by geometrical effects, such as that appearing in the continuity equation (58.5). The fluctuations can now be introduced, and to study the stability of longitudinal vortices we assume the simple form suggested by Görtler (1940a, 1940b):

$$\tilde{u} = \mathbf{u}(y) \cos (\alpha z)\, e^{\beta t}, \tag{58.9}$$

$$\tilde{v} = \mathbf{v}(y) \cos (\alpha z)\, e^{\beta t}, \tag{58.10}$$

$$\tilde{w} = \mathbf{w}(y) \sin (\alpha z)\, e^{\beta t}, \tag{58.11}$$

$$\tilde{p} = \mathbf{p}(y) \cos (\alpha z)\, e^{\beta t}, \tag{58.12}$$

where the boldface symbols are real functions of y.

The equations for the perturbations can now be obtained and linearized for small perturbations. At this stage the terms in $1/R$ corresponding to geometric effects are dropped from the equations. This decision is necessary to avoid complications, and some justification has been given by Smith. The equations then reduce to

$$\mathbf{v}\!\left(U' - \frac{U}{R}\right) = \nu\!\left[\mathbf{u}'' - \left(\alpha^2 + \frac{\beta}{\nu}\right)\mathbf{u}\right], \tag{58.13}$$

$$2\frac{U}{R}\mathbf{u} + \mathbf{p}' = \nu\!\left[\mathbf{v}'' - \left(\alpha^2 + \frac{\beta}{\nu}\right)\mathbf{v}\right], \tag{58.14}$$

$$-\alpha\mathbf{p} = \nu\!\left[\mathbf{w}'' - \left(\alpha^2 + \frac{\beta}{\nu}\right)\mathbf{w}\right], \tag{58.15}$$

$$\alpha\mathbf{w} + \mathbf{v}' = 0. \tag{58.16}$$

Because $R > \delta$, it appears that the term in U/R in (58.13) can be disregarded. It comes from the Coriolis forces. In the free stream the inviscid solution is identically zero (we take β different from zero). Thus the solution does not affect the free stream, except for viscous terms that decay rapidly even when ν is small. This means that our result can easily be extended to flows confined between two walls.

In the boundary layer we see from (58.13) that $\mathbf{u}$ cannot vanish. Otherwise it causes the disappearance of $\mathbf{v}$, followed by that of $\mathbf{w}$ and $\mathbf{p}$ from (58.16) and (58.15). Because the boundary conditions require that both $\mathbf{u}$ and $\mathbf{v}$ vanish at the wall and away from the layer, we make a drastic assumption to cut through the mathematical difficulties. We write

$$\mathbf{v} = a \sin \left(\frac{\pi y}{\delta}\right). \tag{58.17}$$

By using (58.13) with the approximation $U' = U_0/\delta$ we find

$$\mathbf{u} = \frac{-\mathscr{R}}{\pi^2 + \alpha^2\delta^2 + \beta(\delta^2/v)} \, a \, \sin\left(\frac{\pi y}{\delta}\right). \tag{58.18}$$

The corresponding expressions for $\mathbf{w}$ and $\mathbf{p}$ can be obtained from (58.16) and (58.15):

$$\mathbf{w} = -\frac{\pi a}{\alpha\delta} \cos\left(\frac{\pi y}{\delta}\right), \tag{58.19}$$

$$\mathbf{p} = -\frac{v\pi a}{\alpha^2\delta^3}\left(\pi^2 + \alpha^2\delta^2 + \frac{\beta\delta^2}{v}\right)\cos\left(\frac{\pi y}{\delta}\right). \tag{58.20}$$

These functions do not vanish at the outer edge of the layer nor does $\mathbf{w}$ vanish at the wall. This means that (58.17) should be replaced by a more sophisticated expression. Let us introduce (58.20) into (58.14) and simplify. The amplitude a can be dropped, and we find a relation between the function $U(y)$ and various constants. In view of the loose assumptions previously introduced, we simply replace U with the averaged value $\frac{1}{2}U_0$. This leads to the following characteristic equation:

$$\left(\frac{U_0^2\delta^3}{v^2 R}\right)^{1/2} = \left(\frac{\delta}{R}\right)^{1/2}\mathscr{R} = \left(1 + \frac{\pi^2}{\alpha^2\delta^2}\right)^{1/2}\left(\pi^2 + \alpha^2\delta^2 + \frac{\beta\delta^2}{v}\right). \tag{58.21}$$

If α is very large, the right-hand side tends toward $\alpha^2\delta^2$, and if α is very small it tends toward $\pi/\alpha\delta(\pi^2 + \beta^2\delta^2/v)$. This corresponds to the asymptotes shown in Fig. 58.2. For $\alpha\delta = \pi/\sqrt{2}$ we find the minimum value $(\delta/R)^{1/2}\mathscr{R} = 28$.

The results of Görtler and Smith have the same general properties although some quantitative differences are noticeable. Thus our drastic assumption (58.17) is not too far off.

Although (58.21) is only a crude approximation, it already indicates that β cannot be positive unless the factor $(\delta/R)^{1/2}\mathscr{R}$ is large enough. This rules out convex walls and establishes an important equivalence between curvature and friction. Furthermore, in (58.14) the term in $\mathbf{p}'$ is negligible when α is small. Thus the viscosity controls the upper part of the neutral curve and plays a lesser role along the lower part.

Finally, the presence of a factor $\mathscr{R}$ in (58.18) is interesting. It means that $\tilde{u}$ will be larger than $\tilde{v}$ or $\tilde{w}$, as suggested by the slope of the streamlines in Fig. 58.2. This conclusion is supported by Smith's results, which outline the functions $\mathbf{u}$ and $\mathbf{v}$. Thus the vorticity fluctuations are

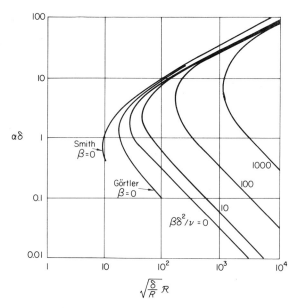

FIG. 58.2. Eigenvalues of a laminar boundary layer along a concave wall. Görtler and Smith used a Blasius layer, $\delta = 100$. Other curves from the highly simplified theory leading to Eq. (58.21).

almost perpendicular to the wall, and it is only their projection along the x-axis that, if considered alone, forms a periodic system of parallel vortices. We surmise that the Coriolis forces should have been retained and that their presence would reduce the ratio $\mathbf{u}/\mathbf{v}$ away from the wall.

Some measurements by M. Clauser and F. Clauser (1937) indicate that transition to turbulence in a laminar boundary layer along a concave wall with $R = 10\delta$ occurs around $\sqrt{\delta/R}\,\mathscr{R} = 300$. A rough estimate indicates that the perturbations of the kind examined in this section could receive an amplification as large as 10^5 or 100 dB.

(c) SECONDARY INSTABILITY

Generally speaking, a secondary flow is one in which there is a vorticity component aligned in the same direction as the mean flow proper. Secondary instability refers to the possible generation of a secondary flow over and above that which may already be present. Clearly, this situation not only exists but is prevalent in all cases in which there are centrifugal force effects.

For these flows (e.g., that over curved boundaries) we do not normally refer to the instability as secondary. On the other hand, the laminar boundary layer over a flat plate presents an ideal problem in which secondary instability is feasible.

When a boundary layer oscillates, according to the Tollmien-Schlichting theory the instantaneous picture of the flow shows curvature of the streamlines. This is the situation where secondary instability can be expected. Secondary instability now becomes a question of stability of the already oscillating flow. The question arises even if the Tollmien-Schlichting waves are neutral or damped. This problem was examined by Görtler and Witting (1958) and Witting (1958).

New axes of reference are taken as ξ, η, ζ along and normal to the streamlines. The stability problem for Tollmien-Schlichting is assumed known, and the stream function resulting from this perturbation can be written as

$$\varepsilon\tilde{\psi}(x, y, t) = \varepsilon \, \text{Re} \, [\phi(y)e^{i\alpha(x-ct)}]. \tag{58.22}$$

The parameter ε is merely an amplitude parameter and is taken as small.

The original velocity $U(y)$ and the corresponding stream function are related as

$$\Psi(y) = \int_0^y U(y') \, dy'. \tag{58.23}$$

The resulting stream function for the combined flow is just

$$\psi(x, y, t) = \Psi(y) + \varepsilon \, q(y) \cos \, [\alpha(x - ct) + p(y)], \tag{58.24}$$

where

$$q(y) = (\phi\phi^*)^{1/2} = (\phi_r^{\,2} + \phi_i^{\,2})^{1/2}, \tag{58.25}$$

and

$$p(y) = \tan^{-1}\left(\frac{\phi_i}{\phi_r}\right). \tag{58.26}$$

Moreover, we are considering a neutral disturbance for (58.22). If ε is small enough, we can expand (58.24) and retain only the terms of order ε to give

$$\psi(x, y, t) = \Psi(y)\left\{y + \varepsilon \frac{q(y)}{U(y)} \cos \, [\alpha(x - ct) + p(y)]\right\}. \tag{58.27}$$

If we define $Q(y) = q(y)/U(y)$, it becomes possible to give the relation between the two coordinate systems as

$$\xi = x - ct + \frac{1}{\alpha} p(y), \tag{58.28}$$

$$\eta = y + \varepsilon Q(y) \cos [\alpha(x - ct) + p(y)], \tag{58.29}$$

$$\zeta = z. \tag{58.30}$$

Now the problem is straightforward. The new flow, as given by (58.27), is perturbed in the new coordinate system. In particular, fluctuating components of the velocity and pressure are taken as

$$u_a = a(\xi, \eta) \cos (\sigma \xi) e^{\gamma(\xi)t}, \tag{58.31}$$

$$u_b = b(\xi, \eta) \cos (\sigma \xi) e^{\gamma(\xi)t}, \tag{58.32}$$

$$u_c = c(\xi, \eta) \sin (\sigma \xi) e^{\gamma(\xi)t}, \tag{58.33}$$

$$p = \hat{p}(\xi, \eta) \cos (\sigma \xi) e^{\gamma(\xi)t}. \tag{58.34}$$

Differential equations are obtained, and only terms up to and including order ε are retained. Boundary conditions require that $a = b = \partial b/\partial \eta = 0$ for $\eta = 0$ and $\eta = \infty$. The details of this work are not important and reference is made to the original. The conclusions found by Görtler and Witting, however, and a sketch of the additional assumptions needed are important.

The streamlines are taken as nearly straight (specifically, $\xi = 0$ is inserted) and changes in the streamwise direction are neglected. As a result, in any locality the flow is similar to the boundary layer flow over a curved surface where the local radius of curvature of the streamlines replaces the radius of curvature of the wall. The analogy is complete if we move with the phase velocity of the Tollmien-Schlichting wave; this is tantamount to having moving walls and the familiar cats-eye structure is found along the critical layer.

Recall that we found instability for the curved wall problem as the velocity increased in the direction of the center of curvature. In this new system the curvature reverses when the critical layer is crossed. Thus secondary instability occurs in the troughs of the streamlines *above* the critical layer, and there is a matching instability at a peak of the streamline *below* the critical layer. The combination consists of paired vortices above and below the layer. Instability is actually found when the Tollmien-Schlichting amplitude is of the order of 10^{-4} times the displacement thickness.

59. Variable Mean Viscosity and Fluctuations

Let us imagine a flow of glycerine over a heated plate. A temperature difference of a few degrees can produce changes in viscosity by several orders of magnitude. Thus the dynamics of the problem are profoundly affected by viscosity gradients, and any study of oscillations will have to include the possibility of fluctuations of the local viscosity. In compressible flows the viscosity depends only on the temperature, and the necessary terms can be included without special difficulties.

In a turbulent flow we have introduced the concept of eddy viscosity and postulated some dependence on the coordinate y. In this case, however, the reality is certainly more complex: the eddy viscosity may depend on upstream conditions and its own history. Therefore v can include a fluctuating part, and we must write $v(y) + \tilde{v}(x, y, t)$. This brings an additional unknown and a supplemental equation is required. If we assume that the viscosity of a fluid particle is preserved during its motion (see Drazin, 1962), we obtain the relation

$$\frac{\partial \tilde{v}}{\partial t} + U \frac{\partial \tilde{v}}{\partial x} + \tilde{v} \frac{dv}{dy} = 0. \tag{59.1}$$

Clearly this equation runs into difficulties at the critical layer. If the viscosity simply depends on the temperature, the diffusion of temperature would remove the singularity; but with an eddy viscosity further assumptions are needed. Perhaps we could assume that this kind of viscosity is self-diffusing.

Similar situations occur in multiphase flows, with or without chemical reactions.

Let us now direct our attention to the Squire theorem, with variable viscosity.

The fluid is Newtonian and the viscosity is a function of y only. After the usual transformation to absorb the case of oblique waves we find that the right-hand side of the Orr-Sommerfeld equation reads as follows:

$$-\frac{iv}{\bar{\alpha}} (\mathbf{v}'''' - 2\bar{\alpha}^2 \mathbf{v}'' + \bar{\alpha}^4 \mathbf{v}) - 2i \frac{v'}{\bar{\alpha}} (\mathbf{v}''' - \bar{\alpha}^2 \mathbf{v}') - i \frac{v''}{\bar{\alpha}} (\mathbf{v}'' + \bar{\alpha}^2 \mathbf{v}). \tag{59.2}$$

This expression shows clearly that the validity of the Squire theorem is not limited by the mean viscosity gradient. If the viscosity fluctuates with a complex amplitude $\mathbf{v}$, the following additional terms appear on the right-hand side:

$$-(\mathbf{v}\overline{U}')'' - \bar{\alpha}^2 \mathbf{v}\overline{U}'. \tag{59.3}$$

Again it can be seen that the Squire theorem remains valid.

60. Viscoelastic Flows and Dusty Gases

(a) BOUNDARY LAYER

A large variety of liquids does not obey the classical Newtonian linear stress-strain law. Most peculiar in this respect are certain solutions of polymers in water, even when very dilute. In general, the stresses are non-linear functions of the rate of strain tensor and its various time derivatives (cf. Wilkinson, 1960). In small oscillations the fluctuating stresses could still be expressed by linearized equations in which the mean flow would contribute to the coefficients. In principle, this would still lead to an Orr-Sommerfeld equation, which would still be amenable to machine computations.

In this section we do not enter into great detail but point out one apparently general property. One of the simplest models of non-Newtonian behavior was proposed by Maxwell who wrote

$$\left(1 + T_M \frac{\partial}{\partial t}\right)\tau_{ij} = \mu\sigma_{ij}, \tag{60.1}$$

where τ_{ij} is the stress tensor and σ_{ij} is the rate of strain tensor. The constant T_M represents the time required by the fluid to build up the equilibrium stress. Another model was proposed by Voigt which corresponds to fluids that require some time before flowing in response to a sudden applied force. With a new constant T_V, this model corresponds to

$$\tau_{ij} = \mu\left(1 + T_V \frac{\partial}{\partial t}\right)\sigma_{ij}. \tag{60.2}$$

A large variety of more sophisticated models can be described by the general equation

$$\tau_{ij}(t) = \int_{-\infty}^{t} K(s - t)\sigma_{ij}(s)\,ds, \tag{60.3}$$

where the kernel K is given by material properties.

Let us now consider the relation between fluctuations of τ_{ij} and of σ_{ij}, proportional to $e^{i\alpha(x-ct)}$. Essentially, the models given by (60.1), (60.2), and (60.3) lead to a relation of the classical type, except that the viscosity becomes complex and frequency-dependent. Finally, in the right-hand side of the Orr-Sommerfeld equation the ordinary viscosity is replaced by $v_0 e^{-i\theta}$, where v_0 is the modulus and $-\theta$ is the argument. In a Maxwellian fluid (60.1) leads to

$$v_0 = \frac{v}{[1 + (\alpha c T_M)^2]^{1/2}}, \tag{60.4}$$

$$\theta = -\arctan(\alpha c T_M). \tag{60.5}$$

Thus a delay in the stress corresponds to a negative angle θ.

The Orr-Sommerfeld equation with complex viscosity was integrated analytically by Wen (1963) and numerically by Betchov (1965). Both authors considered a Blasius boundary layer and obtained the same qualitative results with the machine computations being more accurate. The mean flow is not affected by time-dependent effects such as those of Maxwell or Voigt. Thus it is consistent to use a Blasius profile. Some numerical results are shown in Fig. 60.1, and it can be seen that destabilization occurs when the stresses lag behind the rate of strain. Hence the Maxwell time is destabilizing.

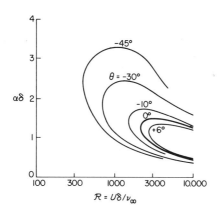

Fig. 60.1. Neutrally stable oscillations of a Blasius boundary layer, with various values of the angle θ accounting for certain viscoelastic effects, according to Eq. (60.5). A delay of the viscous stress is destabilizing.

In Fig. 60.2 the rate of growth and phase velocities are shown for a fixed $\theta = -30°$. The phase velocities exceed 0.3, and therefore we should not use the approximation of a constant mean velocity gradient on which rests the validity of the Tietjens functions.

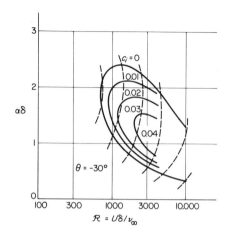

FIG. 60.2. Amplification in a Blasius boundary layer, with a 30° delay between fluctuating stresses and fluctuating rate of strain.

According to (60.1) the stresses are expressed in terms of the local rate of strain tensor and its past values. Intuitively, it seems more satisfactory to take into account the past deformations *of the fluid particle* in place of the past local conditions. Thus the time derivative should be replaced in (60.1) by a total derivative. In (60.4) and (60.5) this leads to the substitution of $U - c$ in place of $-c$. Near the wall this correction is minor, but at the critical layer it restores the usual value of the viscosity. It renders the complex viscosity a function of y and special terms appear, as in Section 61.

Note that the sign of θ will be reversed if $U > c$.

(b) Shear Layers and Analytical Behavior

We have examined the effect of the phase angle of the viscosity on the shear layer $U = \tanh y$ and on the shear layer with a wall at $y = -3$ (see the neutral curve in Fig. 13.4). For low Reynolds numbers we found that c_i tends to grow as θ^2 (see Fig. 60.3). A small linear dependence is also more apparent at low values of α, where the influence of the wall is more pronounced.

However, this linear effect is such that a delay of the stress is stabilizing (see, for example, $\alpha = 0.08$ and $\theta = -10°$ in Fig. 60.3). This is the reverse of the effect observed in a boundary layer (see Fig. 60.1), where a delay is destabilizing.

To examine the effect of large values of θ on a boundary layer flow

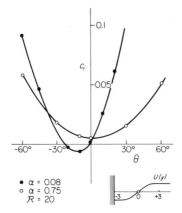

Fig. 60.3. Eigenvalue for a shear layer along a wall. The circles correspond to a large α, near the upper branch of the neutral curve, the dots to a small α, near the lower branch. (See Fig. 13.4.)

(and to satisfy certain curiosities) we determined the eigenvalues for values of θ between 0 and 2π in the Blasius boundary layer for conditions near the neutral curve at the critical Reynolds number. The results, displayed in Fig. 60.4, show an almost linear dependence of c_i on small θ. The periodicity with θ is not surprising and neither is the fact that $c_i(\theta)$ is equal to $-c_i(\pi - \theta)$. These are direct consequences of the structure of the Orr-Sommerfeld equation. In particular, with $\theta = \pi$, we have the equivalent of pure negative friction. Thus, in the sense of the simplified analysis, the diffusion term at the critical layer has reversed polarity as well as the wall diffusion term. As a result, c_i changes sign. In the classical asymptotic theory the sign of the viscosity determines the detour that must be used around the singularity encountered at the critical layer. The two possible detours correspond essentially to $\theta = 0$ and $\theta = \pi$. In the limit $|v| \rightarrow 0$, c_i probably becomes a function of α only, changing sign with $\cos \theta$ at $\theta = \pm \pi/2$.

If $\theta = \pm \pi/2$, the computer cannot converge to a definite eigenvalue, for $v'(-3)$ shows abrupt changes with c. Thus $c_i(\theta)$ is apparently not analytic at these points. An elaborate discussion of the existence of solutions has been given by Lin and Rabenstein (1960).

For a boundary layer c_i is almost an odd function of $\sin \theta$, whereas for a shear layer it seems to be an even function. This is perhaps related to the fact that the viscosity is destabilizing in the boundary layer and always stabilizing in the shear layer problem.

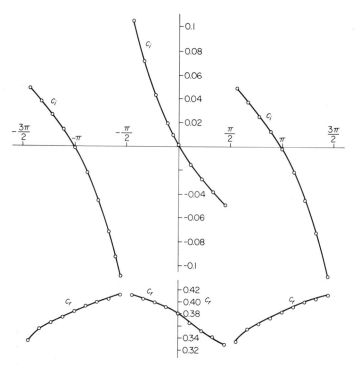

Fig. 60.4. Eigenvalues for a boundary layer with variable phase of the viscous stress fluctuation. Blasius profile, $\nu = \nu_0 e^{-i\theta}$, $U_0\delta/\nu_0 = 2300$, $\alpha\delta = 1.15$.

(c) Dusty Gases

The stability of flows of dusty gases has been examined by Saffman (1962), Michael (1964, 1965), and Liu (1965a, 1965b). These studies have been prompted by the observations that adding dust to a gas flowing in the turbulent regime through pipes can appreciably reduce the energy dissipation. (See Kakevich and Krapivin, 1958; Sproull, 1961.) According to Einstein's formula, the viscosity of a suspension should be increased by the presence of the dust (see Landau and Lifshitz, 1959); this would increase the energy dissipation for a given rate of volume flow. Although the studies of the stability of a laminar dusty flow are not directly related to the properties of turbulent flows, it was hoped that they would bring some clarification.

Saffman simply assumes that the dust particle will follow the motion of the gas with the delays introduced by inertia and the resistance indicated by Stokes's formula.

The gas will simply experience a force equal and opposite to that applied on the dust. With ρ denoting the gas density, m the mass of a dusty particle, and N the number of dust particles per unit volume, it is convenient to define the nondimensional factor of mass concentration,

$$= \frac{mN}{\rho}. \tag{60.6}$$

This factor can be as large as 0.2.

The time constant characteristic for the lag of the dust particles varies between 0.2 and 0.002 sec. The nondimensional form is

$$\tau = \frac{m}{6\pi a \rho v} \frac{U_0}{L_0}, \tag{60.7}$$

where a is the particle radius.

The derivation of the vorticity fluctuation equation leads to the classical Orr-Sommerfeld equation, in which $\bar{U}$ is used in place of U, with the definition

$$\bar{U} = U(y) + \frac{f(U - c)}{1 + i\alpha(U - c)\tau}. \tag{60.8}$$

Two limit cases can be distinguished. If τ is small, we can use the regular profile U and take $\bar{v} = v/(1 + f)$ in place of v. If τ is large, the value of c_i for the dusty gas becomes that of the clean gas, less $f/\alpha\tau$, as we can easily find from (60.8). Thus the dust decreases the effective viscosity when it follows the fluid. This effect is destabilizing. If the lag is large, the effect is stabilizing.

It is also apparent from (60.8) that the effect of the dust disappears at the critical layer. This is not surprising, for the convective processes cease to exist in this region.

61. Turbulent Flows

A promising field of endeavor for stability theory is that of turbulent flow. Let us clarify the concept of the stability of turbulent flow by discussing its analogy with the simple mechanism of a clock. If a small perturbation is applied briefly, the motion is perturbed, but after a while the system will return to its limit cycle. Certain small disturbances, however, such as the ejection of a pin holding the main spring, may have dramatic results. A turbulent flow is nonlinear and aperiodic. It has well-defined average properties, and in principle it is possible to introduce controlled periodic perturbations. The question before us is, "Will this hypothetical perturbation die or grow?"

If the answer to the question is "grow," we can expect that these oscillations already exist as a component of the turbulence and that their amplitude is limited by special nonlinear effects that the linearized theory does not include. In particular, the Reynolds stresses introduced by such oscillations will affect the mean profile. Thus a linearized theory can only uncover part of the structure of turbulent flows. The attention of the reader should be directed to the fact that in a turbulent pipe flow or a turbulent boundary layer the most energetic fluctuations are not isotropic and they are responsible for the turbulent Reynolds stresses, suggesting an analogy with laminar flow oscillations. When the three-dimensional energy spectrum is plotted on a linear scale, it has a clear maximum and a rather limited bandwidth—an apparent resemblance to the c_i versus α curve. These same fluctuations also produce the well-known background of isotropic fluctuations. It is only when the spectrum is plotted on a log-log scale, that it appears to be dominated by the isotropic components, falling as $\alpha^{-5/3}$, until the viscous end. Compared with the energetic fluctuations the isotropic components are small; their effect can also be compared with that of molecular agitation. Thus we meet the concept of eddy viscosity.

There is not even the embryo of a theory for anisotropic turbulent flow, but we may gain some knowledge of the anisotropic components by using the linear stability theory. In this respect we can use the analogy between a turbulent flow and a laminar flow endowed with a special viscosity.

For turbulent flows of the type found in boundary layers, jets, and wakes, it is also observed that the region of random vorticity has a sharply marked boundary known as the "super layer." This special boundary has pronounced undulations, like those of a stormy sea, and is the source of the so-called intermittency of velocity recordings. The length and time scale of the super layer have been measured by Klebanoff (1954) for a turbulent boundary layer. His observations suggest some form of oscillation in which $\alpha\delta \doteq 1$ and $c_r/U_0 = 0.8$. These facts motivated the calculations of Betchov and Criminale (1964).

Betchov and Criminale considered the velocity and viscosity distributions shown in Fig. 61.1. These functions correspond to an unsophisticated representation of a turbulent boundary layer. Using the Orr-Sommerfeld

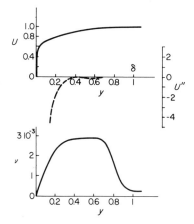

Fig. 61.1. An example of stable flow. These velocity and viscosity profiles simulate a turbulent boundary layer and have been found to be stable.

equation with the terms proportional to gradients (Section 60), they computed the stability of this flow. The machine calculations have shown that this flow is stable. As an experiment, the extra viscous terms were switched off one by one, and it was noted that they have only a modest influence on the eigenvalue.

The next case treated was that of a profile presenting a shear layer in the region of intermittency (see Fig. 61.2). This may correspond to an actual velocity profile when the oscillations of the super layer are under restraint. The amplitude and width of the shear layer represent not more than an educated guess. In regard to this point, we must note that a rapid change in viscosity is likely to be the cause of an increase in velocity. For example, a fluid of low viscosity will tend to slide over a layer of more viscous material. The velocity inflection point will be on the side of the lower viscosity. This formulation leads to instabilities for wavenumbers in the range $0.6 < \alpha < 2.5$, with rather low values of c_i, not exceeding 0.005. (See Fig. 61.3, for $\theta = 0$.) The phase velocity is nearly the velocity of the inflection point.

If the viscous profile is modified so that the viscosity at the inflection point has lower values, c_i is raised to a maximum value of 0.007.

An examination of the eigenfunctions shows that the resulting instabilities resemble those of a free shear oscillation rather than of a boundary layer. An examination of the role of the viscous sublayer at the wall reveals that little transfer of energy from the mean flow to the energetic fluctuation occurs away from the wall.

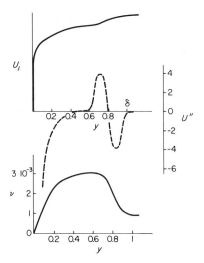

Fig. 61.2. An example of unstable flow. These velocity and viscosity profiles simulate a turbulent boundary layer with an inflection. The amplifications are shown in Fig. 61.3.

Another aspect of the problem was examined briefly by Betchov (1965). In a turbulent flow the eddy-viscosity might be frequency-dependent. Essentially, this is expected if the turbulent stresses that affect the oscillation are given by a more complicated law than a Newtonian relation. Specific examples are indicated in Section 60, which deals with viscoelastic flows. Our intuition suggests that the turbulent stresses will take some time to build up in response to a change in the flow and that the Maxwellian

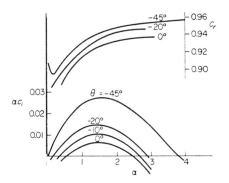

Fig. 61.3. Amplification in a model of a turbulent boundary layer. The case $\theta = 0$ corresponds to a Newtonian viscosity. The other cases simulate delayed stresses.

model is appropriate. If the phase lag amounts to 0, 10, 30, and 45°, the effect on the rate of growth can be significant, as shown in Figure 61.3, which corresponds to the mean flow and viscosity of Fig. 61.2.

In turbulent shear layers the phase angle may also have a significant effect on their stability, as suggested by Fig. 60.3. Thus the response time associated with eddy viscosity could have a notable destabilizing influence.

This sort of analysis invites further investigation of the following points:

1. Does the eddy viscosity travel with fluid or is it self-diffusing?
2. What happens if the basic velocity profile becomes a slowly varying function of the time so that the point of inflection moves about?
3. How do these concepts apply to turbulent channel flows?

Errors in Numerical Procedure

The object of this appendix is to give an example that will illustrate the major forms of error encountered in numerical integration of differential equations. The computer uses binary numbers, and consequently every result must finally be translated from the binary system to the decimal system. This introduces an error that affects the last digit of the results. In a typical computer this error is of the order of 10^{-8}.

If the integration is made by the method proposed by Newton, each function is approximated by a straight line between the points at which it is evaluated and leads to errors of the order of h^2, where h is the step size.

In general, if the function under consideration is approximated by a polynomial of degree n, the errors are of the order of h^{n+1}. The Adams-Moulton method operates with $n = 4$. This value is also employed by the Runge-Kutta method, which uses as intermediate quantities the value of the function at the middle of the interval. Thus integration errors are of the order of h^5. Taking one of the simplest differential equations for demonstration, we have observed the errors shown in Fig. AI.1.

If the calculations are performed with double precision, but with the results given in single precision, the value of h can be so small that the errors are due only to translation. For sufficiently large values of h, the errors grow as h^5. If the calculations are made in single precision, another error becomes apparent. This error is the result of the round-off that takes place after each arithmetic operation, the number of which depends on the method used and the intricacy of the problem. Let us suppose that the computer performs 10^6 operations (which might require a few tenths of a second) and that each operation includes a round-off error of the order of 10^{-8}. These errors are almost random so that the net effect is proportional to the square root of the number of truncations. This means the net error would be 10^{-5}.

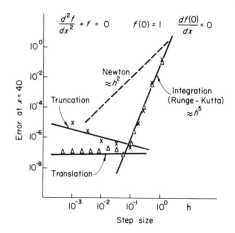

FIG. AI.1. Errors in the integration of a simple equation, with various step sizes: X, simple precision; Δ, computation in double precision, but results given in single precision.

Because the number of operations grows as h^{-1} when the range of integration is kept constant, the truncation errors grow as $h^{-1/2}$. This is apparent in Fig. AI.1 for a single precision program.

These three basic types of error suggest the existence of a range of most favorable values of h. The use of eight significant decimals, together with the choice of a fourth-order method, in single precision generally seems to strike a balance between the contradictory requirements of accuracy and economy.

On Analytic Solutions of the Orr-Sommerfeld Equation

We clearly recognize that the Orr-Sommerfeld equation rests at the center of the problem of stability of parallel flows. From the standpoint of the mathematical formulation, the determination of the stability of any given system involves nothing more than the solutions of this equation combined with the boundary conditions given by the particular flow under scrutiny. Once this information has been obtained, the salient output becomes the eigenvalues, c, the complex phase velocity (or the frequency, $\omega = \alpha c$), as a function of the wavenumber, α (or the wavenumbers α, β), and the Reynolds number, $\mathcal{R}$. The successful accomplishment of this task, however, is surmountable by analytic means only in principle. Indeed, exact solutions cannot be obtained at all and, as a result, approximation methods must be resorted to for analysis. And, this type of result had to wait some sixteen years after the equation was established before it appeared! (Cf. Introduction for a historical review of the development.) A refinement of the method, which corrected the errors of the initial presentation and brought about a significant degree of clarification, had to wait an additional twenty-one years. Fortunately, modern day techniques and tools are available which have facilitated the work even more. The fact still remains, though, that the problem is not in any way simple and can, in some situations, involve a great deal of work. Even the use of the computer, which ultimately produces the most accurate and rapid answers—including the eigenfunctions—must be handled with caution and a good program lest the problem becomes intractable. We propose to make a cursory review of the classical treatment and outline a more recent mode of attack for treating the most commonly encountered situation. This is done by means of what has become known as inner and outer expansions. It is hoped that this presentation will provide both a means of understanding of what is involved in the overall situation and establish a companion to the computer for investigation. Obviously, acute accuracy remains with the computer, for any approximation scheme involves errors of a given order to be established by the particulars involved. On the other hand, functional behavior is to be found by the route of analysis rather than by computation.

Some general form of the Orr-Sommerfeld equation can be obtained when the set of governing linearized equations are combined to produce one equation in terms of one dependent variable. The net result is, by virtue of the level of complexity which we are willing to tolerate, an ordinary, fourth-order, linear and homogeneous differential equation with variable coefficients, expressed best with the fluctuating velocity component which is normal to the direction of the mean flow as the dependent variable. For the most part, this remains true whether or not we are dealing with two-dimensional or three-dimensional disturbances, the spatial or the temporal initial value problem, and even if we are considering the physical complexities of compressibility, stratification, or electrical conduction. Physically, we have seen that this equation should be looked upon as that which governs the fluctuating vorticity component that is normal to the plane of motion of the laminar flow. With two-dimensional disturbances in an incompressible flow, this equation is exactly correct for the vorticity since the operations needed to arrive at the result are tantamount to invoking the curl operator on the momentum equations; the other two components of the vorticity do not even exist. When three-dimensional disturbances for the same system are treated, all three components of vorticity exist, but the equation for the component normal to the motion plane remains the one of principal interest (see Section 23). When actually written in terms of the vorticity, however, we have noted that the Orr-Sommerfeld equation is not necessarily a homogeneous equation, except in those few cases where there is no production. A case in point is to be found in the example of plane Couette flow.

Let us consider the incompressible laminar boundary over a flat plate. By noting that the Squire theorem (Section 19) is valid, we realize that the form of the governing equation is the same for both two- and three-dimensional disturbances. In a somewhat rearranged form, this equation can be written as

$$i\left(\frac{v}{\alpha}\right)\mathbf{v}'''' - \left[2i\left(\frac{v}{\alpha}\right)\alpha^2 - (U - c)\right]\mathbf{v}'' + \left[i\left(\frac{v}{\alpha}\right)\alpha^4 - \frac{d^2U}{dy^2} - \alpha^2(U - c)\right]\mathbf{v} = 0.$$

$$(\text{II.1})$$

All quantities in (II.1) are taken as real except for v and c. We already know that v generally has a small value for the fluids of interest and there exists a critical layer somewhere within the flow where U varies from zero at the plate to the free-stream value of U_0.

The classical method which proved fruitful in solving (II.1) was made by analytically continuing the velocity field in the complex plane. To do this, we can define y as the real axis and let $\xi = y + iz$. Then, the velocity profile $U(y)$ can be assumed as given by $U = U(\xi)$ where $U(\xi)$ is an analytic function. Moreover, $U(\xi) \to U(y)$ as $\xi \to y$ and the equation $U(\xi) = c$ must have a solution. In fact, the point $\xi = \xi_c$ in the complex plane is the location of the critical layer from the relation

$$U(\xi_c) = c. \tag{II.2}$$

Once (II.1) has been transformed, the power of analytic functions becomes accessible. For example, if the mean velocity gradient at the critical layer, U'_c, is positive and the imaginary part of $c = c_i > 0$ too, it can be shown that the point ξ_c lies above the real axis in the complex plane. Likewise, if $U'_c > 0$ but $c_i < 0$, then ξ_c is below the real axis. This can be proved by standard procedure.

Let $\xi_c = y_c + iz$. Now, $U(\xi_c) = U(y_c + iz) = c$. If we separate this relation into real and imaginary parts we will have

$$U(y_c + z) = U_r + iU_i. \tag{II.3}$$

Since $c = c_r + ic_i$, we see that

$$U_r = c_r \tag{II.4}$$

$$U_i = c_i. \tag{II.5}$$

But, $U(\xi)$ is an analytic function and, hence, by the Cauchy-Riemann equations, we can write

$$\partial U_r / \partial y = \partial U_i / \partial z. \tag{II.6}$$

For small z, (II.5) can be taken as

$$c_i = U_i(y_c, 0) + [\partial U_i(y_c, 0)/\partial z]z + \cdots. \tag{II.7}$$

We note that $U'_c > 0$ means that $\partial U_r / \partial y > 0$. In turn, by (II.6) $\partial U_i / \partial z > 0$ must follow. Thus, when $z \to 0$, $U_i \to 0$ and (II.7) shows that c_i and the product $[\partial U_i(y_c, 0)/\partial z]z$ are of the same sign. When this is true, the answer we seek is evident: $c_i > 0$ means $z > 0$ and vice versa. This fact is important to the ultimate solution that can be obtained and will be seen to be compatible with the method we now outline.

Recent mathematical methods lead us to analyze (II.1) in a somewhat different fashion. Inspection of the equation immediately classifies it as one of the singular perturbation variety. This implies that we have an equation where the terms involving the highest derivatives are multiplied by a small coefficient. In this case, the coefficient is (v/α). (Strictly speaking, the coefficient should be on a nondimensional basis. Therefore, (v/α) is really $1/\alpha\mathscr{R}$ with $\mathscr{R}$ the Reynolds number and α is now a nondimensional wavenumber. The concept of small then means that $1/\alpha\mathscr{R} \ll 1$.) Such a situation can be well treated by inner and outer expansions. For reference as to the details, the work of Erdélyi (1963) or the treatise of van Dyke (1964) is recommended.

The basic argument as applied to our problem leads to the physical analogy that viscosity is not important except in special places, such as in the free-stream for example. The most obvious need for the viscosity is at the wall where we must require that there be no slip. In terms of the equation, we can infer this to mean that only a second-order equation must be solved but for the places where we need more than two coefficients to satisfy the boundary conditions. In between, we will need viscosity at the critical layer. From (II.1) it can be concluded that it is impossible to find a solution for **v** that vanishes at $y = y_c$ and there is a singularity in the equation without viscosity. Physically, we have seen that the vorticity tends to increase without limit at the critical layer without viscous diffusion.

The first solution to the equation becomes that due to the outer expansion. We proceed by assuming the expansion

$$\mathbf{v}(y) = \mathbf{v}_0(y) + (v/\alpha)\mathbf{v}_1(y) + (v/\alpha)^2\mathbf{v}_2(y) + \cdots. \tag{II.8}$$

Substitution of (II.8) into (II.1) gives us

$$\mathbf{v}_0'' - \alpha^2\mathbf{v}_0 = [U''/(U - c)]\mathbf{v}_0, \tag{II.9}$$

$$\mathbf{v}_1'' - \alpha^2\mathbf{v}_1 = -i(\mathbf{v}_0'''' - 2\alpha^2\mathbf{v}_0'' + \alpha^4\mathbf{v}_0)/(U - c). \tag{II.10}$$

To the lowest order this simply becomes the Rayleigh equation and expresses a balance between convection of vorticity with production. For the problem of the boundary layer we see that (II.9) is valid in the free stream. Thus, we are assured that $(U - c)$ does not vanish anywhere. The required boundary conditions for **v** as $y \to \infty$ can be easily met.

Far out in the free stream we note that $U = U_0$, $U'' = 0$, and at once a correct solution becomes

$$\mathbf{v}_0 = Ae^{-\alpha y}, \tag{II.11}$$

which satisfies the conditions $\mathbf{v}, \mathbf{v}' \to 0$ as $y \to \infty$. A is a coefficient still to be determined. If we elect to solve (II.9) when U'' has not yet vanished but is reasonably small, it becomes possible by further expansion. Let $U'' = \varepsilon U''$ with $\varepsilon \ll 1$ and assume

$$\mathbf{v}_0(y) = \mathbf{g}_0(y) + \varepsilon \mathbf{g}_1(y) + \varepsilon^2 \mathbf{g}_2(y) + \cdots. \tag{II.12}$$

The use of (II.12) allows us to generate the equations

$$(U - c)(\mathbf{g}_0'' - \alpha^2 \mathbf{g}_0) = 0, \tag{II.13}$$

$$(U - c)(\mathbf{g}_1'' - \alpha^2 \mathbf{g}_1) = U'' \mathbf{g}_0. \tag{II.14}$$

The essential effect of U'' extending into the free stream can now be seen by solving (II.13) together with (II.14). To the order given by these two equations we find

$$\mathbf{v}_0 = A e^{-\alpha y} \left[1 + \left(\frac{U - 1}{1 - c} \right) \right] \tag{II.15}$$

after integration by parts and taking $U \cong 1$.

The next region to be considered occurs at the wall. At this location we know at the outset that the term having the fourth-order derivatives must be retained. A rescaling of the variables is in order to bring this term into the proper perspective and leads to an inner expansion. Let $y = (v/\alpha)^n \eta$ and $\mathbf{v}(y) = (v/\alpha)^s \mathbf{F}(\eta)$ with $n, s > 0$; $U(y)$ is also taken to be expressed in terms of η. Direct substitution of the variable changes into (II.1) leads to

$$i(v/\alpha)^{1 - 4n + s} \mathbf{F}''' - [2i(v/\alpha)\alpha^2 - (U - c)](v/\alpha)^{s - 2n} \mathbf{F}''$$

$$+ [i(v/\alpha)\alpha^4 - U'' - \alpha^2(U - c)](v/\alpha)^s \mathbf{F} = 0, \tag{II.16}$$

where the prime now denotes differentiation with respect to η. A suitable choice of the coefficients which meets the desired goal and which does not produce any terms of higher order than that due to the viscous diffusion requires

$$1 - 4n + s = 0 \tag{II.17}$$

together with

$$s - 2n = 0. \tag{II.18}$$

Thus $n = \frac{1}{2}$ and $s = 1$ and we have the equation

$$i\mathbf{F}''' - [2i(v/\alpha)\alpha^2 - (U - c)]\mathbf{F}''$$
$$+ [i(v/\alpha)\alpha^4 - U'' - \alpha^2(U - c)](v/\alpha)\mathbf{F} = 0. \tag{II.19}$$

As it is to be expected, (II.19) is the governing equation in terms of the viscous wall region which is of order $(v/\alpha)^{1/2}$.

To solve (II.19) we assume the expansion

$$\mathbf{F}(\eta) = \mathbf{F}_0(\eta) + (v/\alpha)^{1/2}\,\mathbf{F}_1(\eta) + (v/\alpha)\,\mathbf{F}_2(\eta) + \cdots. \qquad \text{(II.20)}$$

Along with (II.20) it seems reasonable to take

$$U - c = -c + ky \qquad \text{(II.21)}$$

or

$$U - c = -c + k(v/\alpha)^{1/2}\eta. \qquad \text{(II.22)}$$

Collecting (II.20) and (II.22) together with (II.19) we are able to make use of the lowest order equation. This reads

$$\mathbf{F}_0''' + ic\,\mathbf{F}_0'' = 0. \qquad \text{(II.23)}$$

Equation (II.23) denotes that the convection of the vorticity must be balanced by diffusion due to viscosity. The solution which satisfies the conditions $\mathbf{v} = \mathbf{v}' = 0$ at $y = 0$ is

$$\mathbf{F}_0 = D\sqrt{ic}\left\{\eta + \frac{1}{\sqrt{ic}}[\exp(-\sqrt{ic}\,\eta) - 1]\right\} \qquad \text{(II.24)}$$

with D a coefficient yet to be determined.

We now turn our attention to the intermediate layer between the wall and the free stream where again viscous effects must be retained. This can be done in a manner similar to that at the wall by a rescaling of the variables. The location of the critical layer is taken as y_c from the relation $U - c = 0$ at $y = y_c$. Let $y = y_c + (v/\alpha)^q t$ and $\mathbf{v}(y) = (v/\alpha)^r \mathbf{G}(t)$; $q, r > 0$. With the new quantities, Eq. (II.1) becomes

$$i(v/\alpha)^{1-4q+r}\mathbf{G}''' - [2i(v/\alpha)\alpha^2 - (U - c)](v/\alpha)^{r-2q}\mathbf{G}''$$
$$+ [i(v/\alpha)\alpha^4 - U'' - \alpha^2(U - c)](v/\alpha)^r\mathbf{G} = 0. \qquad \text{(II.25)}$$

This time the prime refers to differentiation with respect to the variable t. We combine (II.25) with the knowledge that $U(y)$ is continuous at $y = y_c$. Accordingly we can expand U near the critical layer as

$$U(y) = U(y_c) + U'(y_c)(y - y_c) + U''(y_c)\frac{(y - y_c)^2}{2!} + \cdots. \qquad \text{(II.26)}$$

Recognizing that $U(y_c) = c$ we have

$$U - c = U'(y_c)(v/\alpha)^q t + U''(y_c)(v/\alpha)^{2q}\frac{t^2}{2!} + \cdots. \qquad \text{(II.27)}$$

Substituting (II.27) into (II.25) yields the approximate leading equation

$$i(v/\alpha)^{1-4q+r}\mathbf{G}'''' - [2i(v/\alpha)^{1+r-2q}\alpha^2 - U'(y_c)(v/\alpha)^{r-q}t]\mathbf{G}''$$

$$+ [i(v/\alpha)^{1+r}\alpha^4 - U''(y_c)(v/\alpha)^{2q+r} - \alpha^2 U'(y_c)(v/\alpha)^{q+r}]\mathbf{G} = 0. \quad \text{(II.28)}$$

The relations for the coefficients are

$$1 - 4q + r = 0 \qquad\qquad\qquad\qquad \text{(II.29)}$$

and

$$r - q = 0 \qquad\qquad\qquad\qquad \text{(II.30)}$$

or $r = q = \tfrac{1}{3}$. In other words, the critical layer has a thickness different from that at the wall and is of order $(v/\alpha)^{1/3}$. The complete transformed equation is

$$i\mathbf{G}'''' - [2i(v/\alpha)^{2/3}\alpha^2 - U'(y_c)t]\,\mathbf{G}''$$

$$+ [i(v/\alpha)^{4/3}\alpha^4 - U''(y_c)(v/\alpha) - \alpha^2 U'(y_c)(v/\alpha)^{2/3}]\mathbf{G} = 0. \quad \text{(II.31)}$$

An expansion for obtaining solutions to (II.31) is generated by taking

$$\mathbf{G}(t) = \mathbf{G}_0(t) + (v/\alpha)^{1/3}\,\mathbf{G}_1(t) + (v/\alpha)^{2/3}\,\mathbf{G}_2(t) + \cdots. \quad \text{(II.32)}$$

To lowest order, then, we must have

$$\mathbf{G}_0'''' - iU'(y_c)\,t\mathbf{G}_0'' = 0, \qquad\qquad \text{(II.33)}$$

which is again a balance between convection and diffusion of vorticity. Equation (II.33) can be recognized as the well-known Airy equation when $\mathbf{G}_0''$ is considered as the dependent variable; the properties and the solutions are well reviewed in almost any book which deals with Bessel functions. Generally speaking, there are four solutions for $\mathbf{G}_0$: two correspond to solutions which grow as t (or y) increases and the other two correspond to solutions that decay as t (or y) increases. Since we desire the overall solution to be bounded as $y \to \infty$, the two coefficients belonging to the growing solutions must be set equal to zero. This leaves us with but two undetermined coefficients. On either side of the critical layer, we have a solution with only one unknown coefficient, making a net total of four, exactly the number we should have for a solution to the Orr-Sommerfeld equation.

The final step in the inner and outer expansion procedure is to match the solution so that we have a relation over the entire flow region. This operation is done by rewriting the adjacent solutions in terms of the neighbor's independent variable and matching in the limit. In this way the coefficients are evaluated and a uniformly valid solution for **v** is determined. For example, let us consider the matching between the outer solution, **v**^{o} say, which is found from Eq. (II.9) and that at the critical layer which we shall call **v**^{i} from Eq. (II.33). The composite result will be a uniform expansion from the critical layer to infinity which can be symbolically given as

$$\mathbf{v} = \mathbf{v}^o + \mathbf{v}^i - (\mathbf{v}^i)^o,$$

where $(\mathbf{v}^i)^o$ implies that an outer expansion of the inner solution has been made. An identical result can be had if we prefer to write ,

$$\mathbf{v} = \mathbf{v}^o + \mathbf{v}^i - (\mathbf{v}^o)^i.$$

In short, the inner and outer expansions are related by the stretching or scale transformations that we have systematically made with the variables. A similar process can be used between the solution at the wall and the critical layer.

If we refer to the classical method of analysis we will find that all computations have been made in the complex plane. Moreover, two pairs of solutions are considered: one set is attributed to viscosity and a second is inviscid. In order to make a uniformly valid solution at the end, some consideration must be given to the integration involved in the answers so that the singular point (critical layer) is properly taken into account. Hence, the reason for the discussion of the location of this point. It turns out that the search for the proper path of integration provides the link of similarity with this approach to that of the inner and outer expansion method. The determination of the proper path is the same as making the correct matching of the respective inner and outer solutions.

Use of the integration path can be made with the computer too. We could actually integrate the equation with $v = 0$ by first moving along the real axis to the critical layer and then switch to the imaginary axis until we made the crossing. Once on the other side, the real axis can be followed to the wall.

A few last remarks should be recorded. The outline of the inner-outer expansions has been made for a particular case. In essence we have treated the situation of the boundary layer where the critical layer is clearly distinct from the wall and the free stream. Should the critical layer occur very close to the wall or further out toward the free stream, then the entire procedure must be reconsidered. And, of course, other laminar flows besides the boundary layer type must be examined for other points or different behavior. A more general treatment of the problem as to what might be expected for practically every possibility and combination has been given by Graebel (1966). Finally, it should be emphasized that, although we did not consider such possibilities in this review, the method of inner and outer expansions can treat with equal success any of the important variations in the problem due to changes in the operating bases. This would include the spatial initial value problem where α is complex and ω is real. Also a variable or complex viscosity can likewise be incorporated.

Technique of Orthogonalization

The Kaplan method is related to a much more general and powerful procedure which prevents a set of vectors from becoming parallel. Let us begin with a simple differential equation:

$$f'' + a(y)f' + b(y)f = 0, \tag{III.1}$$

where all quantities are real.

At any point along the y axis, the solution can be expressed in terms of the local modes

$$f = A(y)e^{py} + B(y)e^{qy}, \tag{III.2}$$

where $p(y)$ and $q(y)$ are the roots of

$$p^2 + ap + b = 0. \tag{III.3}$$

If the coefficients a and b are constants, A and B will also be constants. We can suppose that p is much larger than q so that we have the typical difficulty which is encountered throughout this book. As for boundary conditions, we shall only say that the emphasis is placed on the B mode, in the range $0 \leq y \leq 1$.

Following the classical method, we make two integration passes, starting from $y = 0$ with $A = 1$, $B = 0$ and $A = 0$, $B = 1$, respectively. As we know, both solutions tend to grow because the A mode becomes dominant.

The quantities Ae^{py} and Be^{qy} can be treated as the coordinates of a point in a mode space. The first pass then corresponds to a vector which varies with y. Initially the vector is of unit length and along the A axis. The second pass corresponds to another vector, initially along the B axis. As y grows, this vector rotates toward the A axis and becomes very large. The application of boundary conditions at the end of the range depends crucially upon the small angular difference between these two large vectors. This is where the accuracy of the computer is very insufficient.

The method of Kaplan amounts to reconsidering the initial conditions of the second pass at several stations. The initial conditions are not actually modified because that would be below the level of accuracy of the computer. This gives a set of piecewise solutions, which stay in the vicinity of the B axis.

If the amplification is very large, the computer may become unable to handle the first pass. Furthermore, a procedure that follows strictly the B axis during the second pass would be more elegant and perhaps more accurate.

The function f and f' are related to Ae^{py} and Be^{qy} by a linear transformation, well defined at every point. Thus each vector in the mode space corresponds to a vector in the phase space, whose coordinates are f and f'. In the phase space, the two passes tend to run away along a line whose direction is given by the vector $f = 1, f' = p(y)$. It is in this space that the procedure of orthogonalization is generally carried out, but the difference is not essential. In phase space, the initial conditions are usually $f = 1$, $f' = 0$ for one pass and $f = 0, f' = 1$ for the other. Let V_1 and V_2 denote the two vectors obtained from the two initial points, at the first station, say at $y = 0.1$. Following Bellman and Kalaba (1965), we shall now define two new vectors V_1' and V_2' which specify the two solutions that will be integrated between $y = 0.1$ and the next station, say $y = 0.2$. At $y = 0.1$ we look for relations of the following type:

$$V_i' = a_{ij}V_j \qquad\qquad (\text{III.4})$$

such that V_1', V_2' will be normalized and orthogonal. The desired properties impose three conditions on the four coefficients a_{ij}. In the Kaplan method, the conditions are $a_{11} = a_{22} = 1$ instead of normalization, $a_{12} = 0$, with some suitable choice of a_{21}. In the Gram-Schmidt procedure the fourth condition is also $a_{12} = 0$.

The procedure is repeated at each station, with the location of the station being arbitrary. At the end of the range of integration, the boundary conditions are applied as usual.

With the Orr-Sommerfeld equation, there are four complex modes, but two must be zero in the free stream, such as for boundary layers or similar flows. The search for a general solution formally requires four passes, and (III.4) must be generalized. However two solutions are identically zero through the entire range, so that each station is reached with two complex vectors only and there is only one arbitrary coefficient a_{12}.

In more complicated problems, we have N modes with $N = 4, 6, 8$, etc. Thus there are N^2 coefficients. Normalization imposes N equations, and orthogonalization adds $N(N-1)/2$ relations. Let us now select $N/2$ vectors, from the set $V_1' V_2' \cdots V_N'$ and decide that, as we leave each station, these vectors will again be parallel to their original direction at $y = 0$. This means that $N - 1$ components of each vector must vanish, and this gives $N(N-1)/2$ additional relations. This leaves no arbitrary coefficient a_{ij}. In selecting the $N/2$ modes, it might be advantageous, for accuracy, to give priority to those less likely to grow. Thus the pressure comes first, next the electric current if the magnetic Reynolds number is low. If the Prandtl number is large, vorticity should be selected rather than entropy, etc.

The procedure of orthogonalization has been used by Wazzan *et al.* (1966) for incompressible two-dimensional flows with Reynolds numbers as large as 500,000 (see Fig. 14.6).

References Cited

Abernathy, F. H., and Kronauer, R. E. (1962). The formation of vortex sheets. *J. Fluid Mech.* **13**, 1–20.

Alder, B., and Fernbach, S., eds. (1964). "Methods in Computational Physics Vol 3: Fundamental Methods in Hydrodynamics." Academic Press. New York.

Alekseyev, Y. N., and Korotkin, A. I. (1966). Influence of transverse velocity on the stability of the laminar flow (in Russian). *Izv. Akad. Nauk SSSR Mekhan. Zhid kosti i Gaza* **1**, 32–36.

Batchelor, F. K., and Gill, A. E. (1962). Analysis of the stability of axisymmetric jets. *J. Fluid Mech.* **14**, 529–551.

Bellman, R. E., and Kalaba, R. E. (1965)."Quasilinearization and Boundary-Value Problems." American Elsevier, New York.

Bénard, H. (1901). Les tourbillons cellulaires dans une nappe liquide transportant de la chaleur par convection en régime permanent. *Ann. Chimie (Paris)* **23**, 62–144.

Bénard, G. (1908). Formation des centres de giration à l'arrière d'un obstacle en mouvement. *Compt. Rend.* **147**, 839–842.

Benjamin, T. B. (1960). Effects of a flexible boundary on hydrodynamic stability. *J. Fluid Mech.* **9**, 4, 513–532.

Benjamin, T. B. (1962). Development of three-dimensional disturbances in an unstable film flowing down an inclined plane. *J. Fluid Mech.* **10**, 401–419.

Benjamin, T. B. (1963). The threefold classification of unstable disturbance in flexible surfaces bounding inviscid flows. *J. Fluid Mech.* **16**, 436–450.

Benney, D. J. (1961). A non-linear theory for oscillations in a parallel flow. *J. Fluid Mech.* **10**, 2, 209–236.

Benney, D. J. (1964). Finite amplitude effects in an unstable boundary layer. *Phys. Fluids* **7**, 319–326.

Benney, D. J., and Lin, C. C. (1960). On the secondary motion induced by oscillations in shear flow. *Phys. Fluids* **3**, 656–657.

Betchov, R. (1960a). On the mechanism of turbulent transition. *Phys. Fluids* **3**, 1026–1027.

Betchov, R. (1960b). Simplified analysis of boundary layer oscillations. *J. Ship Res.* **4**, 2, 37–54.

Betchov, R. (1961). Thermal agitation and turbulence. *Proc. Second Intern. Symp. Rarefied Gas Dynamics* pp 307–321. Academic Press, New York.

Betchov, R. (1965). Stability of parallel flows with frequency-dependent viscosity. *Phys. Fluids* **8**, 1910–1911.

Betchov, R., and Criminale, W. O. (1964). Oscillations of a turbulent flow. *Phys. Fluids* **7**, 1915–1919.

Betchov, R., and Criminale, W. O. (1966). Spatial instability of the inviscid jet and wake. *Phys. Fluids* **9**, 359–362.

Betchov, R., and Szewczyk, A. B. (1963). Stability of a shear layer between parallel streams. *Phys. Fluids* **6**, 1391–1396.

Bickley, W. (1937). The plane jet. *Phil. Mag.* **23**, 727–731.

Birdsall, C. K., Brewer, G. R., and Haeff, A. V. (1953). The resistive wall amplifier. *Proc. I. R. E.* **41**, 865–875.

Birkoff, G., and Fisher, J. (1959). Do vortex sheets roll up? *Circolo Mat. Palermo Rendi.* **8**, 77–90.

Blasius, H. (1908). Grenzschichten in Flüssigkeiten mit kleiner Reibung. *Z. Angew. Math. Phys.* **56**, (Transl.: Boundary layers in fluids of small viscosity. N. A. C. A. Tech. Memo. 1256, February 1950).

Briggs, R. J. (1964). "Electron-Stream Interaction with Plasmas." M. I. T. Press, Cambridge, Massachusetts (Research monograph No. 29).

Brown, W. B. (1959). Numerical calculation of the stability of cross flow profiles in laminar boundary layers on a rotating disc and on a swept back wing and an exact calculation of the stability of the Blasius velocity profile. Northrop Aircraft Inc., Rep. NAI 59–5.

Brown, W. B. (1961a). A stability criterion for three-dimensional laminar boundary layers. *In* "Boundary Layer and Flow Control" (G. V. Lachmann, ed.), Vol. 2, pp. 913–923. Pergamon, London.

Brown, W. B. (1961b). Exact solution of the stability equations for laminar boundary layers in compressible flow. *In* "Boundary Layer and Flow Control" (G. V. Lachmann, ed.), Vol. 2, pp. 1033–1048. Pergamon, London.

Brown, W. B. (1962). Exact numerical solutions of the complete linearized equations for the stability of compressible boundary layers. Northrop Aircraft Inc., Norair Division Rep. NOR–62–15.

Brown, W. B. (1965). Stability of compressible boundary layers including the effects of two-dimensional linear flows and three dimensional disturbances. Northrop Aircraft Inc., Norair Division Rep.

Burgers, J. M. (1948). A mathematical model illustrating the theory of turbulence. *Advan. Appl. Mech.* **1**, 171–199.

Chandrasekhar, S. (1961), "Hydrodynamic and Hydromagnetic Stability." Oxford Univ. Press, London and New York.

Cheng, S. I. (1953). On the stability of laminar boundary layer flow. *Quart. Appl. Math.* **11**, 346–350.

Chiarulli, P., and Freeman, J. C. (1948). Stability of the boundary layer. Tech. Rep. No. F–TR/1197–1A. Headquarters, Air Material Command, Dayton, Ohio.

Chu, B. T., and Kovásznay, L. S. G. (1958). Non-linear interactions in a viscous heat-conducting compressible gas. *J. Fluid Mech.* **3**, 494.

Clauser, F. H. (1963). Concept of field modes and the behavior of the magnetohydrodynamic field. *Phys. Fluids* **6**, 231–253.

Clauser, F. H., and Clauser, M. U. (1937). The effect of curvature on the transition from laminar to turbulent boundary layer. N. A. C. A. Tech. Note 613.

Cŏlak-Antić, P. (1964a). Hitzdrahtmessungen des laminar-turbulenten Umschlags bei freier Konvektion. *WGLR Jahrbuch (Wiss. Ges. Luftfahrt- und Raumfahrtforschung)*, pp. 172–176.

Cŏlak-Antić, P. (1964b). Dreidimensionale Instabilitätserscheinungen des laminar-turbulenten Umschlages bei freier Konvection längs einer vertikalen geheizten Platte. *Heidelberger Akad. Wiss. Math. Naturwiss. Kl. Sitzungsber.* No. 6.

Conrad, P. W., and Criminale, W. O. (1965a). The stability of time-dependent laminar flow: parallel flows. *Z. angew. Math. Phys.* **16**, 233–254.

Conrad, P. W., and Criminale, W. O. (1965b). The stability of time-dependent laminar flow: flow with curved streamlines. *Z. angew. Math. Phys.* **16**, 569–582.

Corcos, G. M., and Sellars, J. R. (1959). On the stability of fully developed flow in a pipe. *J. Fluid Mech.* **5**, 97–113.

Cowling, T. G. (1957). "Magnetohydrodynamics." Wiley (Interscience), New York.

Criminale, W. O. (1960). Three-dimensional laminar instability. AGARD Rep. No. 266.

Criminale, W. O., and Kovásznay, L. S. G. (1962). The growth of localized disturbances in a laminar boundary layer. *J. Fluid Mech.* **14**, 59–80.

Demetriades, A. (1958). An experimental investigation of the stability of the hypersonic laminar boundary layer. California Institute of Technology, Guggenheim Aeronautical Laboratory, Hypersonic Research Project, Memo. No. 43.

Drazin, P. G. (1958). The stability of a shear layer in an unbounded heterogeneous inviscid fluid. *J. Fluid Mech.* **4**, 214–224.

Drazin, P. G. (1962). On stability of parallel flow of an incompressible fluid of variable density and viscosity. *Proc. Cambridge Phil. Soc.* **58**, 646–661.

Drazin, P. G., and Howard, L. N. (1962). The instability to long waves of unbounded parallel inviscid flow. *J. Fluid Mech.* **14**, 257–283.

Dryden, H. L. (1948). Recent advances in the mechanics of boundary-layer flow. *Advan. Appl. Mech.* **1**, 1–40.

Dunn, D. W., and Lin, C. C. (1955). On the stability of the laminar boundary layer in a compressible fluid. *J. Aeron. Sci.* **22**, 455–477.

Eckhaus, W. (1962a). Problèmes non linéaires dans la théorie de la stabilité. *J. Mécan.* **1**, 49–77.

Eckhaus, W. (1962b). Problèmes non linéaires de stabilité dans un espace à deux dimensions. I. Solutions péridoques. *J. Mécan.* **1**, 413–438

Eckhaus, W. (1963). Problèmes non linéaires de stabilité dans un espace à deux dimensions. II. Stabilité des solutions péridoques. *J. Mécan.* **2**, 153–172.

Eckhaus, W. (1965). "Studies in Non-linear Stability Theory." Springer, Berlin.

Eichhorn, R. (1962). Measurement of low speed gas flows by particle trajectories: a new determination of free convection velocity profiles. *Intern. J. Heat Mass Transfer* **5**, 915.

Erdélyi, A. (1963). Singular perturbations of boundary value problems involving ordinary differential equations. *J. Soc. Ind. Appl. Math.* **11**, 105–116.

Esch, R. E. (1957). The instability of a shear layer between two parallel streams. *J. Fluid Mech.* **3**, 289–303.

Evans, M. W., and Harlow, F. H. (1958). Calculation of supersonic flow past an axially symmetric cylinder. *J. Aeronaut. Sci.* **25**, 269–270

Fromm, J. E. (1964). The time dependent flow of an incompressible viscous fluid. *In* "Methods in Computational Physics" (B. Alder, ed), Vol. 3, pp. 346–382. Academic Press, New York.

Fromm, J. E., and Harlow, F. H. (1963). Numerical solution of the problem of vortex street development. *Phys. Fluids* **6**, 975–982.

Gallagher, A. P., and Mercer, A. M. (1962). On the behaviour of small disturbances in plane Couette flow. I. *J. Fluid Mech.* **13**, 91–100.

Gallagher, A. P., and Mercer, A. M. (1964). On the behaviour of small disturbances in plane Couette flow. II. The higher eigenvalues. *J. Fluid Mech.* **18**, 350–352.

Gaster, M. (1963). A note on a relation between temporally increasing and spatially increasing disturbances in hydrodynamic stability. *J. Fluid Mech.* **14**, 222–224.

Gaster, M. (1965a). The role of spatially growing waves in the theory of hydrodynamic stability. *Prog. Aeron. Sci.* **6**, 251–270.

Gaster, M. (1965b). On the generation of spatially growing waves in a boundary layer. *J. Fluid Mech.* **22**, 433–441.

Gill, A. E. (1965). On the behaviour of small disturbances to Poiseuille flow in a circular pipe. *J. Fluid Mech.* **21**, 145–172.

Görtler, H. (1940a). Über eine dreidimensionale Instabilität laminarer Grenzschichten an konkaven Wänden. *Nachr. Akad. Wiss, Goettingen Math-Physik Kl. IIa, Math-Physik-Chem. Abt.* **2**, 1–26 (Transl.: On the three-dimensional instability of laminar boundary layers on concave walls. N. A. C. A. Tech. Memo. 1375, June 1954).

Görtler, H. (1940b). Über den Einfluss der Wandkrümmung auf die Enstehung der Turbulenz. *Z. Angew. Math. Mech.* **20**, 138–147.

Görtler, H., and Witting, H. (1958). Theorie der sekundären Instabilität der laminaren Grenzschichten. International Union of Theoretical and Applied Mechanics, *Grenzschichtforschung* (*Boundary Layer Research Symposium*), *Freiburg, 1957* pp. 110–126. Springer, Berlin.

Gold, H. (1963). Stability of laminar wakes. Ph. D. Thesis, California Institute of Technology.

Gotoh, K. (1965). The damping of the large wave number disturbance in a free boundary layer flow. *J. Phys. Soc. Japan* **20**, 164–169.

Graebel, W. P. (1966). On determination of the characteristic equations for the stability of parallel flows. *J. Fluid Mech.* **24**, 497–508.

Greenspan, H. P., and Carrier, G. F. (1959). The magnetohydrodynamic flow past a flat plate. *J. Fluid Mech.* **6**, 77–96.

Gregory, N., Stuart, J. T., and Walker, W. S. (1955). On the stability of three dimensional boundary layers with application to the flow due to a rotating disk. *Phil. Trans. Roy. Soc.* (*London*) **A248**, 155–199.

Grohne, R. (1954). Über das Spectrum bei Eigenschwingungen ebener Laminar-Strömungen. *Z. Angew. Math. Mech.* **35**, 355–357.

Hagen, G. (1855). Über den Einfluss der Temperatur auf die Bewegung des Wassers in Rohren. *Math. Abh. Akad. Wiss.* (aus dem Jahr 1854), 17–98.

Hama, F. R. (1962). Streakline in a perturbed shear flow. *Phys. Fluids* **5**, 644–650.

Hama, F. R., and Burke, E. R. (1960). On the rolling-up of a vortex sheet. University of Maryland, Institute for Fluid Dynamics and Applied Mathematics, Tech. Note BN-220.

Hamming, R. W. (1962). "Numerical Methods for Scientists and Engineers." McGraw Hill, New York.

Harlow, F. H., and Fromm, J. E. (1964). Dynamics and heat transfer in the von Kármàn wake of a rectangular cylinder. *Phys. Fluids* **7**, 1147–1156.

Hartmann, J. (1937). Theory of the laminar flow of an electrically conductive liquid in a homogeneous magnetic field. *Danske Videnskab. Selskab. Math.-Fys. Medd.* **15**, No. 6.

Harvard Computation Laboratory (1945). "Tables of the Modified Hankel Functions of Order One-Third and of Their Derivatives." Harvard Univ. Press., Cambridge, Massachusetts.

Hasimoto, H. (1960). Magnetohydrodynamic wakes in a viscous conducting fluid. *Rev. Mod. Phys.* **32**, 860–866.

Heisenberg, W. (1924). Über Stabilität und Turbulenz von Flüssigkeitsströmen. *Ann. Physik* **74**, 577–627.

Helmholtz, H. (1868). Über discontinuirliche Flüssigkeits-Bewegungen. *Akad. Wiss., Berlin, Monatsber.* 215.

Høiland, E. (1953). On two-dimensional perturbations of linear flow. *Geofys. Publikasjoner, Norske Videnskaps-Akad. Oslo* **18**, No. 9, 1–12.

Holstein, H. (1950). Über die äussere und innere Reibungsschicht bei Störungen laminar Strömungen. *Z. Angew. Math. Mech.* **30**, 25–49.

Howard, L. N. (1964). The number of unstable modes in hydrodynamic stability problems. *J. Mécan.* **3**, 433–443.

Kakevich, F. P., and Krapivin, A. M. (1958). Investigation of heat transfer and of the aerodynamical resistance in tube assemblies when the flow of gas is dust-laden (in Russian). *Isv. Vysshikh Uchebn. Zavedenii, Energ.* No. 1, 101–107.

Kaplan, R. E. (1964). The stability of laminar incompressible boundary layers in the presence of compliant boundaries. Massachusetts Institute of Technology, Aero-Elastic and Structures Research Laboratory, ASRL–TR 116-1.

Kelvin, Lord. (1880). On a disturbance in Lord Rayleigh's solution for waves in a plane vortex stratum. *In* "Mathematical and Physical Papers," Vol 4, pp. 186–187. Cambridge Univ. Press, London and New York.

Kelvin, Lord. (1887a). Rectilinear motion of a viscous fluid between parallel plates. *In* "Mathematical and Physical Papers," Vol. 4, pp. 321–330. Cambridge Univ. Press, London and New York.

Kelvin, Lord. (1887b). Broad river flowing down an inclined plane bed. *In* "Mathematical and Physical Papers," Vol. 4, pp. 330–337. Cambridge Univ. Press, London and New York.

Kendall, J. M. (1966). Supersonic boundary layer stability. (Abstract only.) 1966 Divisional Meeting of the Division of Fluid Dynamics, *Bull. Am. Phys. Soc.*

Klebanoff, P. S. (1954). Characteristics of turbulence in a boundary layer with zero pressure gradient. N. A. C. A. Tech. Note 3178. (Also N. A. C. A. Rep. 1247.)

Klebanoff, P. S., and Tidstrom, K. D. (1959). Evolution of amplified waves leading to transition in a boundary layer with zero pressure gradient. N. A. C. A. Tech. Note D–195.

Klebanoff, P. S., Tidstrom, K. D., and Sargent, L. M. (1962). The three dimensional nature of boundary layer instability. *J. Fluid Mech.* **12**, 1–34.

Kochin, N. E., Kibel, I. A., and Roze, N. V. (1964). "Theoretical Hydrodynamics" (translated from the fifth Russian edition). Wiley (Interscience), New York.

Korotkin, A. I. (1966). Stability of the laminar boundary layer of an incompressible fluid at an elastic surface (in Russian). *Izv. Akad. Nauk SSSR Mekhan.* **1966**, 39–44.

Kovásznay, L. S. G. (1949). Hot-wire investigation of the wake behind cylinders at low Reynolds numbers. *Proc. Roy. Soc. (London)* **A198**, 174–190.

Kovásznay, L. S. G. (1953). Turbulence in supersonic flow. *J. Aeron. Sci.* **20**, 657–675.

Kovásznay, L. S. G., Komoda, H., and Vasudeva, B. R. (1962). Detailed flow field in transition. *Proc. Heat Transfer and Fluid Mech. Inst., Univ. of Washington, 1962* pp. 1–26. Stanford Univ. Press, Stanford, California.

Krylov, N. M., and Bogliubov, N. (1943). "Introduction to Non-linear Mechanics" (Translated by S. Lefschetz). Princeton Univ. Press, Princeton, New Jersey.

Kuethe, A. M., and Raman, K. R. (1959). Some details of the transition to turbulent flow in Poiseuille flow in a tube. Michigan University, Department of Aeronautical and Astronautical Engineering, AFOSR TR 59–84.

Kurtz, E. F., Jr., and Crandall, S. H. (1962). Computer-aided analysis of hydrodynamic stability. *J. Math. Phys.* **41**, 264–279.

Kuwabara, S. (1966). Nonlinear instability of the plane Couette flow. I. U. T. A. M.–I. U. G. G. Symposium on Boundary Layers and Turbulence Including Geophysical Applications, Kyoto, Japan. (To appear in Proceedings as special supplement to *Physics of Fluids*, 1967).

Lamb, H. (1932). "Hydrodynamics," Chap. 9, Sect. 272. Dover, New York.

Lanchon, H., and Eckhaus, W. (1964). Sur l'analyse de la stabilité des écoulements faiblement divergents. *J. Mécan.* **3**, 445–459.

Landahl, M. T. (1962). On the stability of a laminar incompressible boundary layer over a flexible surface. *J. Fluid Mech.* **13**, 4, 609–632.

Landahl, M. T. (1966). A time-shared program system for the solution of the stability problem for parallel flows over rigid or flexible surfaces. Rep. No. ASRL 116–4, Massachusetts Institute of Technology Aeroelastic and Structures Research Laboratory, Cambridge, Massachusetts.

Landahl, M. T., and Kaplan, R. E. (1965). Effect of compliant walls on boundary layer stability and transition. AGARDograph 97–1–353, May 1965.

Landau, L. D. (1944). Stability of tangential discontinuities in compressible fluid. *Dokl. Akad. Nauk. SSSR* **44**, 139–141.

Landau, L. D., and Lifshitz, E. M. (1953). "Mechanics of Continuous Media" (in Russian). Gostekhizdat, Moscow.

Landau, L. D., and Lifschitz, E. M. (1959). "Fluid Mechanics." Pergamon, London.

Laufer, J., and Vrebalovich, T. (1957). Experiments on the instability of a supersonic boundary layer. *Proc. 9th. Intern. Cong. Appl. Mech.* **4**, 121–131.

Laufer, J., and Vrebalovich, T. (1958). Stability of a supersonic laminar boundary layer on a flat plate. California Institute of Technology, Jet Propulsion Laboratory Rep. 20–116.

Laufer, J., and Vrebalovich, T. (1960). Stability and transition of a supersonic laminar boundary layer on an insulated flat plate. *J. Fluid Mech.* **9**, 257–299.

Lees, L. (1947). The stability of the laminar boundary layer in a compressible fluid. N. A. C. A. Tech. Rept. No. 876.

Lees, L., and Gold, H. (1964). Stability of laminar boundary layers and wakes at hypersonic speeds. I. Stability of laminar wakes. *Fundamental Phenomena in Hypersonic Flow, Proc. Internat. Symp. Buffalo, N. Y., 1964* pp. 310–339. Cornell Univ. Press, Ithaca, N. Y.

Lees, L., and Lin, C. C. (1946). Investigation of the stability of the laminar boundary layer in a compressible fluid. N. A. C. A. Tech. Note No. 1115.

Lees, L., and Reshotko, E. (1962). Stability of the compressible laminar boundary layer. *J. Fluid Mech.* **12**, 555–590.

Leite, R. J. (1956). An experimental investigation of the stability of axially symmetric poiseuille flow. Ph. D. Thesis, University of Michigan.

Leite, R. J. (1959). An experimental investigation of the stability of Poiseuille flow. *J. Fluid Mech.* **5**, 81–96.

Lessen, M. (1952). Note on a sufficient condition for the stability of general plane parallel flows. *Quart. Appl. Math.* **10**, 184–186.

Lessen, M., Fox, J. A., Bhat, W. V., and Liu, T. (1964). Stability of Hagen-Poiseuille flow. *Phys. Fluids* **7**, 1384–1385.

Lessen, M., Fox, J. A., and Zien, H. M. (1965). The instability of inviscid jets and wakes in compressible fluid. *J. Fluid Mech.* **21**, 129–143.

Liepmann, H. W. (1943). Investigation of laminar boundary layer stability and transition on curved boundaries. *N.A.C.A. Advisory Conference Rept.* No. 31730.

Lin, C. C. (1944). On the stability of two-dimensional parallel flows. *Proc. Nat. Acad. Sci. U. S.* **30**, 316–323.

Lin, C. C. (1945). On the stability of two-dimensional parallel flows, Parts I, II, III. *Quart. Appl. Math.* **3**, 117–142, 218–234, 277–301.

Lin, C. C. (1955). "The Theory of Hydrodynamic Stability." Cambridge Univ. Press., London and New York.

Lin, C. C. (1961). Some mathematical problems in the theory of the stability of parallel flows. *J. Fluid Mech.* **10**, 430–438.

Lin, C. C., and Rabenstein, A. L. (1960). On the asymptotic solutions of a class of ordinary differential equations of the fourth order. *Trans. Amer. Math. Soc.* **94**, 24–57.

Liu, J. T. C. (1965a). Some physical considerations of the hydrodynamic stability of parallel flows of a dusty gas. Princeton University, Department of Aerospace and Mechanical Sciences, Rept. No. 745.

Liu, J. T. C. (1965b). On the hydrodynamic stability of a parallel dusty gas flows. *Phys. Fluids* **8**, 1939–1945.

Lock, R. C. (1955). The stability of flow of an electrically conducting fluid between parallel planes under a transverse magnetic field. *Proc. Roy. Soc. (London)* **A233**, 105–125.

Mack, L. M. (1960). Numerical calculation of the stability of the compressible, laminar boundary layer. California Institute of Technology, Jet Propulsion Laboratory Rept. No. 20-122.

Mack, L. M. (1965a). Computation of the stability of the laminar compressible boundary layer. *In* "Methods in Computational Physics" (B. Alder, ed.), Vol. 4, pp. 247–299. Academic Press, New York.

Mack, L. M. (1965b). Stability of the compressible layer according to a direct numerical solution. Recent Developments in Boundary Layer Research (AGARDograph 97, part I), 329–362.

Mack, L. M. (1966). Viscous and inviscid amplification rates of two and three-dimensional disturbances in a compressible boundary layer. *Space Prog. Summary* **42**, IV, November.

Meksyn, D., and Stuart, J. T. (1951). Stability of viscous motion between parallel flows for finite disturbances. *Proc. Roy. Soc. (London)* **A208**, 517–526.

Michael, D. H. (1953). Stability of plane parallel flows of electrically conducting fluids. *Proc. Cambridge Phil. Soc.* **49**, 166–168.

Michael, D. H. (1964). The stability of plane Poiseuille flow of a dusty gas. *J. Fluid Mech.* **18**, 19–32.

Michael, D. H. (1965). Kelvin-Helmholtz instability of a dusty gas. *Proc. Cambridge Phil. Soc.* **61**, 569–572.

Michalke, A. (1964). On the inviscid instability of the hypersonic tangent velocity profile. *J. Fluid Mech.* **19**, 543–556.

Michalke, A. (1965). Vortex formation in a free boundary layer according to stability theory. *J. Fluid Mech.* **22**, 371–383.

Michalke, A., and Schade, H. (1963). Zur Stabilität von freien Grenzschichten. *Ing. Arch.* **33**, 1–23.

Miles, J. W. (1957). On the generation of surface waves by shear flows. *J. Fluid Mech.* **3**, 185–204.

Minorsky, N. (1947). "Introduction to Nonlinear Mechanics." Edwards, Ann Arbor.

Morawetz, C. S. (1952). The eigenvalues of some stability problems involving viscosity. *J. Rational Mech. and Anal.* **1**, 579–603.

Nachtsheim, P. R. (1963). Stability of free-convection boundary-layer flows. N. A. S. A. Tech. Note D–2089.

Nachtsheim, P. R. (1964). An initial value method for the numerical treatment of the Orr-Sommerfeld equation for the case of plane Poiseuille flow. N. A. S. A. Tech. Note D–2414.

Nachtsheim, P. R., and Reshotko, E. (1965). Role of conductivity in hydromagnetic stability of parallel flows. N. A. S. A. Tech. Note D–3144.

Nosova, L. N., and Tumarkin, S. A. (1965). "Tables of Generalized Airy Functions." Pergamon, London.

Orr, W. McF. (1907). The stability or instability of the steady motions of a liquid. *Proc. Roy. Irish Acad.* **A27**, 9–138.

Ostrach, S. (1953). Analysis of laminar free-convection flow and heat transfer about a flat plate. N. A. C. A. Rept. 1111 (supercedes N. A. C. A. Tech. Note 2635).

Perkis, C. L. (1948). Stability of a laminar flow through a straight pipe of circular cross-section to infinitesmal disturbances which are symmetrical about the axis of the pipe. *Proc. Natl. Acad. Sci. U. S.* **34**, 285–295.

Powers, J. O., Heiche, G., and Shen, S. F. (1963). The stability of selected boundary-layer profiles. U. S. Naval Ordnance Laboratory, Aerodynamics Res. Rept. No. 186.

Prandtl, L. (1921–1926). Bemerkungen über die Entstehung der Turbulenz. *Z. Angew. Math. Mech.* **1**, 431–436 (1921); *Physik. Z.* **23**, 19–23 (1922). Discussion after Solberg's paper (1924) and with F. Noether, *Z. Angew. Math. Mech.* **6**, 339, 428 (1926).

Prandtl, L. (1929). Einfluss stabilisierender Kräfte auf die Turbulenz. *Vorträge aus dem Gebiete der Aerodynamik und verwandter Gebiete*, Aachen, 1930, 1–17. Springer, Berlin.

Prandtl, L. (1935). *In* "Aerodynamic Theory" (W. F. Durand, ed.), Vol. 3, pp. 178–190. Springer, Berlin.

Pretsch, J. (1941). Über die Stabilität einer Laminarströmung in einem geraden Rohr mit kreisförmingen Querschnitt. *Z. Angew. Math. Mech.* **21**, 204–217.

Pretsch, J. (1952). Excitation of unstable perturbations in a laminar layer. N. A. C. A. Tech. Memo. No. 1343.

Raetz, G. S. (1959). Northrop Aircraft Inc., Norair Division Rept. NOR–59–383 (BLC–121).

Raetz, G. S. (1964). Calculation of precise proper solutions for the resonance theory of transition. I. Theoretical investigations. Contract AF 33 657–11618 Final Rept. Tech. Document Rept. ASD–TDR, Northrop Aircraft Inc., Norair Division, Hawthorne, California.

Rayleigh, Lord (1878). On the stability of jets. *In* "Scientific Papers," Vol. 1, pp. 361–371. Cambridge Univ. Press, London and New York.

Rayleigh, Lord (1880). On the stability or instability of certain fluid motions. *In* "Scientific Papers," Vol. 1, pp. 474–484. Cambridge Univ. Press, London and New York.

Rayleigh, Lord (1887). On the stability or instability of certain fluid motions. II. *In* "Scientific Papers," Vol. 3, pp. 17–23. Cambridge Univ. Press, London and New York.

Rayleigh, Lord (1892a). On the question of the stability of the flow of fluids. *In* "Scientific Papers," Vol. 3, pp. 575–584. Cambridge Univ. Press, London and New York.

Rayleigh, Lord (1892b). On the instability of a cylinder of viscous liquid under capillary force. *In* "Scientific Papers," Vol. 3, pp. 2–23. Cambridge Univ. Press, London and New York.

Rayleigh, Lord (1892c). On the instability of cylindrical fluid surfaces. *In* "Scientific Papers," Vol. 3, pp. 594–596. Cambridge Univ. Press, London and New York.

Rayleigh, Lord (1895). On the stability or instability of certain fluid motions. III. *In* "Scientific Papers," Vol. 4, pp. 203–209. Cambridge Univ. Press, London and New York.

Rayleigh, Lord (1911). Hydrodynamical notes. *Phil. Mag.* **21**, 177–195.

Rayleigh, Lord (1913). On the stability of the laminar motion of an inviscid fluid. *In* "Scientific Papers," Vol. 6, pp. 197–204. Cambridge Univ. Press, London and New York.

Rayleigh, Lord (1914). Further remarks on the stability of viscous fluid motion. *In* "Scientific Papers," Vol. 6, pp. 266–275. Cambridge Univ. Press, London and New York.

Rayleigh, Lord (1915). On the stability of the simple shearing motion of a viscous incompressible fluid. *In* "Scientific Papers," Vol. 6, pp. 341–349. Cambridge Univ. Press, London and New York.

Rayleigh, Lord (1916a). On convection currents in a horizontal layer of fluid when the higher temperature is on the other side. *In* "Scientific Papers," Vol. 6, pp. 432–446. Cambridge Univ. Press, London and New York.

Rayleigh, Lord (1916b). On the dynamics of revolving fluids. *In* "Scientific Papers," Vol. 6, pp. 447–453. Cambridge Univ. Press, London and New York.

Rayleigh, Lord (1917). On the dynamics of revolving fluids. *Proc. Roy. Soc. (London)* **A93**, 148–159.

Reid, W. H. (1965). The stability of parallel flows. *In* "Basic Developments in Fluid Dynamics" (M. Holt, ed.), Vol. 1, pp. 249–307. Academic Press, New York.

Reshotko, E. (1960). Stability of the compressible laminar boundary layer. California Institute of Technology, Guggenheim Aeronautical Laboratory, GALCIT Memo. No. 52.

Reynolds, O. (1883). An experimental investigation of the circumstances which determine whether the motion of water shall be direct or sinuous, and of the law of resistance in parallel channels. *In* "Scientific Papers," Vol. 2, pp. 51–105. Cambridge Univ. Press, London and New York.

Rosenhead, L. (1931). The formation of vortices from a surface of discontinuity. *Proc. Roy. Soc. (London)* **A134**, 170–192.

Roshko, A. (1954). On the development of turbulent wakes from vortex streets. N. A. C. A. Rept. No. 1191.

Rossow, V. J. (1958a). On flow of electrically conducting fluids over a flat plate in the presence of a transverse magnetic field. N. A. C. A. Rept. No. 1358.

Rossow, V. J. (1958b). Boundary-layer stability diagrams for electrically conducting fluids in the presence of a magnetic field. N. A. C. A. Tech. Note No. 4282.

Saffman, P. G. (1962). On the stability of laminar flow of a dusty gas. *J. Fluid Mech.* **13**, 120–128.

Sato, H., and Kuriki, K. (1961). Mechanism of transition in the wake of a thin flat plate placed parallel to a uniform flow. *J. Fluid Mech.* **11**, 321–352.

Schade, H. (1964a). Introduction to the nonlinear hydrodynamical stability theory (in German). *Deut. Versuchsantalt für Luft- und Raumfahrt.* DVL–379.

Schade, H. (1964b). Contribution to the nonlinear stability theory of inviscid shear layers. *Phys. Fluids* **7**, 623–628.

Schensted, I. V. (1960). Contributions to the theory of hydrodynamic stability. Ph. D. Thesis, University of Michigan.

Schlichting, H. (1932a). Über die Stabilität der Couetteströmung. *Ann. Physik (Leipzig)* **14**, 905–936.

Schlichting, H. (1932b). Über die Enstehung der Turbulenz bei der Plattenströmung. *Gessellschaft der Wissenschaften. Göttingen. Mathematisch–Naturwissenschaftliche Klasse.* 160–198.

Schlichting, H. (1933a). Zur Entstehung der Turbulenz bei der Plattenströmung. *Gesellschaft der Wissenschaften. Göttingen. Mathematisch–Physikalische Klasse. Nachrichten,* 181–208.

Schlichting, H. (1933b). Berechnung der Anfachung kleiner Störungen bei der Plattenströmung. *Z. Angew. Math. Mech.* **13**, 171–174.

Schlichting, H. (1933c). Laminar spread of a jet. *Z. Angew. Math. Mech.* **13**, 260–263.

Schlichting, H. (1934). Neuere Untersuchungen über die Turbulenzenstehung. *Naturwiss.* **22**, 376–381.

Schlichting, H. (1935). Amplitudenverteilung und Energiebilanz der kleinen Störungen bei der Plattengrenzschicht. *Gesellschaft der Wissenschatten. Göttingen. Mathematisch–Naturwissenschattliche Klasse,* 1, 47–78.

Schlichting, H. (1960). "Boundary Layer Theory," 4th. ed. McGraw-Hill, New York.

Schmidt, E., and Beckmann, W. (1930). Das Temperatur-und Geschwindigkeitsfeld von einer Wärme abgebenden senkrechten Platte bei natürlicher Konvektion. *Tech. Mech. und Thermodynamik* **1**, Nos. 10 and 11.

Schubauer, G. B., and Klebanoff, P. S. (1955). Contributions on the mechanics of boundary-layer transition. N. A. C. A. Tech. Note 3489.

Schubauer, G. B., and Klebanoff, P. S. (1956). Contributions on the mechanics of boundary-layer transition. N. A. C. A. Rept. 1289.

Schubauer, G. B., and Skramstad, H. K. (1943). Laminar boundary layer oscillations and transition on a flat plate. N. A. C. A. Tech. Rept. No. 909 (Originally issued as N. A. C. A. A. C. R., April 1943).

Serrin, J. (1959). Mathematical principles of classical fluid mechanics. "Handbuch der Physik," Vol. 8, P. 1, pp. 125–263. Springer, Berlin.

Sexl, T. (1927a). Zur Stabilitätsfrage der Poiseuilleschen und Couetteschen Strömung. *Ann. Physik* **83**, 835–848.

Sexl, T. (1927b). Über dreidimensionale Störungen der Poiseuilleschen Strömung. *Ann. Physik* **84**, 807–822.

Shen, S. F. (1952). On the boundary layer equations in hypersonic flow. *J. Aeron. Sci.* **19**, 500–501.

Shen, S. F. (1954). Calculated amplified oscillations in plane Poiseuille and Blasius flows *J. Aeron. Sci.* **21**, 62–64.

Shen, S. F. (1954). Stability of Laminar flows. "Theory of Laminar Flows" (High Speed Aerodynamics and Jet Propulsion, Vol. 4), Section G. Princeton Univ. Press, Princeton, New Jersey.

Singh, K., Lumley, J. L., and Betchov, R. (1963). Modified Hankel functions and their integrals to argument 10. Pennsylvania State University, Engineering Res. Bull. B–87.

Smirnov, A. D. (1960). "Tables of Airy Functions and Special Confluent Hypergeometric Functions." Pergamon, London.

Smith, A. M. O. (1955). On the growth of Taylor-Görtler vortices along highly concave walls. *Quart. Appl. Math.* **13**, 233–262.

Smith, A. M. O. (1957). Transition, pressure gradient and stability theory. *Proc. Ninth Intern. Congr. Appl. Mech.* Vol. 4, pp. 234–244.

Sommerfeld, A. (1908). Ein Beitrag zur hydrodynamischen Erklärung der turbulenten Flüssigkeitsbewegung. *Proc. Fourth Intern. Congr. Mathematicians, Rome* pp. 116–124.

Sproull, W. T. (1961). Viscosity of dusty gases, *Nature* **190**, 976–978.

Squire, H. B. (1933). On the stability of three-dimensional disturbances of viscous flow between parallel walls. *Proc. Roy. Soc. (London)* **A142**, 621–628.

Stuart, J. T. (1954). On the stability of viscous flow between parallel planes in the presence of a co-planar magnetic field. *Proc. Roy. Soc. (London)* **A221**, 189–206.

Stuart, J. T. (1958). On the non-linear mechanics of hydrodynamic stability. *J. Fluid Mech.* **4**, 1–21.

Stuart, J. T. (1960). On the non-linear mechanics of wave disturbances in stable and unstable parallel flows. I. The basic behaviour in plane Poiseuille flow. *J. Fluid Mech.* **9**, 353–370.

Stuart, J. T. (1962). Non-linear effects in hydrodynamic stability. *Proc. Tenth Intern. Congr. Appl. Mech. Stressa* pp. 63–97. Elsevier, Amsterdam.

Stuart, J. T. (1963). Hydrodynamic stability. *In* "Laminar Boundary Layers" (L. Rosenhead, ed.), pp. 629–670. Oxford Univ. Press (Clarendon), London and New York.

Stuart, J. T. (1965). The production of intense shear layers by vortex stretching and convection. Great Britain, National Physical Laboratory, Teddington, England, Aerodynamics Division, NPL Aero Rept. 1147.

Swift, J. (1726). "Travels into Several Remote Nations of the World, by *Lemuel Gulliver.*" Printed for B. Motte, London.

Synge, J. L. (1938). Hydrodynamic stability. "Semi-centennial Publications of the American Mathematical Society," Vol. 2 (Addresses), pp. 227–269. Am. Math. Soc., Providence, Rhode Island.

Szewczyk, A. B. (1962). Stability and transition of the free convection layer. *Intern. J. Heat Mass Transfer* **5**, 903.

Tani, I., and Komoda, H. (1962). Boundary layer transition in presence of streamwise vortices. *J. Aeron. Sci.* **29**, 440–444.

Tatsumi, T. (1952). Stability of the laminar inlet-flow prior to the formation of Poiseuille region. *J. Phys. Soc. Japan* **7**, 489–502.

Tatsumi, T., and Gotoh, K. (1960). The stability of free boundary layers between two uniform streams. *J. Fluid Mech.* **7**, 433–441.

Tatsumi, T., and Kakutani, T. (1958). The stability of a two-dimensional jet. *J. Fluid Mech.* **4**, 261–275.

Tatsumi, T., Gotoh, K., and Ayukawa, K. (1964). The stability of a free boundary layer at large Reynolds numbers. *J. Phys. Soc. Japan* **19**, 1966–1980.

Taylor, G. I. (1923). Stability of a viscous liquid contained between two rotating cylinders. *Phil. Trans. Roy. Soc. (London)* **A223**, 289–343.

Thoman, D. C., and Szewczyk, A. B. (1966). Numerical solutions of time dependent two-dimensional flow of a viscous incompressible fluid over stationary and rotating cylinders. Heat Transfer and Fluid Mechanics Laboratory Tech. Rept. 66–14, University of Notre Dame.

Thomas, L. H. (1953). The stability of plane Poiseuille flow. *Phys. Rev.* **91**, 780–783.

Thompson, W. B. (1962). "An Introduction to Plasma Physics." Addison-Wesley, Reading, Massachusetts.

Tietjens, O. (1925). Beiträge zur Entstehung der Turbulenz. *Z. Angew. Math. Mech.* **5**, 200–217.

Tollmien, W. (1929). Über die Entstehung der Turbulenz. *Gesellschaft der Wissenschaften. Göttingen. Mathematisch-Naturwissenschaftliche Klasse. Nachrichten,* 21–44.

Tollmien, W. (1935). Ein allgemeines Kriterium der Instabilität laminarer Geschwindig-keitsverteilungen. *Gesellschaft der Wissenschaften. Göttingen. Mathematisch-Naturwissenschaftliche Klasse. Nachrichten,* 50, 79–114.

Tollmien, W. (1947). Asymptotische Integration der Störungsdifferentialgleichung ebener laminarer Strömungen bei hohen Reynoldschen Zahlen. *Z. Angew. Math. Mech.* **25/27**, 33–50, 70–83.

Townsend, A. A. (1956). "Structure of Turbulent Shear Flow." Cambridge Univ. Press, London and New York.

Van Dyke, M. (1964). "Perturbation Methods in Fluid Mechanics." Academic Press, New York.

Velikhov, E. P. (1959). Stability of a plane Poiseuille flow of an ideally conducting fluid in a longitudinal magnetic field. *Soviet Phys. JETP (English Transl.)* **4**, 848–855.

Von Kármàn, T. (1912). Über den Mechanismus des Flüssigkeits-und Luftwiderstandes. *Physik. Z.* **13**, 49–59.

Wang, H. (1960). Toward mechanical mathematics. *IBM J. Res. Develop.*

Watson, J. (1960). Three-dimensional disturbances in flow between parallel planes. *Proc. Roy. Soc. (London)* **A254**, 562–569.

Watson, J. (1962). On spatially-growing finite disturbances in plane Poiseuille flow. *J. Fluid Mech.* **14**, 211–221.

Wazzan, A. R., Okamura, T., and Smith, A. M. O. (1966). Spatial stability study of some Falkner-Skan similarity profiles. *Proc. Fifth U. S. Natl. Congr. Appl. Mech.* p. 836. A. S. M. E., University of Minnesota, June 1966.

Wen, K. S. (1963). On the stability of a laminar boundary layer for a Maxwellian fluid. General Electric Missile and Space Division, Tech. Information Division Rept. R63SD102.

Wilkinson, W. L. (1960). "Non-Newtonian Fluids." Pergamon, London.

Witting, H. (1958). Über den Einfluss der Strömlinienkrümmung auf die Stabilität laminarer Strömungen. *Arch. Rational Mech. and Anal.* **2**, 243–283.

Additional Bibliography

Compiled and edited by J. K. Lucker, Assistant Librarian for Science and Technology, Princeton University, Princeton, New Jersey

In general, the following bibliography represents a survey of the literature from 1954 through 1966. For the period prior to 1954 the reader is recommended to refer to the Bibliography on pages 141–153 of C. C. Lin's "The Theory of Hydrodynamic Stability' (Cambridge University Press, London and New York, 1955). None of the references listed in Lin are duplicated here but some items published prior to 1954 and not listed in Lin have been included.

Alishaev, M. G. (1962). Couette flow with a pulsating boundary (in Russian). *Moscow. Universitet. Vestnik. Seriia Matematiki, mekhanika* No. 2, 59–62.

Alterman, Z. (1961). Effect of magnetic field and rotation on Kelvin-Helmholtz instability. *Phys. Fluids* **4**, 1207–1210.

Alterman, Z. (1961). Kelvin-Helmholtz instability in media of variable density. *Phys. Fluids* **4**, 1177–1179.

American Mathematical Society (1962). "Hydrodynamic Stability" (Proc. Symp. Appl. Math.), 13. Amer. Math. Soc. Providence, Rhode Island.

Amsden, A. A., and Harlow, F. H. (1964). Slip instability. *Phys. Fluids* **7**, 327–334.

Andreev, A. F. (1964). Stability of the laminar flow of thin liquid layers. *Soviet Phys. JETP (English Transl.)* **18**, 519–521.

Anliker, M., and Shih Pi, W. –Y. (1963). Effects of geometry and unidirectional body forces on the stability of liquid layers. *Astronaut. Acta* **9**. 325–350.

Arkhipov, V. N. (1957). Stability of a laminar accompanying region (in Russian). *Moscow. Universitet. Vestnik. Seriia Matematiki, Mekhaniki, Astronomii, Fiziki, Khimii* No. 4, 41–44.

Arkhipov, V. N. (1959). Influence of a magnetic field on boundary layer stability (in Russian). *Dokl. Akad. Nauk SSSR* **129**, 751–753 (*English Transl.: Soviet Phys. "Doklady"* **4** (1960), 1199–1201).

Arnold, V. I. (1965). Conditions for nonlinear stability of the plane stationary curvilinear flow of an ideal fluid (in Russian). *Dokl. Akad. Nauk SSSR* **162**, 975–978.

Arnold, V. I. (1965). Variational principle for three-dimensional stationary flows of an ideal fluid (in Russian). *Prikl. Mat. i Mekh.* **29**, 848–851.

Arnold, V. I. (1966). On one apriori estimation of the theory of hydrodynamic stability (in Russian). *Izv. Vysshikh Uchebn. Zavedenii, Mat.* No. 5, 3–5.

Arnold, V. I. (1966). Sur la géometrie différentielle des groupes de Lie de dimension infinite et ses applications à l'hydrodynamique des fluides parfaits. *Ann. Inst. Fourier* **16**, 319–361.

Arnold, V. I. (1966). Sur un principe variationnel pour les écoulements stationnaires des liquides parfaits et ses applications aux problèmes de stabilité non linéaires. *J. Mécan.* **5**, 29–43.

Artiushkov, E. V., and Morozov, A. I. (1964). The longitudinal stability of a one-dimensional conducting gas flow (in Russian). *Teplofiz. Vysokikh Temperatur, Akad. Nauk SSSR* **2**, 525–534.

Axford, W. I. (1960). The stability of plane current-vortex sheets. *Quart. J. Mech. and Appl. Math.* **13**, 314–324.

Becker, E. (1960). Die laminare inkompressible Grenzschicht an einer durch laufende Wellen deformierten ebenen Wand. *Z. Flugwiss.* **8**, 308–316; *Deut. Versuchsanstalt Luftfahrt* **132**.

Bellman, R. E., and Wing, G. M. (1956). Hydrodynamical stability and Poincaré-Lyapunov theory. Rand Corporation Paper P-960.

Ben Daniel, D. J., and Hurwitz, H. (1964). Class of multiply periodic magnetic field configurations exhibiting hydromagnetic stability. *Phys. Fluids* **7**, 1874–1875.

Benjamin, T. B. (1957). Wave formation in laminar flow down an inclined plane. *J. Fluid Mech.* **2**, 554–574.

Benney, D. J., and Greenspan, H. P. (1962). Remarks on transition and the stability of time-dependent shear layers. *Phys. Fluids* **5**, 862–863.

Benney, D. J., and Rosenblat, S. (1964). Stability of spatially varying and time-dependent flows. *Phys. Fluids* **7**, 1385–1386.

Betz, A. (1955). Calculation of the transition boundary layer in outer-flow. *In* "Fifty Years of Boundary Layer Research" (W. Görtler and W. Tollmien, eds.), pp. 63–70. Vieweg, Braunschweig.

Birikh, R. V. (1965). Spectrum of small perturbations of a plane-parallel Couette flow (in Russian). *Prikl. Mat. i Mekh.* **29**, 798–800.

Birikh, R. V., Gershuni, G. Z., and Zhukhovitskii, E. M. (1965). Perturbation spectrum of plane-parallel flows with small Reynolds numbers (in Russian). *Prikl. Mat. i Mekh.* **29**, 88–98.

Bismut, M. (1963). Ecoulement perturbé entre plaques parallèles–recherche per calcul analogique des valeurs propres des coefficients de l'équation Orr-Sommerfeld. Association Internationale pour le Calcul Analogique, *Proc. Intern. Techniques de Calcul Analogique et Numérique en Aeronautique. Liège*, pp. 64–68.

Bjørgum, O. (1956). On the passage from laminar to turbulent flow. *Bergen, Norway. Universitet. Aarbok: Naturvitenskapelig Rekke*, no. 7, 10 pp.

Bloxsom, D. E., Jr. (1965). Transition of air laminar boundary layers. *AIAA J.* **3**, 982–984.

Bodoia, J. R., and Osterle, J. F. (1961). Finite difference analysis of plane Poiseuille and Couette flow developments. *Appl. Sci. Res.* **10A**, 265–276.

Bourrieres, F. –J. (1953). Self-excited oscillations and the intrinsic equations of viscous fluids (in French). Ministère de l'Air, Publications Scientifique et Technique No. 279.

Boyanovitch, D. (1965). On the application of hydrodynamics to the study of the stability of singular points of differential equations–autonomous systems. 1965 *Joint Automatic Control Conference (Preprints of Technical Papers)*, pp. 620–629. Amer. Soc. of Mech. Eng., New York.

Bradshaw, P., Stuart, J. T., and Watson, J. (1960). Flow stability in the presence of finite initial disturbances. AGARD Rept. No. 264.

Brand, R. S. (1964). On the stability of plane parallel flows. *Acta Polytech. Scand. Phys.* No. 31.

Braslow, A. L. (1966). A review of factors affecting boundary-layer transition. N. A. S. A. Tech. Note D–3384.

Braslow, A. L., Knox, E. C., and Horton, E. (1959). Effect of distributed three-dimensional roughness and surface cooling on boundary-layer transition and lateral spread of turbulence at supersonic speeds. N. A. S. A. Tech. Note D–53.

Brillouin, L. (1960). "Wave Propagation and Group Velocity." Academic Press, New York.

Brown, F. T., Graber, S. D., and Wallhagen, R. (1964). Investigation of stability predictions of fluid-jet amplifier systems; semi-annual report. N. A. S. A. CR–54244, Massachusetts Institute of Technology, Engineering Projects Laboratory.

Brushlinskaia, N. N. (1965). On behavior of solutions of hydrodynamical equations when the Reynolds number crosses the critical one (in Russian). *Dokl. Akad. Nauk SSSR* **162**, 431–434.

Bushmanov, V. K. (1961). Hydrodynamic stability of a liquid layer on a vertical wall. *Soviet Phys. JETP (English Transl.)* **12**, 873–877.

Caldwell, D. R. (1964). Oscillating boundary layer in magnetohydrodynamics. *Phys. Fluids* **7**, 1062–1070, 1338–1348.

Carrier, G. F. (1954). Boundary problems in applied mathematics. *Commun. Pure Appl. Math.* **7**, 11–17.

Carrier, G. F., and Chang, C. T. (1959). On an initial value problem concerning Taylor instability of incompressible fluids. *Quart. Appl. Math.* **16**, 436–439.

Carrier, G. F., and Di Prima, R. C. (1957). On the unsteady motion of a viscous fluid past a semi-infinite flat plate. *J. Math. Phys.* **35**, 359–383.

Carrier, G. F., and Greenspan, H. P. (1960). The time-dependent magnetohydrodynamic flow past a flat plate. *J. Fluid Mech.* **7**, 22–32.

Carstens, M. R. (1957). Transition from laminar to turbulent flow during unsteady flow in a smooth pipe. *Proc. 9th Intern. Congr. Appl. Mech.* **3**, pp. 370–378.

Case, K. M. (1960). Edge effects and the stability of plane Couette flow. *Phys. Fluids* **3**, 432–435.

Case, K. M. (1960a). Stability of inviscid plane Couette flow. *Phys. Fluids* **3**, 143–148.

Case, K. M. (1960b). Stability of an idealized atmosphere. I. Discussion of results. *Phys. Fluids* **3**, 149–154.

Case, K. M. (1961). Hydrodynamic stability and the inviscid limit. *J. Fluid Mech.* **10**, 420–429.

Cercignani, C. (1965). Plane Poiseuille flow according to the method of elementary solutions. *J. Math. Anal. Appl.* **12**, 254–262.

Chan, S. K., and Kurtz, E. F., Jr. (1965). On using an analog computer to study hydrodynamic stability. *IEEE Trans. Electron. Computers* **EC-14**, 233–238.

Chang, I. D., and Russell, P. E. (1965). Stability of a liquid layer adjacent to a high-speed gas stream. *Phys. Fluids* **8**, 1018–1026.

Chang, T. S. (1965). An exact analysis of magnetohydrodynamic stability of non-dissipative Couette flow. *Quart. J. Mech. Appl. Math.* **18**, 491–500.

Cheng, S. I. (1956). The unsteady laminar boundary layer on a flat plate. 1956 Heat Transfer and Fluid Mechanics Institute, Preprint No. 14.

Cheng, S. I. (1957). Some aspects of unsteady laminar boundary layer flows. *Quart. Appl. Math.* **14**, 337–352.

Chudov, L. A., and Kuskova, T. V. (1963). Application of difference schemes to calculating nonstationary flows of a viscous incompressible liquid (in Russian). *In* "Numerical Methods in Gas Dynamics" (G. S. Rosliakov and L. A. Chudov, eds.), Vol 2, pp. 190–207. Izdalel'stvo Moskovskogo Universiteta, Moscow. (English translation published by Israel Program for Scientific Translation, Jerusalem, 1966.)

Clay, W. G., Labitt, M., and Slattery, R. E. (1965). Measured transition from laminar to turbulent flow and subsequent growth of turbulent wakes. *AIAA J.* **3**, 837–841.

Clenshaw, C. W., and Elliott, D. A. (1960). A numerical treatment of the Orr-Sommerfeld equation in the case of a laminar jet. *Quart. J. Mech. Appl. Math.* **13**, 300–313.

Conte, S. D., and Miles, J. W. (1959). On the numerical integration of the Orr-Sommerfeld equation. *J. Soc. Ind. Appl. Math.* **7**, 361–366.

Coppi, B. (1963). On the stability of hydromagnetic systems with dissipation. Princeton University, Plasma Physics Laboratory, MATT–143.

Covert, E. E. (1957). The stability of binary boundary layers. Massachusetts Institute of Technology, Naval Supersonic Laboratory, Tech. Rept. 217 (AFOSR–TN–57–200; AD–126 497).

Cox, B. G. (1964). An experimental investigation of the streamlines in viscous fluid expelled from a tube. *J. Fluid Mech.* **20**, 193–200.

Crow, S. C. (1966). The spanwise perturbation of two-dimensional boundary layers. *J. Fluid Mech.* **24**, 153–164.

Curle, N. (1956). Hydrodynamic stability of the laminar mixing region between parallel streams. Great Britain, Aeronautical Research Council, unpublished Rept. 18426.

Curle, N. (1956). On hydrodynamic stability of unlimited antisymmetrical velocity profiles. Great Britain, Aeronautical Research Council, unpublished Rept. 18564.

Curle, N. (1957). On hydrodynamic stability in unlimited fields of viscous flow. *Proc. Roy. Soc. (London)* **A238**, 489–501.

Curle, N. (1958). Hydrodynamic stability of laminar wakes. *Phys. Fluids* **1**, 159–160.

Datta, S. K. (1964). Note on the stability of an elasticoviscous liquid in Couette flow. *Phys. Fluids* **7**, 1915–1919.

Davies, M. H. (1965). A note on the stability of elastico-viscous liquids. *Appl. Sci. Res.* **15A**, 253–260.

Deardorff, J. W. (1963). On the stability of viscous plane Couette flow. *J. Fluid Mech.* **15**, 623–631.

Deardorff, J. W. (1965). Gravitational instability between horizontal plates with shear. *Phys. Fluids* **8**, 1027–1030.

Debler, W. R. (1966). On the analogy between thermal and rotational hydrodynamic stability. *J. Fluid Mech.* **24**, 165–176.

Debye, P., and Daen, J. (1959). Stability considerations on non-viscous jets exhibiting surface or body tension. *Phys. Fluids* **2**, 416–421.

Deem, R. E. (1963). Bibliography on boundary layer transition. Douglas Aircraft Company, Missile and Space Systems Division, SM–43059.

Demetriades, A. (1960). An experiment on the stability of hypersonic laminar boundary layers. *J. Fluid Mech.* **7**, 385–396.

De Santo, D. F., and Keller, H. B. (1962). Numerical studies of transition from laminar to turbulent flow over a flat plate. *J. Soc. Ind. Appl. Math.* **10**, 569–595.

Devan, L. (1965). Approximate solution of the shear flow boundary layer on a flat plate. *Phys. Fluids* **8**, 2211–2215.

Dhawan, S., and Narasimha, R. (1958). Some properties of boundary layer flow during the transition from laminar to turbulent motion. *J. Fluid Mech.* **3**, 418–436.

Dikii, L. A. (1960). On the stability of plane parallel flows of an inhomogeneous fluid (in Russian). *Prikl. Mat. i Mekh.* **24**, 249–257 (*English Transl.: J. Appl. Math. Mech.* **24**, 357–369).

Dikii, L. A. (1960). The stability of plane-parallel flows of an ideal fluid (in Russian). *Dokl. Akad. Nauk SSSR* **135**, 1068–1071 (*English Transl.: Soviet Phys. "Doklady"* **5** (1961), 1179–1182).

Dikii, L. A. (1964). On the stability of plane-parallel Couette flow (in Russian). *Prikl. Mat. i Mekh.* **28**, 389–392 (*English Transl.: J. Appl. Math. Mech.* **28**, 479–483).

Dikii, L. A. (1965). Contribution to the nonlinear theory of hydrodynamic stability (in Russian). *Prikl. Mat. i Mekh.* **29**, 852–855.

Dikii, L. A. (1965). On non-linear theory of stability of zonal flows (in Russian). *Fiz. Atmosfery i Okeana* No. 1, 1117–1122.

Di Prima, R. C. (1954). A note on the asymptotic solutions of the equation of hydrodynamic stability. *J. Math. Phys.* **33**, 249–257.

Di Prima, R. C. (1955). Application of the Galerkin method to problems in hydrodynamic stability. *Quart. Appl. Math.* **13**, 55–62.

Di Prima, R. C., and Dunn, D. W. (1956). The effect of heating and cooling on the stability of the boundary-layer flow of a liquid over a curved surface. *J. Aeron. Sci.* **23**, 913–916.

Di Prima, R. C., and Sani, R. (1965). The convergence of the Galerkin method for the Taylor-Dean stability problem. *Quart. Appl. Math.* **23**, 183–187.

Dolaptschiev, B. (1957). Bemerkungen über die Stabilitätsuntersuchungen der Wirbelstrassen. *Akademie der Wissenschaften, Berlin. Forschungsinstitut für Mathematik. Schriften*, no. 4, 28 pp.

Dolidze, D. E. (1957). Unsteady flow of viscous fluid between parallel porous walls (in Russian). *Dokl. Akad. Nauk SSSR* **117**, 380–383.

Dolph, C. L., and Lewis, D. C. (1958). On the application of infinite systems of ordinary differential equations to perturbations of plane Poiseuille flow. *Quart. Appl. Math.* **16**, 97–110.

Domm, U. (1955). The stability of vortex sheets with consideration of the spread of vorticity of individual vortices. *J. Aeron. Sci.* **22**, 750–754.

Domm, U. (1956). Über die Wirbelstrassen von geringster Instabilität. *Z. Angew. Math. Mech.* **36**, 367–371.

Donnelly, R. J., and Glaberson, W. (1966). Experiments on the capillary instability of a liquid jet. *Proc. Roy. Soc.* (*London*) **A290**, 547–556.

Drazin, P. G. (1960). Stability of parallel flow in a parallel magnetic field at small magnitude Reynolds numbers. *J. Fluid Mech.* **8**, 130–142.

Drazin, P. G. (1961). Discontinuous velocity profiles for the Orr-Sommerfeld equations. *J. Fluid Mech.* **10**, 571–583.

Drazin, P. G. (1961). Stability of a broken-line jet in a parallel magnetic field. *J. Math. Phys.* **39**, 49–53.

Drazin, P. G., and Howard, L. N. (1961). Stability in a continuously stratified fluid. *J. Am. Soc. Civil Eng, Eng. Mech. Div.* **87**, EM6 101–116.

Drazin, P. G., and Howard, L. N. (1966). Hydrodynamic stability of parallel flow of inviscid fluid. *Advan. Appl. Mech.* **9**, 1–89.

Dunn, D. W. (1958). On the theory of the stability of the boundary layer in a compressible fluid. National Research Council of Canada, NAE Rept. (unpublished).

Dunn, D. W. (1960). Stability of laminar flows. National Research Council of Canada, Quarterly Bulletin of the Division of Mechanical Engineering and the National Aeronautical Establishment, July 1–September 30, 1960, 15–58; *Can. Aeron. J.* **7**, (1961) 153–170.

Eagles, P. M. (1966). The stability of a family of Jeffery-Hamel solutions for divergent channel flow. *J. Fluid Mech.* **24**, 191–207.

Eckart, C. (1963). Extension of Howard's circle theorem for adiabatic jets. *Phys. Fluids* **6**, 1042–1047.

Eisler, T. J. (1965). Stability of a shear flow in an unstable layer. Catholic University of America, Department of Space Science and Applied Physics, N. A. S. A. CR–57874; *Phys. Fluids* **8**, 1635–1640.

Eisler, T. J. (1966). Taylor instability in a stratified flow. N. A. S. A. CR–569.

Enig, J. W. (1961). Stability criteria for numerical solutions in unsteady two-dimensional cylindrical Lagrangian flow. *J. Math. Phys.* **40**, 23–32.

Ershin, S. A., and Sakipov, Z. B. (1959). A study of the initial section of a turbulent jet of compressed gas. *Soviet Physics—Tech. Phys. (English Transl.)* **4**, 43–49.

Esch, R. E. (1962). Stability of the parallel flow of a fluid over a slightly heavier fluid. *J. Fluid Mech.* **12**, 192–208.

Faller, A. J. (1963). An experimental study of the instability of the laminar Ekman boundary layer. *J. Fluid Mech.* **15**, 560–576.

Faller, A. J., and Kaylor, R. E. (1965). Investigations of stability and transition in rotating boundary layers. Maryland University, Institute for Fluid Dynamics and Applied Mathematics, Tech. Note BN–427.

Faller, A. J., and Kaylor, R. E. (1966). A numerical study of the instability of the laminar Ekman boundary layer. *J. Atmospheric Sci.* **23**, 466–480.

Fejer, J. A. (1964). Hydromagnetic stability at a fluid velocity discontinuity between compressible fluids. *Phys. Fluids.* **7**, 499–503.

Fejer, J. A., and Miles, J. W. (1963). On the stability of a plane vortex sheet with respect to three-dimensional disturbances. *J. Fluid Mech.* **15**, 335–336.

Feldman, S. (1957). On the hydrodynamic stability of two viscous incompressible fluids in parallel uniform shearing motion. *J. Fluid Mech.* **2**, 343–370.

Ferrio, A., and Vaglio-Lanrin, R. (1956). A note on the effect of centrifugal forces and accelerated motion on the instability of the laminar boundary layer about highly cooled bodies. Brooklyn Polytechnic Institute, Department of Aeronautical Engineering and Applied Mechanics, PIBAL Rept. 313.

Fettis, H. E. (1955). On the integration of a class of differential equations occurring in boundary-layer and other hydrodynamic problems. *Proc. 4th Midwestern Conf. Fluid Mech.* pp. 93–114.

Filler, L., and Ludloff, H. F. (1961). Stability analysis and integration of the viscous equations of motion. *Math. Computation* **15**, 261–274.

Flax, A. H., Treanor, C. E., and Curtis, J. T. (1953). Stability of flow in air induction systems for boundary layer suction. U.S. Air Force, Wright Air Development Center, Tech. Rept. 53–189.

Freeman, N. C. (1955). A theory of the stability of plane shock waves. *Proc. Roy. Soc. (London)* **A228**, 341–362.

Fromm, J. E. (1963). A method for computing nonsteady, incompressible, viscous fluid flows. Los Alamos Scientific Laboratory Rept. LA–2910.

Fussell, D. D., and Hellums, J. D. (1965). The numerical solution of boundary-layer problems. *Am. Inst. Chem. Eng. J.* **11**, 733–739.

Gallagher, A. P., and Mercer, A. M. (1965). On the behaviour of small disturbances in plane Couette flow with a temperature gradient. *Proc. Roy. Soc. (London)* **A286**, 117–128.

Genensky, S. M. (1960). A general theorem concerning the stability of a particular non-Newtonian fluid. *Quart. Appl. Math.* **18**, 245–250.

Gershuni, G. Z., and Zhukhovitskii, E. M. (1963). On parametric excitation of convective instability (in Russian). *Prikl. Mat. i Mekh.* **27**, 779–783 (*English Transl.: J. Appl. Math. Mech.* **27**, 1197–1204).

Gheorghitza, S. I. (1961). The marginal stability in porous inhomogeneous media. *Proc. Cambridge Phil. Soc.* **57**, 871–877.

Gill, A. E. (1962). On the occurrence of condensations in steady axisymmetric jets. *J. Fluid Mech.* **14**, 557–567.

Gill, A. E. (1965a), Instabilities of " top-hat " jets and wakes in compressible fluids. *Phys. Fluids* **8**, 1428–1430.

Gill, A. E. (1965b). A mechanism for instability of plane Couette flow and of Poiseuille flow in a pipe. *J. Fluid Mech.* **21**, 503–511.

Gill, A. E., and Drazin, P. G. (1965). Note on instability of compressible jets and wakes to long-wave disturbances. *J. Fluid Mech.* **22**, 415.

Görtler, H. (1959). On an analogy between instabilities of laminar boundary-layer flows over concave walls and heated walls (in German). *Ingen.-Arch.* **28**, 71–78.

Görtler, H. (1960). Einige neuere Ergebnisse zur hydrodynamischen Stabilitätstheorie. *Z. Flugwiss.* **8**, 1–8.

Görtler, H., and Tollmien, W. (1955). " Fifty Years of Boundary-Layer Research (*Fünfzig Jahre Grenzschichtforschung*)." Vieweg, Braunschweig.

Goldstein, S. (1960). " Lectures on Fluid Mechanics." (Lectures in Applied Mathematics, Vol. 2; Proc. Summer Seminar, Boulder, Colorado, 1957). Wiley (Interscience), New York.

Goldstine, H. H., and Gillis, J. (1955). On the stability of two superposed compressible fluids. *Ann. Mat. Pura ed Applicata* **40**, 261–267.

Goren, S. L. (1962). The instability of an annular thread of a fluid. *J. Fluid Mech.* **12**, 309–319.

Gotoh, K. (1961). Effect of a magnetic field upon the stability of free boundary layers between two uniform streams. *J. Phys. Soc. Japan* **16**, 559–570.

Gourdine, M. C. (1961). Magnetohydrodynamic channel flow of a rotating fluid. " Magnetohydrodynamics" (Proc. 4th Gas Dynamics Symp.), pp. 19–24. North-Western Univ. Press, Evanston, Illinois.

Graebel, W. P. (1960). The stability of a stratified flow. *J. Fluid Mech.* **8**, 321–336.

Graebel, W. P. (1961). Stability of a Stokesian fluid in Couette flow. *Phys. Fluids* **4**, 362–368.

Graebel, W. P. (1964). The hydrodynamic stability of a Bingham fluid in Couette flow. "Second-Order Effects in Elasticity, Plasticity and Fluid Dynamics" (Intern. Symp., Haifa, 1962), pp. 636–649. Pergamon, London.

Grant, R. P., and Middleman, S. (1966). Newtonian jet stability. *J. Am. Inst. Chem. Eng.* **12**, 669–678.

Greene, J. M., and Johnson, J. L. (1962). Stability criterion for arbitrary hydromagnetic equilibria. *Phys. Fluids* **5**, 510–517.

Greenspan, H. P., and Benney, D. J. (1963). On shear-layer instability, breakdown and transition. *J. Fluid Mech.* **15**, 133–153.

Gregory, N., and Love, E. M. (1962). Progress report on an experiment on the effect of surface flexibility on the stability of laminar flow. Great Britain, Aeronautical Research Council, Current Paper 602.

Gregory, N., and Walker, W. S. (1958). Experiments on the use of suction through perforated strips for maintaining laminar flow: transition and drag measurements. Great Britain, Aeronautical Research Council, Reports and Memo. 3083.

Gribov, V. N., and Gurevich, L. E. (1957). On the theory of the stability of a layer located at a superadiabatic temperature gradient in a gravitational field. *Soviet Phys. JETP (English Transl.)* **4**, 720–729.

Grinberg, G. A. (1964). The steady flow of a viscous conducting fluid along rectilinear tubes in a tranverse external magnetic field (in Russian). *Zh. Tekhn. Fiz.* **34**, 1721–1731.

Gupalo, Y. P. (1960). Stability of laminar motion of a fluid with heavy impurities (in Russian). *Izv. Akad. Nauk SSSR Otd. Tekhn. Nauk Mekhan. i Mashinostr.* No. 6, 38–46.

Gurevich, M. I. (1959). On the instability of certain jet flows with free surface. *Soviet Phys. "Doklady"* **4**, 54–56.

Gurevich, V. L. (1964). Growth of fluctuations associated with instability of a system (in Russian). *Zh. Eksperim. i Teor. Fiz.* **46**, 354 (*English Transl.: Soviet Phys. JETP* **19**, 242).

Hämmerlin, G. (1955). Über das Eigenwertproblem der dreidimensionalen Instabilität laminarer Grenzschichten an konkaven Wanden. *J. Rational Mech. and Anal.* **4**, 279–321.

Hämmerlin, G. (1956). Zur Theorie der dreidimensionalen Instabilität laminarer Grenzschichten. *Z. Angew. Math. Phys.* **7**, 156–164.

Hämmerlin, G. (1958). Die Stabilität der Strömung in einem gekrümmten Kanal. *Arch. Rational Mech. and Anal.* **1**, 212–224.

Hagerty, W. W., and Shea, J. F. (1955). A study of the stability of plane fluid sheets. *J. Appl. Mech.* **22**, 509–514.

Hains, F. D. (1963). Comparison of the stability of Poiseulle flow and the Blasius profile for flexible walls. Boeing Scientific Research Laboratories Rept. No. 75.

Hains, F. D. (1964). Additional modes of instability for Poiseuille flow over flexible walls. *AIAA J.* **2**, 1147–1148.

Hains, F. D. (1964). On the stability of MHD channel flow. Boeing Scientific Research Laboratories, D1–82–0331.

Hains, F. D., and Price, J. F. (1962). Stability of plane Poiseuille flow between flexible walls. *Proc. 4th Nat. Congr. Appl. Mech.* Vol. 2, pp. 1263–1268. Amer. Soc. Mech. Eng., New York.

Hains, F. D., and Price, J. F. (1962). Effect of a flexible wall on the stability of Poiseuille flow. *Phys. Fluids* **5**, 365.

Hall, G. R. (1966). Interaction of the wake from bluff bodies with an initially laminar boundary layer. American Institute of Aeronautics and Astronautics, Aerospace Sciences Meeting, 3rd, New York, January 24–26, 1966, Paper 66–126.

Hama, F. R. (1957). Note on the boundary-layer instability on a flat plate stopped suddenly. *J. Aeron. Sci.* **24**, 471–472.

Harris, L. P. (1960). "Hydromagnetic Channel Flows." M. I. T. Press, Cambridge, Massachusetts.

Hasimoto, H. (1963). Magneto-fluid-dynamic wakes. *Appl. Mech. Rev.* **16**, 253–257.

Hayama, S. A. (1963). A study on the hydrodynamic instability in boiling channels. I. The instability in a single boiling channel. *Bull. J. SME* **6**, 549–556.

Hinze, J. O. (1959). "Turbulence," pp. 376–447. McGraw-Hill, New York.

Hirsch, R. (1961). Essai de calcul de la couche limite turbulente et de la transition sur une plaque plane. I. Étude stationnaire. *Technique et Science Aéronautiques* July–August 1961, 273–283.

Hocking, L. M. (1964). The instability of a non-uniform vortex sheet. *J. Fluid Mech.* **18**, 177–186.

Hocking, L. M. (1965). Note on the instability of a non-uniform vortex sheet. *J. Fluid Mech.* **21**, 333–336.

Holman, J. P., Gartrell, H. E., and Soehngen, E. E. (1960). An interferometric method of studying boundary layer oscillations. *J. Heat Transfer* (ASME Trans. Part C) **3**, 264.

Hoult, D. P. (1965). Effect of an axial magnetic field on the stability of an axisymmetric jet or wake. *Phys. Fluids* **8**, 1456–1460.

Howard, L. N. (1959). Hydrodynamic stability of a jet. *J. Math. Phys.* **37**, 283–298.

Howard, L. N. (1963). Neutral curves and stability boundaries in stratified flow. *J. Fluid Mech.* **16**, 333–342.

Howard, L. N., and Drazin, P. G. (1964). On instability of parallel flow of inviscid fluid in a rotating system with variable Coriolis parameter. *J. Math. Phys.* **43**, 83–99.

Howarth, L. (1959). Laminar boundary layers. "Handbuch der Physik," Vol. 8, p. 1 (Strömungsmechanik I, edited by C. Truesdell), pp. 264–350. Springer Verlag, Berlin.

Hruska, A. (1964). Hydromagnetic instability of contra-streaming plasmas. *Czech. J. Phys.* **14B**, 586–593.

Hsieh, D. –Y. (1964). Stability of helium II flow down an inclined plane. *Phys. Fluids* **7**, 1755–1761.

Hsieh, D. –Y. (1965). Stability of a conducting fluid flowing down an inclined plane in a magnetic field. *Phys. Fluids* **8**, 1785–1791.

Hughes, T. H., and Reid, W. H. (1965). On the stability of the asymptotic suction boundary-layer profile. *J. Fluid Mech.* **23**, 715–735.

Hughes, T. H., and Reid, W. H. (1965). The stability of laminar boundary layers at separation. *J. Fluid Mech.* **23**, 737–747.

Huhnt, D. (1959). Stable forms of flow and transition phenomena in a layer of water on a slightly inclined flat plate (in German). *Deut. Versuchsanstalt Luftfahrt* **92**.

Hunt, J. N. (1961). Stable vortex wakes near rigid boundaries. *J. Math. Phys.* **40**, 33–40.

Iagodkin, V. I. (1960). On the theory of stability of flow of viscous fluids in channels. *PMM J. Appl. Math. Mech.* **24**, 1304–1315.

Ingersoll, A. P. (1966). Convective instabilities in plane Couette flow. *Phys. Fluids* **9**, 682–689.

Iudovich, V. I. (1965). An example of generation of the secondary stationary or periodic flow while loosing the stability by laminar flow of viscous incompressible fluids (in Russian). *Prikl. Mat. i Mekh.* **29**, 453–467 (*English Transl.: J. Appl. Math. Mech.* **29**, 527–544).

Iudovich, V. I. (1965). Stability of steady flows of viscous incompressible fluids (in Russian). *Dokl. Akad. Nauk SSSR* **161**, 1037–1040 (*English Transl.: Soviet Phys. "Doklady"* **10**, 293–295).

Iudovich, V. I. (1966). On instability of parallel flows of viscous incompressible fluids relative to space-periodic disturbances (in Russian). *Vychislitel'naia Mat. i Mat. Fiz.* (Suppl.) 242–249.

Ivanilov, I. P. (1960). On the stability of plane-parallel flow of a viscous fluid over an inclined bottom. *J. Appl. Math. Mech.* **24**, 549–552.

Ivanilov, I. P., and Pashinina, L. V. (1965). Stability of long waves in a flow of a viscous incompressible fluid (in Russian). *Izv. Akad. Nauk SSSR Mekh.* Jan.–Feb. 1965, 29–33.

Jaffe, P., and Prislin, R. H. (1966). Effect of boundary-layer transition on dynamic stability. *J. Spacecraft and Rockets* **3**, 46–52.

Jarvis, S., Jr. (1965). Stability of two-phase annular flow in a vertical pipe. U. S. National Bureau of Standards. Tech. Note No. 314.

Jeffrey, A., and Taniuti, T. (1964). "Non-Linear Wave Propagation; with Applications to Physics and Magnetohydrodynamics." Academic Press, New York.

Johnson, J. A. (1963). The stability of shearing motion in a rotating fluid. *J. Fluid Mech.* **17**, 337–352.

Joseph, D. D. (1964). Variable viscosity effects on the flow and stability of flow in channels and pipes. *Phys. Fluids* **7**, 1761–1771.

Joseph, D. D. (1965). Stability of frictionally-heated flow. *Phys. Fluids* **8**, 2195–2200.

Jungclaus, G. (1957). On the stability of laminar flow with three-dimensional disturbances. University of Maryland, Institute of Fluid Dynamics and Applied Mathematics, AFOSR TN 57–577.

Kaganov, S. A. (1962). Stability of laminar flow of an incompressible fluid in a plane channel and circular tube, with consideration of the heat of friction and dependence of viscosity on temperature (in Russian). *Zh. Prikl. Mekhan. i Tekh. Fiz.* No. 3, 96–99.

Kakutani, T. (1964). The hydromagnetic stability of the modified plane Couette flow in the presence of a transverse magnetic field. *J. Phys. Soc. Japan* **19**, 1041–1057.

Kao, T. W. (1964). Stability of two-layer viscous stratified flow down an inclined plane. Catholic University of America, Department of Space Science and Applied Physics, N. A. S. A. CR–57878; *Phys. Fluids* **8**, (1965) 812–820.

Karplus, R., and Watson, K. M. (1958). On the problem of Helmholtz instability. General Dynamics Corporation, ZPh–019.

Kaskel, A. (1961). Experimental study of the stability of pipe flow. California Institute of Technology. Jet Propulsion Laboratory, Tech. Rept. No. 32–138.

Kawamura, R., and Tsien, F. –H. (1956). On the stability of two-dimensional transonic potential flows. *Proc. 6th Japan Nat. Congr. Appl. Mech.* pp. 249–252.

Kelly, R. E. (1965). The stability of an unsteady Kelvin-Helmholtz flow. *J. Fluid Mech.* **22**, 547–560.

Kharin, V. T. (1965). On calculation of eigenvalues by the Boubnov and Galerkin method and its application to the theory of hydrodynamic stability (in Russian). *Prikl. Mat. i Mekhan.* **29**, 1111–1115.

Kiselev, M. I. (1966). On magneto-elastic flutter (in Russian). *Magnitnaya Gidrodinamika* No. 1, 51–59.

Klebanoff, P. S. (1966). The effect of a two-dimensional roughness element on boundary-layer transition. *Proc. Eleventh Intern. Congr. Appl. Mech.* pp. 803–805. Springer, Berlin.

Klebanoff, P. S., Schubauer, G. B., and Tidstrom, K. D. (1955). Measurements of the effect of two-dimensional and three-dimensional roughness elements on boundary-layer transition. *J. Aeron. Sci.* **22**, 803–804.

Kobashi, Y., and Onji, A. (1964). An experimental investigation of stability characteristics of unsteady laminar boundary layer. National Aerospace Laboratory, Tokyo, NAL–TR–65.

Kohlman, D. L., and Mollo-Christensen, E. (1965). Measurement of drag of cylinders and spheres in a Couette-flow channel. *Phys. Fluids* **8**, 1013–1017.

Kolomy, J. (1960). An application of Galerkin's method to stability problems of a stream line viscous flow (in Czech). *Aplikace Matematiky* **5**, 40–44.

Komarov, A. M. (1959). Application of a method of Galerkin type for investigation of the development of perturbed flow of a viscous fluid in a plane channel (in Russian). *Moscow. Univeristet. Vestnik. Seriia Matematika, Mekhanika, Astronomii, Fiziki, Khimii* No. 2, 55–59.

Komarov, A. M. (1964). Development of perturbations in a flow of a viscous fluid through a plane channel (in Russian). *Izv. Akad. Nauk SSSR Mekhan. i Mashinost.* Nov.–Dec. 1964, 142–145.

Korkegi, R. H. (1956). Transition studies and skin-friction measurements on an insulated flat plate at a Mach number of 5.8. *J. Aeron. Sci.* **23**, 97–107.

Korotkin, A. I. (1965). Stability of plane Poiseuille flow in the presence of elastic boundaries (in Russian). *Prikl. Mat. i Mekhan.* **29**, 1122–1127.

Korotkin, A. I. (1966). Stability of a laminar boundary layer in an incompressible liquid with variable physical properties (in Russian). *Izv. Akad. Nauk SSSR Mekhan. Zhidkosti i Gaza* July–Aug. 1966, 76–80.

Koslov, L. F. (1962). The calculation of the transition from a laminar boundary to a turbulent layer under the action of an incident flow (in Russian). *Inzh. Fiz. Zh. Akad. Nauk Belorussk. SSR* **5**, 103–106.

Kováסznay, L. S. G. (1960). A new look at transition. *Proc. Durand Centennial Conf. Aeron. Astron.* pp. 161–172. Pergamon, London.

Kramer, M. O. (1960). Boundary layer stabilization by distributed damping. *J. Amer. Soc. Naval Eng.* **72**, 25–33.

Kreith, F. (1965). Reverse transition in radial source flow between two parallel planes. *Phys. Fluids* **8**, 1189–1190.

Kropik, K. (1964). Contribution to the stability problem of Poiseuille flow (in German). *Acta. Phys. Austraica* **17**, 351–377.

Krylov, A. L. (1963). A proof of the instability of one flow of a viscous incompressible fluid (in Russian). *Dokl. Akad. Nauk SSSR* **153**, 787–789.

Krylov, A. L. (1964). On the stability of a Poiseuille flow in a shallow channel (in Russian). *Dokl. Akad. Nauk SSSR* **159**, 978–981.

Krzywoblocki, M. Z. V. (1956). Jets—review of literature. *Jet Propulsion* **26**, 760–778.

Kudryashev, L. I., and Golovin, V. M. (1961). On the solution of stability of the laminar flow of viscous fluids flowing between flat parallel walls (in Russian). *Izv. Vysshikh Uchebn. Zavedenii, Aviats. Tekhn.* No. 1, 13–18.

Kuethe, A. M. (1956). Some features of boundary layers and transition to turbulent flow. *J. Aeron. Sci.* **23**, 444–452.

Kuo, H. L. (1963). Perturbations of plane Couette flow in stratified fluid and origin of cloud streets. *Phys. Fluids* **6**, 195–211.

Kurtz, E. F., Jr. (1961) Study of the stability of laminar parallel flows. Ph. D. Thesis, Massachusetts Institute of Technology.

Kurzrock, J. W., and Mates, R. E. (1966). Exact numerical solutions of the time-dependent compressible Navier-Stokes equations. American Institute of Aeronautics and Astronautics, Aerospace Sciences Meeting, 3rd, New York, Jan. 24–26, 1966, Paper 66–30.

Kurzweg, U. H. (1963). A note on the stability of generalized Couette flow. *Z. Angew. Math. Phys.* **14**, 380–383.

Kurzweg, U. H. (1965). Convective instability of a hydromagnetic fluid within a rectangular cavity. *Intern. J. Heat Mass Transfer* **8**, 35–41.

Kuznetsov, A. V. (1964). The problem of jet flow past a contour performing small oscillations (in Russian). *Prikl. Mat. i Mekhan.* **28**, 567–571 (*English Transl.: J. Appl. Math. Mech.* **28**, 700–705).

Kuznetsov, V. M. (1965). On a possible oscillating mechanism in magnetohydrodynamics (in Russian). *Inzh. Zh.* **5**, 346–348.

Kwoh-l, A. D. (1960). Small perturbations in the unsteady flow of a rarefied gas based on Grad's thirteen-moment approximation. *Proc. 2nd Intern. Symp. Rarefied Gas Dynamics* pp. 345–367. Academic Press, New York.

Lachmann, G. V. (1954). Laminarization through boundary-layer control. *Aeron. Eng. Rev.* **13**, No. 8, 37–51.

Lachmann, G. V. (1956). Practical application of boundary layer control (in German). *Z. Flugwiss.* **4**, 9–14.

Ladyzhenskaia, O. A. (1963). "The Mathematical Theory of Viscous Incompressible Flow." Gordon and Breach, New York.

Landahl, M. T. (1965). A wave-guide model for turbulent shear flow. McDonnel Aircraft Corporation, N. A. S. A. CR–317.

Lariontsev, E. G. (1964). Some problems of the hydrodynamic and hydromagnetic stability of a cylindrical jet (in Russian). *Prikl. Mat. i Mekhan.* **28**, 962–964.

Laufer, J. (1956). Experimental observation of laminar boundary-layer oscillations in supersonic flow. *J. Aeron. Sci.* **23**, 184–185.

Laval, G. (1963). Équilibres de bifurcation et la stabilité hydromagnétique. *Nucl. Fusion* **3**, 99.

Lehnert, B. (1955). An instability of laminar flow of mercury caused by an external magnetic field. *Proc. Roy. Soc. (London)*, **A233**, 299–302.

Leslie, F. M. (1964). The stability of Couette flow of certain anisotropic fluids. *Proc. Cambridge Phil. Soc.* **60**, 949–955.

Lessen, M. (1957). On the hydrodynamic stability of curved laminar compressible flows. *Proc. 5th Midwestern Conf. Fluid Mech.* pp. 22–28. Univ. of Michigan Press, Ann Arbor, Michigan.

Lessen, M. (1958). On the hydrodynamic stability of curved laminar flows. *Z. Angew. Math. Mech.* **38**, 95–99.

Lessen, M., and Fox, J. A. (1955). The stability of boundary layer type flows with finite boundary conditions. *In* " Fifty Years of Boundary Layer Research " (H. Görtler and W. Tollmien, eds.), pp. 122–126. Vieweg, Braunschweig.

Lessen, M., and Ko, S. –H. (1966). Viscous instability of an incompressible fluid half-jet flow. *Phys. Fluids* **9**, 1179–1183.

Lessen, M., *et al.* (1954). Hydrodynamic stability. Pennsylvania State University, Departments of Aeronautical Engineering and Engineering Research, Tech. Rept. No. 2.

Lessen, M., Fox, J. A., and Zien, H. M. (1965). On the inviscid stability of the laminar mixing of two parallel streams of a compressible fluid. *J. Fluid Mech.* **23**, 355–367.

Lessen, M., Fox, J. A., and Zien, H. M. (1966). Stability of the laminar mixing of two parallel streams with respect to supersonic disturbances. *J. Fluid Mech.* **25**, 737–742.

Lew, H. G. (1955). On the stability of the axially symmetric laminar jet. *Quart. Appl. Math.* **13**, 310–314.

Lick, W. (1965). The instability of a fluid layer with time-dependent heating. *J. Fluid Mech.* **21**, 565–576.

Lighthill, M. J. (1957). The fundamental solution for small steady three-dimensional disturbances to a two-dimensional parallel shear flow. *J. Fluid Mech.* **3**, 113.

Lighthill, M. J. (1963). Boundary layer theory. *In* " Laminar Boundary Layers " (L. Rosenhead, ed.), Chap. 2. Clarendon Press, Oxford.

Liley, B. S. (1962). A generalization of a sufficient condition for hydromagnetic stability. *J. Nucl. Energy: P. C* **4**, 325–328.

Lilly, D. K. (1966). On the instability of Ekman boundary flow. *J. Atmospheric Sci.* **23**, 481–494.

Lin, C. C. (1954). Hydrodynamic stability. "Hydrodynamic Stability " (Proc. 13th Symp. Appl. Math.), pp. 1–18. Amer. Math. Soc., Providence, Rhode Island.

Lin, C. C. (1956). Aspects of the problem of turbulent motion. *J. Aeron. Sci.* **23**, 453–461.

Lin, C. C. (1957). On the stability of the laminar boundary layer. *Proc. Symp. Naval Hydrodynamics* pp. 353–373. National Research Council, Washington, D. C.

Lin, C. C. (1957). On uniformly valid asymptotic solutions of the Orr-Sommerfeld equation. *Proc. 9th Intern. Congr. Appl. Mech.* Vol. 2, pp. 136–148.

Lin, C. C. (1957). Stability of laminar flows. *Appl. Mech. Rev.* **10**, 1–3.

Lin, C. C. (1958). On the instability of laminar flow and its transition to turbulence. International Union of Theoretical and Applied Mechanics, *Grenzschichtforschung Symposium* pp. 144–160. Springer, Berlin.

Lin, C. C., and Benney, D. J. (1966). On the instability of shear flows and their transition to turbulence. *Proc. 11th Intern. Congr. Appl. Mech.* pp. 797–802. Springer, Berlin.

Lingren, E. R. (1957). The transition process and other phenomena in viscous flow. *Arkiv Fysik* **12**, 1–169.

Listrov, A. T. (1965). Flow stability of a viscoelastic fluid flowing down an inclined plane (in Russian). *Zh. Prikl. Mekhan. i Tekhn. Fiz.* Sept.–Oct. 1965, 102–105.

Listrov, A. T. (1965). Stability of parallel flows of non-Newtonian media (in Russian). *Dokl. Akad. Nauk SSSR* **164**, 1001–1004.

Lock, R. C. (1954). Hydrodynamic stability of the flow in the laminar boundary layer between parallel streams. *Proc. Cambridge Phil. Soc.* **50**, 105–124.

Loitsianskii, L. G. (1955). Theory of laminar and turbulent channel flows (in Russian). *Tr. Leningr. Politekhn. Inst.* **176**, 101–114.

Low, F. E. (1961). Persistence of stability in Lagrangian systems. *Phys. Fluids* **4**, 842–846.

Low, G. M. (1955). Stability of compressible laminar boundary layer with internal heat sources or sinks. *J. Aeron. Sci.* **22**, 329–336.

Ludwieg, H. (1964). Experimentelle Nachprüfung der Stabilitätstheorien für reibungsfreie Strömungen mit schraubenlinienförmigen Stromlinien. *Z. Flugwiss.* **12**, 304–309.

Ludwieg, H. (1965). Erklärung des Wirbelaufplatzens mit Hilfe der Stabilitätstheorie für Strömungen mit schraubenlinienförmigen Stromlinien. *Z. Flugwiss.* **13**, 437–442.

Lyshevskii, A. S. (1959). The breakdown of a stream of viscous liquid under the action of unsymmetrical perturbations (in Russian). *Izv. Vysshikh Uchebn. Zavedenii, Energ.* No. 3, 114–123.

Malkus, W. V. R. (1956). Outline of a theory of turbulent shear flow. *J. Fluid Mech.* **1**, 521–539.

Mallick, D. D. (1963). Instability of a surface of discontinuity of velocity in a parallel uniform magnetic field. *J. Fluid Mech.* **16**, 187–196.

Manohar, M. (1960). Stability of flow near an oscillating surface. *J. Sci. Eng. Res.* (*Indian Inst. Technol., Kharagpur*) **4**, 217–238.

Maurin, J. A. (1963). Un modèle simple de la transition en écoulement de Poiseuille. *Compt. Rend.* **256**, 587–589.

Maurin, L. N., and Sorokin, V. S. (1962). On the wave motion of thin layers of a viscous fluid (in Russian). *Zh. Prikl. Mekhan. i Tekhn. Fiz.* No. 4, 60–67.

Mawardi, O. K. (1959). Magnetohydrodynamics—a survey of the literature. *Appl. Mech. Rev.* **12**, 443–446.

McClure, J. D. (1962). On perturbed boundary layer flows. Massachusetts Institute of Technology, Fluid Dynamics Research Laboratory Rept. No. 62–2.

McCracken, D. D. (1965). "A Guide to FORTRAN IV Programming." Wiley, New York.

McKeon, C. H. (1956). On the stability of a laminar wake. Great Britain, Aeronautical Research Council, Current Papers No. 303.

Medvedev, V. A. (1964). On the application of the Bubnov-Galerkin method in the theory of hydrodynamic stability (in Russian). *Prikl. Mat. i Mekhan.* **28**, 780–782 (*English Transl: J. Appl. Math. Mech.* **28**, 948–951).

Meister, E. (1965). Zur Theorie der ebenen, instationären Unterschallströmung um ein schwingendes Profil im Kanal. *Z. Angew. Math. Phys.* **16**, 770–780.

Meksyn, D. (1958). The boundary layer equations of compressible flow seperation. *Z. Angew. Math. Mech.* **38**, 372–379.

Meksyn, D. (1961). "New Methods in Laminar Boundary-Layer Theory." Pergamon, London.

Meksyn, D. (1964). Stability of laminar flow between parallel planes for two- and three-dimensional finite disturbances. *Z. Physik* **178**, 159–172.

Meksyn, D. (1966). The foundations of turbulent motion flow between parallel planes. *Z. Physik* **195**, 485–488.

Menkes, J. (1959). On the stability of a shear layer. *J. Fluid Mech.* **6**, 518–522.

Menkes, J. (1961). On the stability of a heterogeneous shear layer subject to a body force. *J. Fluid Mech.* **11**, 284–290.

Menkes, J. (1962). Stability of a heterogeneous shear layer in a magnetic field. *Phys. Fluids* **5**, 1424–1427.

Menzel, K. (1964). Three-dimensional instability of the boundary layer of a plate. (As reported by W. Velte, Beiträge zur hydrodynamischen Stabilität. *Proc. 11th Intern. Congr. Appl. Mech. Munich, 1964* pp. 806–814. Berlin: Springer, Berlin, 1966.)

Merrill, E. W., Mickley, H. S., and Ram, A. (1962). Instability in Couette flow of solutions of macromolecules. *J. Fluid Mech.* **13**, 86–90.

Meshalkin, L. D., and Sinai, Ia. G. (1961). Investigations of the stability of a stationary solution for the plane movement of an incompressible viscous liquid. *PMM J. Appl. Math. Mech.* **25**, 1700–1705.

Michael, D. H. (1961a). Energy considerations in the instability of a current-vortex sheet. *Proc. Cambridge Phil. Soc.* **57**, 628–637,

Michael, D. H. (1961b). Note on the stability of plane parallel flows. *J. Fluid Mech.* **10**, 525–528.

Michael, D. H. (1962). The stability of an incompressible jet with an aligned magnetic field. *Mathematika* **9**, 162–169.

Michalke, A. (1964). Instability and nonlinear development of a disturbed shear layer (in German). *Ing. Arch.* **33**, 264–276.

Michalke, A. (1965). On spatially growing disturbances in an inviscid shear layer. *J. Fluid Mech.* **23**, 521–544.

Michalke, A., and Wille, R. (1966). Flow processes in the laminar-turbulent transition region of freestream boundary layers (in German). *Proc. 11th Intern. Congr. Appl. Mech.* pp. 962–972. Springer, Berlin.

Michel, R., and Sirieix, M. (1956). A new method of triggering boundary-layer transition (in French). *Compt. Rend.* **243**, 231–233.

Miles, J. W. (1956). On the aerodynamic instability of thin panels. *J. Aeron. Sci.* **23**, 771–780.

Miles, J. W. (1958). On the disturbed motion of a plane vortex sheet. *J. Fluid Mech.* **4**, 538–552.

Miles, J. W. (1959). On the generation of surface waves by shear flows. *J. Fluid Mech.* **6**, 568–582, 583–598.

Miles, J. W. (1960). The hydrodynamic stability of a thin film of liquid in uniform shearing motion. *J. Fluid Mech.* **8**, 593–610.

Miles, J. W. (1960). On the generation of surface waves by turbulent shear flows. *J. Fluid Mech.* **7**, 469–478.

Miles, J. W. (1961). On the stability of heterogeneous shear flows. *J. Fluid Mech.* **10**, 496–508.

Miles, J. W. (1962). A note on the inviscid Orr-Sommerfeld equation. *J. Fluid Mech.* **13**, 427–432.

Miles, J. W. (1963). On the stability of heterogeneous shear flows. II. *J. Fluid Mech.* **16**, 209–227.

Miles, J. W., and Howard, L. N. (1964). Note on a heterogeneous shear flow. *J. Fluid Mech.* **20**, 331-336.

Moeckel, W. E. (1956). Some effects of bluntness on boundary-layer transition and heat transfer at supersonic speeds. N. A. C. A. Tech. Note 3653.

Monaghan, R. J. (1955). On the behaviour of boundary layers at supersonic speeds. *Proc. 5th Intern. Aeron. Conf., Los Angeles, June 20–23, 1955* pp. 277–315. Inst. Aeron. Sci., New York.

Monin, A. S., and Iaglom, A. M. (1965). "Statistical Hydromechanics" (in Russian). Nauka, Moscow.

Moore, F. K. (1956). Three-dimensional boundary layer theory. *Advan. Appl. Mech.* **4**, 159–228.

Moore, F. K. (1964). "Theory of Laminar Flows," Vol. 4 of "High Speed Aerodynamics and Jet Propulsion." Princeton Univ. Press, Princeton, New Jersey.

Moore, R. L. (1964). A general theory of stability of the motion of a fluid with heat transport and arbitrary force fields. Douglas Aircraft Company Rept. 31964.

Morduchow, M., and Grape, R. G. (1955). Separation, stability, and other properties of compressible laminar boundary layer with pressure gradient and heat transfer. N. A. C. A. Tech. Note 3296.

Morduchow, M., Grape, R. G., and Shaw, R. P. (1957). Stability of laminar boundary layer near a stagnation point over an impermeable wall and a wall cooled by normal fluid injection. N. A. C. A. Tech Note 4037.

Murphy, W. D. (1966). Numerical analysis of boundary layer problems. New York University, Courant Institute of Mathematical Sciences, AEC Computing and Applied Mathematics Center, Rept. NYO–1480–63.

Nayfeh, A. H. (1964). Stability of dust laden two-dimensional laminar wakes. Heliodyne Corporation Rept. RR–6 (AD–608712).

Nihoul, J. C. J. (1964). Stability of a current-vortex sheet created by the motion of a semi-infinite plate at right angle to a uniform magnetic field. *Revue Roumaine Math. Pures et Appl.* **9**, 157–165.

Noh, W. F., and Protter, M. H. (1963). Difference methods and the equations of hydrodynamics. *J. Math. Mech.* **12**, 149–191.

Nonweiler, T. (1963). Qualitative solution of the stability equation for a boundary layer in contact with various forms of flexible surface. Great Britain, Aeronautical Research Council, Current Paper 622.

Ostrach, S., and Maslen, S. H. (1961). Stability of laminar viscous flows with a body force. *Intern. Develop. in Heat Transfer*, 1017–1023. Amer. Soc. Mech. Eng., New York.

Ostrach, S., and Thornton, P. E. (1962). Stability of compressible boundary layers induced by a moving wave. *J. Aerospace Sci.* **29**, 289–296.

Oswatitsch, K. (1966). Flow surfaces in weakly perturbed parallel flows (in German). *Z. Flugwiss.* **14**, 14–18.

Ovchinnikov, V. V. (1966). Inducing turbulence to lower the resistance of some fish in motion (in Russian). *Biofizika* **11**, No. 1, 186–188.

Pai, S. I. (1954). On a generalization of Synge's criterion for sufficient stability of plane parallel flows. *Quart. Appl. Math.* **12**, 203–206.

Pai, S. I. (1955). On the stability of parallel flows with respect to periodic disturbances. *J. Franklin Inst.* **259**, 197–208.

Pai, S. I. (1956). "Viscous Flow Theory. I. Laminar Flow." Van Nostrand, Princeton, New Jersey.

Pai, S. I. (1962). On the stability of axisymmetrical flows. General Electric Company, Missile and Space Division, R62SD75.

Pasta, J. R., and Ulam, S. (1959). Heuristic numerical work in some problems of hydrodynamics. *Mathematical Tables and Other Aids to Computation* **13**, 1–12.

Pavlov, B. M., Rosliakov, G. S., and Chudov, L. A. (1965). "Numerical Methods in Gas Dynamics," Vol. 4 (in Russian). Izdalel'stvo Moskovskogo Universiteta, Moscow.

Pavlov, K. B. (1962). Stability of plane Couette flow in the presence of a magnetic field (in Russian). *Vopr. Magnitn. Gidrodinam. i Dinam. Plazmy, Tr. Konf. Akad. Nauk Latv. SSRS, Inst. Fiz.* **2**, 99–106

Pavlov, K. B., and Tarasov, Yu. A. (1960). On the stability of the flow of a viscous conducting liquid between parallel planes in a vertical magnetic field (in Russian). *Prikl. Mat. i Mekhan.* **24**, No. 4, 723–725.

Pezzoli, G. (1964). The stability of vortext streets—the double row of Bérnard-Kármàn (in Italian). *Atti Accad. Naz. Lincei, Rend. Classe Sci, Fis. Mat. Nat.* **36**, 145–152.

Pierce, D. (1964). A brief note on periodic disturbances in a laminar shear layer. Great Britain, Royal Aircraft Establishment, TN–AERO–2970.

Pierce, J. G. (1964). Stability of potential flow. *Phys. Fluids* **7**, 1109–1113.

Plapp, J. E. (1957). The analytic study of laminar boundary-layer stability in free convection. *J. Aeron. Sci.* **24**, 318–319.

Plesset, M. S., and Hsieh, D. –Y. (1964). General analysis of the stability of superposed fluids. *Phys. Fluids* **7**, 1099–1108.

Pocinki, L. S. (1955). The stability of a simple baroclinic flow with horizontal shear. U.S. Air Force, Cambridge Research Center, Geophysics Research Directorate, Geophysics Research Paper No. 38.

Potter, M. C. (1966). Stability of plane Couette-Poiseuille flow. *J. Fluid Mech.* **24**, 609–619.

Princeton University, Plasma Physics Laboratory (1964). Plasma Physics Summer Institute Lecture Notes, Vol. 5: General stability theory in plasma physics.

Quinn, B. (1964). Fréquence de la formation des tourbillons dans un sillage laminaire. *Compt. Rend.* **258**, 5356–5358.

Rabenstein, A. L. (1959). The determination of the inverse matrix for a basic reference equation for the theory of hydrodynamic stability. *Arch. Rational Mech. Anal.* **2**, 355–366.

Raja Gopal, E. S. (1963). Motion and stability of vortices in a finite channel: application to liquid helium II. *Ann. Phys. N.Y.* **25**, 196–220.

Rajapps, N. R., and Chang, I. D. (1966). Interfacial instability of a liquid with constant shear. *Phys. Fluids* **9**, 616–617.

Reid, W. H. (1961). The effects of surface tension and viscosity on the stability of two superposed fluids. *Proc. Cambridge Phil. Soc.* **57**, 415–425.

Reid, W. H. (1961). Inviscid modes of instability in flow over a concave wall. *J. Math. Anal. and Appl.* **2**, 419–427.

Reid, W. H. (1965). Asymptotic approximations in hydrodynamic stability. *In* "Non-Equilibrium Thermodynamics, Variation Techniques and Stability" (R. J. Donnelly and others, eds.), pp. 115–123. Univ. of Chicago Press, Chicago, Illinois.

Reshotko, E. (1958). Experimental study of the stability of pipe flow. I. Establishment of an axially symmetric Poiseuille flow. California Institute of Technology, Jet Propulsion Laboratory, Progress Rept. 20–364.

Reshotko, E. (1962). Stability of three-dimensional compressible boundary layers. N. A. S. A. Tech. Note D–1220.

Reshotko, E. (1963). Transition reversal and Tollmien-Schlichting instability. *Phys. Fluids* **6**, 335–342

Reshotko, E., and Monnin, C. F. (1965). Stability of two-fluid wheel flows. N. A. S. A. Tech. Note D–2696.

Rheinboldt, W. (1956). External boundary values in boundary-layer equations (in German). *Z. Angew. Math. Mech.* **36**, 153–154.

Richtmyer, R. D., and Morton, K. W. (1964). Stability studies for difference equations. I: Non-linear stability. II: Coupled sound and heat flow. New York University, Courant Institute of Mathematical Sciences Rept. NYO–1480–5.

Riis, E. (1962). The stability of Couette-flow in non-stratified and stratified viscous fluids. *Geofys. Publikasjoner, Norske Videnskaps-Akad. Oslo* **23**, No. 4.

Roberts, K. V., and Taylor, J. B. (1965). Gravitational resistive instability of an incompressible plasma in a sheared magnetic field. *Phys. Fluids* **8**, 315–322.

Roberts, P. H. (1964). The stability of hydromagnetic Couette flow. *Proc. Cambridge Phil. Soc.* **60**, 635–651.

Robinson, A. R. (1962). The instability of a thermal ocean circulation. *J. Marine Res.* **20**, 189–200

Rom (Rabinowicz), J. (1962). Dynamic stability of a curved laminar compressible flow. *Amer. Rocket Soc. J.* **32**, 952–954.

Rosenbluth, M. N., and Simon, A. (1964). Necessary and sufficient condition for the stability of plane parallel inviscid flow. *Phys. Fluids* **7**, 557.

Rosenhead, L. (1963). "Laminar Boundary Layers." Oxford Univ. Press, London and New York.

Rosliakov, G. S., and Chudov, L. A. (1963). "Numerical Methods in Gas Dynamics," Vol. 2 (in Russian). Izdalel'stvo Moskovskogo Universiteta, Moscow. (Translated by Israel Program for Scientific Translation, Jerusalem, 1966.)

Rowe, P. W. (1957). An introduction to aerodynamic stability. Sandia Corporation Tech. Memo. 220–56(51).

Rozenshtok, I. L. (1965). An unsteady thermal boundary layer on a semi-infinite plate in a viscous flow (in Russian). *Izv. Akad. Nauk SSSR Mekhan.* Nov.–Dec. 1965, 20–26.

Rubin, S. G. (1963). On the propagation of small amplitude waves in magneto-gasdynamic channel flows. Cornell University, Graduate School of Aerospace Engineering.

Rudraiah, N. (1964). Magnetohydrodynamic stability of heterogeneous incompressible nondissipative conducting liquids. *Appl. Sci. Res.* **11B**, 105–117.

Rudraiah, N. (1964). Magnetohydrodynamic stability of heterogeneous dissipative conducting liquids. *Appl. Sci. Res.* **11B**, 118–133.

Rusbridge, M. G., and Saunders, P. A. (1963). Report of the Culham Laboratory study group on plasma instabilities. Culham Laboratory (England) Rept. CLM–M21.

Sani, R. (1964). Note on flow instability in heated ducts. *Z. Angew. Math. Phys.* **15**, 381–387.

Sani, R. L. (1966). Remarks on the Bénard and related stability problems. *Z. Angew. Math. Phys.* **17**, 642–645.

Sato, H. (1956). Experimental investigation on the transition of laminar separated layer. *J. Phys. Soc. Japan* **11**, 702–709.

Sato, H. (1960). The stability and transition of a two-dimensional jet. *J. Fluid Mech.* **7**, 53–80.

Schade, H. (1961). Zur Stabilitätstheorie axialsymmetrischer Parallel-Strömungen. *Z. Angew. Math. Mech.* **41**, T154–T157.

Schade, H. (1962). Hydrodynamic stability of two-dimensional axially symmetric laminar flow (in German). Deutsche Versuchsanstalt für Luft- und Raumfahrt, D. V. L. R. Rept. No. 190.

Schade, H. (1962). Stability theory of axisymmetric parallel-flow (in German). *Ing.-Arch.* **31**, 301–316.

Schade, H., and Michalke, A. (1962). Formation of vortices in a free boundary layer (in German). *Z. Flugwiss.* **10**, 147–154.

Schubauer, G. B. (1958). Mechanics of transition at subsonic speeds. International Union of Theoretical and Applied Mechanics, *Grenzschichtforschung Symposium*, pp. 85–109. Springer, Berlin.

Sen, A. K. (1963). Stability of hydromagnetic Kelvin-Helmholtz discontinuity. *Phys. Fluids* **6**, 1154–1163.

Sergienko, A. A., and Gretsov, V. K. (1959). Transition from a turbulent into a laminar boundary layer. (*English Transl.: Soviet Phys. "Doklady"* **4**, 275–276).

Serrin, J. (1959). A note on the existence of periodic solutions of the Navier-Stokes equations. *Arch. Rational Mech. and Anal.* **3**, 120–122.

Serrin, J. (1959). On the stability of viscous fluid motions. *Arch. Rational Mech. and Anal.* **3**, 1–13.

Sexl, T., and Spielberg, K. (1958). Zum Stabilitätsproblem der Poiseuille-Strömung. *Acta. Phys. Austraica* **12**, 9–28.

Shair, F. H. (1964). Interaction of an oscillating magnetic field with fluid in Couette flow. *AIAA J.* **2**, 1510–1511.

Shen, S. F. (1959). Some considerations of the laminar stability of incompressible time-dependent basic flows. U.S. Naval Ordnance Laboratory, NAVORD Rept. 6654.

Shen, S. F. (1960). Some considerations of the laminar stability of incompressible, parallel time-dependent flows. Institute of the Aeronautical Sciences, Paper No. 60–33.

Shen, S. F. (1961). Some considerations on the laminar stability of time-dependent basic flows. *J. Aerospace Sci.* **28**, 397–404, 417.

Shen, S. T. (1957). The theory of stability of compressible laminar boundary layers with injection of a foreign gas. U. S. Naval Ordnance Laboratory, NAVORD Rept. 4467.

Shirely, L. K. (1966). Asymptotic expansions in the theory of hydrodynamic stability. *Proc. Fifth U. S. Natl. Congr. Appl. Mech.* pp. 705–713. Amer. Soc. Mech. Eng., New York.

Sinha, K. D., and Choudhary, R. C. (1965). Flow of a viscous incompressible-fluid between two parallel plates, one in uniform motion and the other at rest, with suction at the stationary plate. *Proc. Indian Acad. Sci.* **61A**, 308–318.

Sirazetdinov, T. K. (1965). Contribution to the theory of the stability of motion of a fluid subject to continuous perturbations (in Russian). *Aviatsionnaia Tekhn.* **8**, No. 4, 62–74.

Sizonenko, V. L. (1964). The stability of tangential discontinuities in magnetohydrodynamics. (in Russian). *Zh. Prikl. Mekan. i Tekhn. Fiz.* **6**, 23–60.

Slezkin, N. A., and Shustov, S. N. (1954). On the stability of particles suspended in a laminar flow (in Russian). *Dokl. Akad. Nauk SSSR* **96**, 933–936.

Smith, M. H. (1955). Bibliography on boundary layer control. Princeton University, James Forrestal Research Center Library, Literature Search No. 6.

Sorokin, V. S., and Suskin, I. V. (1960). Stability of equilibrium of a conducting liquid heated from below in a magnetic field (in Russian). *Zh. Eksperim. i Teor. Fiz.* **38**, 612–620 (*English Transl.: Soviet Phys. JETP* **11**, 440–445).

Spangenberg, W. G., and Rowland, W. R. (1960). Optical study of boundary layer transition processes in a supersonic air stream. *Phys. Fluids* **3**, 667–684.

Sparrow, E. M., Goldstein, R. J., and Jonsson, V. K. (1964). Thermal instability in a horizontal fluid layer—effect of boundary conditions and non-linear temperature profile. *J. Fluid Mech.* **18**, 513–528.

Sparrow, E. M., Tsou, F. K., and Kurtz, E. F., Jr. (1965). Stability of laminar free-convection flow on a vertical plate. *Phys. Fluids* **8**, 1559–1561.

Speidel, L. (1957). Influencing the laminar boundary layer through periodic disturbances by injections (in German). *Z. Flugwiss.* **5**, 270–275.

Spence, D. A., and Randall, D. G. (1953). The influence of surface waves on the stability of a laminar boundary layer with uniform suction. Great Britain, Aeronautical Research Council, Current Paper No. 161.

Spiegel, E. A. (1960). The convective instability of a radiating fluid layer. *Astrophys. J.* **132**, 716–728.

Spielberg, K., and Timan, H. (1960). On three and two-dimensional disturbances of pipe flow. *J. Appl. Mech.* **27**, 381–389.

Srinivasan, S. K. (1966). A novel approach to the kinetic theory and hydrodynamic turbulence. *Z. Physik* **193**, 394; **197**, 435–439.

Stepanov, K. N., and Khomeniuk, V. V. (1962). On the stability conditions of equilibriun magnetohydrodynamic configurations (in Russian). *Prikl. Mat. i Mekhan.* **26**, 466–470 (*English Transl.: J. Appl. Math. Mech.* **26**, 695–703).

Stern, M. E. (1962). Relative stability of the solutions in non-linear convection. *Phys. Fluids* **5**, 120–121.

Stern, M. E. (1963). Joint instability of hydromagnetic fields which are separately stable. *Phys. Fluids* **6**, 636–642.

Sternberg, J. (1962). A theory for the viscous sublayer of a turbulent flow. *J. Fluid Mech.* **13**, 241–271.

Sternling, C. V., and Scriven, L. E. (1959). Interfacial turbulence: hydrodynamic instability and the Marangoni effect. *Am. Inst. Chem. Eng. J.* **5**, 514–523.

Steurer, J. W. (1960). Instability of the laminar layer and the mechanisms of transition. Institute of the Aeronautical Sciences, Minta Martin Aeronautical Student Fund, First Award Student Papers, 185–198.

Stewartson, K. (1960). The theory of unsteady laminar boundary layers. *Advan. Appl. Mech.* **6**, 1–37.

Struminskii, V. V. (1963). On the non-linear theory of aerodynamic stability (in Russian). *Dokl. Akad. Nauk SSSR* **151**, 1046–1049.

Stuart, J. T. (1956). On the effects of the Reynolds stress on hydrodynamic stability. *Z. Angew. Math. Mech.* S32–S38.

Stuart, J. T. (1956). On the role of Reynolds stresses in stability theory. *J. Aeron. Sci.* **23**, 86–88.

Stuart, J. T. (1960). On three-dimensional non-linear effects in the stability of parallel flows. *Advan. Aeron. Sci. (Proc. 2nd Intern. Congr. Aeron. Sci., Zürich, 1960)* **3**, 121–142.

Stüper, J. (1956). The effect of surface excrescencies on the transition of the boundary layer over a plate (in German). *Z. Flugwiss.* **4**, 30–34.

Stuetzer, O. M. (1959). Instability of certain electrohydrodynamic systems. *Phys. Fluids* **2**, 642–648.

Sundaram, A. K. (1965). Oscillations and stability of a plasma jet. *Phys. Fluids* **8**, 195–199.

Suprunenko, I. P. (1965). Stability of jet flows (in Russian). *Izv. Akad. Nauk SSSR Mekhan.* July–Aug. 1965, 31–35.

Szablewski, W. (1962). Behavior of wave-forming disturbances of a plane Poiseuille flow with small parameter αRe (in German). *Z. Angew. Math. Mech.* **42**, 194–202.

Szablewski, W. (1964). Longitudinal vortices in a plane Poiseuille flow (in German). *Z. Angew. Math. Mech.* **44**, 253–263.

Talwar, S. P. (1959). Hydromagnetic stability of a conducting, inviscid, incompressible fluid of variable density. *Z. Astrophys.* **47**, 161–168.

Talwar, S. P. (1961). Helmholtz instability in hydromagnetics. *Proc. Natl. Inst. Sci. India* **A27**, 263–269.

Talwar, S. P. (1965). Hydromagnetic stability. N. A. S. A. Tech. Note D–2791.

Taneda, S. (1963). The stability of two-dimensional laminar wakes at low Reynolds numbers. *J. Phys. Soc. Japan* **18**, 288–296.

Taneda, S. (1965). Experimental investigations of vortex streets. *J. Phys. Soc. Japan* **20**, 1714–1721.

Tarasov, Yu. A. (1959). Stability of a plane Poiseuille flow of a finite conductivity plasma in a magnetic field (in Russian). *Zh. Eksperim. i Teor. Fiz.* **3**, No. 6, 1708–1713.

Tarasov, Yu. A. (1960). On the stability of plane Poiseuille flow of a plasma of finite conductivity in a magnetic field. *Soviet Phys. JETP* (*English Transl.*) **10**, 1209–1212.

Tchen, C. –M. (1956). Approximate theory on the stability of interfacial waves between two streams. *J. Appl. Phys.* **27**, 1533–1536.

Tchen, C. –M. (1956). Stability of oscillations of superposed fluids. *J. Appl. Phys.* **27**, 760–767.

Teddington, England, National Physical Laboratory (1955). *Proc. Boundary Layer Effects in Aerodynamics* H. M. Stationery Office, London.

Tetervin, N. (1957). A discussion of cone and flat-plate Reynolds numbers for equal ratios of the laminar shear to the shear caused by small velocity fluctuation in a laminar boundary layer. N. A. C. A. Tech. Note 4078.

Tetervin, N. (1958). Theoretical distribution of laminar boundary-layer thickness, boundary-layer Reynolds number and stability limit, and roughness Reynolds number for a sphere and disk in incompressible flow. N. A. C. A. Tech. Note 4350.

Tetervin, N. (1959). Charts and tables for estimating the stability of the compressible laminar boundary layer with heat transfer and arbitrary pressure gradient. N. A. S. A. Memo. 5–4–59L.

Tetervin, N. (1961). An estimate of the minimum Reynolds number for transition from laminar to turbulent flow by means of energy considerations. *J. Aerospace Sci.* **28**, 160–161.

Thompson, W. B. (1960). A stability survey. Denmark, Atomic Energy Commission Research Establishment, Risö Rept. No. 18, 237–248.

Timman, R., Zaat, J. A., and Burgenhout, T. J. (1956). Stability diagrams for laminar boundary-layer flow. Amsterdam, Nationaal Luchtvaart Laboratorium, Tech. Note F–193.

Tollmien, W. (1962). Aspects of the physics of flow (in German). *Z. Flugwiss.* **10**, 403–413.

Tollmien, W., and Grohne, D. (1961). The nature of transition. *In* "Boundary Layer and Flow Control" (G. V. Lachmann, ed.), Vol. 1, pp. 602–636. Pergamon, London.

Toong, T. -Y. (1963). Combustion instability in laminar boundary layers. American Institute of Aeronautics and Astronautics, Preprint 63–225.

Trehan, S. K., Malik, S. K., and Dikshit, M. S. (1965). Oscillations of a capillary jet in the presence of a uniform magnetic field. *Phys. Fluids* **8**, 1461–1466.

Trilling, L. (1958). Oscillating shock boundary-layer interaction. *J. Aero/Space Sci.* **25**, 301–304.

Tserkovnikov, Yu. A. (1960). On the problem of convective instability of a plasma (in Russian). *Dokl. Akad. Nauk SSSR* **130**, 295–298.

Tsou, F. K., Sparrow, E., M. and Kurtz, E. F., Jr. (1966). Hydrodynamic stability of the boundary layer on a continuous moving surface. *J. Fluid Mech.* **26**, 145–161.

Uberoi, M. S., and Chow, C. -Y. (1963). Instability of a current-carrying fluid jet issuing from a nozzle. *Phys. Fluids* **6**, 1237–1241.

Van Driest, E. R. (1965). Recent studies in boundary layer transition. *Intern. J. Eng. Sci.* **3**, 341–353.

Van Driest, E. R., and Blumer, C. B. (1962). Boundary layer transition-roughness and free-stream turbulence effects. *Proc. 4th Intern. Symp. Space Technol. and Sci., Tokyo, August 27–31, 1962* pp. 203–208.

Vedenov, A. A., and Ponomarenko, I. B. (1964). Onset of turbulence (in Russian). *Zh. Eksperim. i Teor. Fiz.* **46**, 2247–2250.

Velte, W. (1962). Stability of flow in a horizontal tube in an unevenly heated wall (in German). *Z. Angew. Math. Phys.* **13**, 591–600.

Viilu, A. (1962). An experimental determination of the minimum Reynolds number for instability in a free jet. *Trans. Amer. Soc. Mech. Eng.* **84E**, 506–508.

Von Baranoff, A. (1960). Sur une méthode de résolution de l'équation d'Orr-Sommerfeld régissant l'apparition de la transition dans un couche limite. *AGARD Rept.* 270.

Voronin, V. I. (1954). On the asymptotic solution of the equations of the laminar boundary layer of a flat plate (in Russian). *Tr. Voronezhsk. Gos. Univ.* No. 33, 63–69.

Wang, K. C. (1962). Instability of a vortex sheet in nonequilibrium flows. *Phys. Fluids* **5**, 1368–1373.

Wang, K. C., and Maslen, S. H. (1964). Hydromagnetic stability of a vortex sheet in compressible fluids. *Phys. Fluids* **7**, 1780–1784.

Watson, J. (1958). A solution of the Navier-Stokes equations illustrating the response of a laminar boundary layer to a given change in the external stream velocity. *Quart. J. Math. Appl. Math.* **11**, 302–325.

Watson, J. (1960). On the nonlinear mechanics of wave disturbances in stable and unstable parallel flows. II. The development of a solution for plane Poiseuille flow and for plane Couette flow. *J. Fluid Mech.* **9**, 371–389.

Wedemeyer, E. (1965). Stability of a flow between two curved parallel walls (in German). *Z. Flugwiss.* **13**, 459–462.

Wen, K. S. (1964). Wake transition. *AIAA J.* **2**, 956–957.

Whitham, G. B. (1960). A note on group velocity. *J. Fluid Mech.* **9**, 347.

Wille, R. (1960). Modellvorstellungen zum Übergang laminar-turbulent. *Arbeitsgemeinschaft Forsch. Landes Nordrhein–Westfalen* **72**; *DVL Bericht* No. 113.

Willson, A. J. (1965). On the stability of two superposed fluids. *Proc. Cambridge Phil. Soc.* **61**, 595–607.

Wooding, R. A. (1959). The stability of a viscous liquid in a vertical tube containing porous material. *Proc. Roy. Soc.* (*London*) **A252**, 120–134.

Wooding, R. A. (1960). Instability of a viscous liquid of variable density in a vertical Hele-Shaw cell. *J. Fluid Mech.* **7**, 501–515.

Wooding, R. A. (1960). Rayleigh instability of a thermal boundary layer in flow through a porous medium. *J. Fluid Mech.* **9**, 183–192.

Wooding, R. A. (1962). The stability of an interface between miscible fluids in a porous medium. *Z. Angew. Math. Phys.* **13**, 255–266.

Wooler, P. T. (1961). Instability of flow between parallel planes with a coplanar magnetic field. *Phys. Fluids* **4**, 24–27.

Wortmann, F. X. (1955). Investigation of unstable boundary layer oscillations in a water channel by the tellurium method (in German). *In* "Fifty Years of Boundary Layer Research" (H. Görtler and W. Tollmien, eds.), pp. 460–471. Vieweg, Braunschweig.

Wortmann, F. X. (1966). Experimental investigation of laminar boundary layers in unstable stratification (in German). *Proc. 11th Intern. Congr. Appl. Mech.* pp. 815–825. Springer, Berlin.

Wright, J. P. (1964). General relativistic instability. *Phys. Rev.* **136**, B288–B289.

Yamada, H. (1960). Stability of laminar shear layer between parallel uniform streams. I. *Rept. Res. Inst. Appl. Mech. Kyushu Univ.* **8**, No. 30, 1–18.

Yamada, H. (1964). On the stability to long waves of two-dimensional laminar jet. *Rept. Res. Inst. Appl. Mech. Kyushu Univ.* **12**, No. 43, 1–15.

Yamada, H. (1965). Stability of laminar shear layer between parallel uniform streams. II. *Rept. Res. Inst. Appl. Mech. Kyushu Univ.* **13**, No. 45, 1–15.

Yang, K. -T., and Jerger, E. W. (1964). First-order perturbations of laminar free-convection boundary layers on a vertical plate. *Trans. Amer. Soc. Mech. Eng.* **86C**, 107–115.

Yang, T. -T., and Kelleher, M. D. (1964). On hydrodynamic stability of two-dimensional unsteady incompressible laminar boundary layers. University of Notre Dame, Department of Mechanical Engineering, Tech. Note 64–22.

Yih, C. -S. (1955). Stability of parallel laminar flow with a free surface, *Proc. 2nd U.S. Natl. Congr. Appl. Mech.* pp. 623–628.

Yih, C. -S. (1955). Stability of two-dimensional parallel flows for three-dimensional disturbances. *Quart. Appl. Math.* **12**, 434–435.

Yih, C. -S. (1957). Stability of laminar flow in curved channels. *Phil. Mag.* **2**, series 8, 305–310.

Yih, C. -S. (1959). Inhibition of hydrodynamic instability by an electric current. *Phys. Fluids* **2**, 125–130.

Yih, C. -S. (1959). Thermal instability of viscous fluids. *Quart. Appl. Math.* **17**, 25–42.

Yih, C. -S. (1961). Dual role of viscosity in the instability of revolving fluids of variable density. *Phys. Fluids* **4**, 806–811.

Yih, C. -S. (1963). Stability of liquid flow down an inclined plane. *Phys. Fluids* **6**, 321–334.

Yih, C. -S. (1965). Hydrodynamic stability. "Dynamics of Nonhomogeneous Fluids," pp. 141–195. Macmillan, New York.

Yih, C. -S. (1965). Gravitational instability of a viscous fluid in a magnetic field. *J. Fluid Mech.* **22**, 579–586.

Zaat, J. A. (1958). Numerical contribution to the stability theory of boundary layers (in German). International Union of Theoretical and Applied Mechanics, *Grenzschichtforschung Symp.* pp. 127–138. Springer, Berlin.

Zaitsev, A. A. (1960). The stability of a viscous layer on a solid body in a flow of gas (in Russian). *Dokl. Akad. Nauk SSSR* **130**, 1228–1231 (*English Transl.: Soviet Phys. "Doklady"* **5**, 49–53).

Zaslavskii, G. M., Moiseev, S. S., and Sagdeev, R. Z. (1964). Asymptotic method of solving a fourth-order differential equation with two small parameters in hydrodynamic stability theory (in Russian). *Dokl. Akad. Nauk SSSR* **158**, 1295–1298 (*English Transl.: Soviet Phys. "Doklady"* **9**, 863–866).

Zeiberg, S. L. (1962). The wake behind an oscillating vehicle. *J. Aerospace Sci.* **29**, 1344–1346, 1357.

Zumartbaev, M. T. (1961). Stability of magnetic tangential discontinuities in relativistic hydrodynamics. (*English Transl.: Soviet Phys. JETP* **13**, 1006–1009).

Zysina-Molozhen, L. M. (1955). Certain quantitative characteristics of the transition from laminar to turbulent flow in the boundary layer (in Russian). *Zh. Tekhn. Fiz.* **25**, 1280–1287.

Author Index

Numbers in italics refer to pages on which the references are listed.

Subject Index

A

Acoustic fluctuations, equation, 207
Acoustic instability, 179
Acoustic power, 181, 182
Acoustic radiation, 178, 181
Acoustic resonance, 187
Acoustic waves, 178, 179, 182
Adams-Moulton method, 264
Airy equation, 272
Alfvén waves, 206-209
Algebraic methods, 35-36, 197
Amplification, see growth factor
Analytical continuation, 33, 67, 78, 268
Axisymmetric flows, 228-233

B

Bernoulli theorem, 19
Bessel functions, 51, 230
Bjørkness effect, 176
Blasius boundary layer
  eigenvalues, 63, 84, 89, 90
  experimental and theoretical comparison, 131, 132
  in non-Newtonian fluid, eigenvalues, 255, 256, 258
  magnetic, 210, 211
  stability, 64
  three-dimensional eigenvalues, 112-116
Blasius functions, 46
Boundary conditions
  magnetic, 201-202
  wavy wall, 244-245
Boundary layer
  boundary conditions, 88-89
  compressible, eigenvalues, 187
  compressible inviscid, eigenvalues, 180
  compressible neutral stability, 187
  curved wall
    eigenvalues, 246, 250
    stability, 246

decaying solution, 73
definition, 45
displacement thickness, 130
experimental eigenfunctions, 132
experimental eigenvalues, 132
experiments, 129-134
  confirming theory, 129-133
final results, 62-65
hypersonic, 174
in adverse pressure gradient, 64, 65
inviscid oscillations, 49
inviscid solutions, 46-50
nonlinear effects, 140-141
numerical integration, 83-91
numerical introduction, 83-84
separation, 64
simplified analysis, 45-73
spatial instability, 222-224
stability
  deformable wall, 242
  history of experimental results, 3, 129
turbulent eigenvalues, 262
unstable, 66-73
unstable solutions, 66-68
viscoelastic, 254-256
with suction, 64, 65
Boussinesq approximation, 189
Brownian motion, 228
Buoyancy
  across mean flow, 189-191
  along mean flow, 191-194
  forces, 188-194
  definition, 189

C

Cats eyes, 152-153, 165
Cauchy-Riemann equations, 218, 268
Cauchy theorem, 66, 72
Centrifugal forces, effects, 191, 245-253

1-MONTH